Integrating Modern Mathematics and Sensor Technologies in Civil Engineering

Abdulsattar Abdullah Hamad
https://orcid.org/0000-0003-0497-5115
University of Samarra, Iraq

IGI Global
Scientific Publishing
Publishing Tomorrow's Research Today

Vice President of Editorial Melissa Wagner
Director of Acquisitions Mikaela Felty
Director of Book Development Jocelynn Hessler
Production Manager Mike Brehm
Cover Design Jose Rosado

Published in the United States of America by
IGI Global Scientific Publishing
701 East Chocolate Avenue
Hershey, PA, 17033, USA
Tel: 717-533-8845 | Fax: 717-533-7115
Website: https://www.igi-global.com E-mail: cust@igi-global.com

Library of Congress Cataloging-in-Publication Data

Names: Hamad, Abdulsattar Abdullah, 1987- editor
Title: Integrating modern mathematics and sensor technologies in civil engineering / edited by: Abdulsattar Hamad.
Description: Hershey PA : IGI Global Scientific Publishing, [2026] | Includes bibliographical references. | Summary: "This book aspires to be a comprehensive scientific reference that explores how the integration of modern mathematics-including differential equations, numerical analysis, and probabilistic modeling-with advanced sensor technologies can help address complex challenges in civil engineering"-- Provided by publisher.
Identifiers: LCCN 2025029107 (print) | LCCN 2025029108 (ebook) | ISBN 9798337352480 hardcover | ISBN 9798337352497 paperback | ISBN 9798337352503 ebook
Subjects: LCSH: Civil engineering--Mathematics | Detectors
Classification: LCC TA331 .I58 2026 (print) | LCC TA331 (ebook)
LC record available at https://lccn.loc.gov/2025029107
LC ebook record available at https://lccn.loc.gov/2025029108

British Cataloguing in Publication Data
A Cataloguing in Publication record for this book is available from the British Library.

Table of Contents

Preface xix

Acknowledgment xxiv

Chapter 1
A Haar Wavelet Collocation Method for Inverse Problem: Application to Black-Scholes Model 1
Muhammad Ahsan, Harbin Institute of Technology, Shenzhen, China & University of Swabi, Pakistan
Muhammad Riaz, University of Swabi, Pakistan
Masood Ahmed, School of Mathematics and Statistics, Central South University, Changsha, China
Maher Alwuthaynani, Taif University, Saudi Arabia
Weidong Lei, Harbin Institute of Technology, Shenzhen, China

Chapter 2
Applications of Complex Intuitionistic Fuzzy Sets in Engineering Data Analytics and Smart Infrastructure Systems 21
R. Myna, Erode Sengunthar Engineering College, India
J. Haritha, Bannari Amman Institute of Technology, India
S. Ramya, Vellalar College of Engineering and Technology, India
C. V. Mythily, Erode Sengunthar Engineering College, India

Chapter 3
Artificial Intelligence and Machine Learning 61
Bipin Sule, Vishwakarma Institute of Technology, India
Durga Meshram, Datta Meghe Institute of Management Studies, India
Subramaniam Seshan Iyer, Datta Meghe Institute of Management Studies, India
Ganesh Pandit Pathak, Sri Balaji University, Pune, India
Nilesh Anute, Sri Balaji University, Pune, India
P. Selvakumar, Department of Science and Humanities, Nehru Institute of Technology, Coimbatore, India
T. C. Manjunath, Rajarajeswari College of Engineering, India

Chapter 4
Comparative Study Between YOLOv7, YOLOv8, and YOLOv9 for Pothole Detection on Indian Roads 89
Pranav Jaysankar, Anna University, Chennai, India
Deepak Rajeshkumar, Anna University, Chennai, India
Balaji Chandrasekaran, Anna University, Chennai, India
G. S. Smrithy, VIT Chennai, India

Chapter 5
Design and Implementation of IoT-Based Sensing Systems for Infrastructure Monitoring 123
Zaid A. Mohammed, Technical College of Engineering, Al-Bayan University, Baghdad, Iraq
G. Vijayakumar, V.S.B. Engineering College, Karur, India
Mohanad Hatem Shadhar, Department of Civil Engineering, College of Engineering, Al-Iraqia University, Baghdad, Iraq
Balaji Sheshara o Ingole, Independent Researcher, USA
S. B. Sridhara, Amruta Institute of Engineering and Management Sciences, Bidadi, India

Chapter 6
Designing Adaptive and Self-Responsive Structures 149
Moqdad Ibrahim Jassem, Tikrit University, Iraq
Ahmed Ali Abdullah, Tikrit University, Iraq
Hameed Hasan Khalaf, University of Tikrit, Iraq
M. O. Noor, Independent Researcher, Iraq

Chapter 7
Development of Autonomous Systems for Infrastructure Inspection and Maintenance Using ML and IoT 171
Mohanad Hatem Shadhar, Department of Civil Engineering, College of Engineering, Al-Iraqia University, Baghdad, Iraq
K. P. Nagapushpa, Acharya Institute of Technology, Karnataka, India.
Zaid A. Mohammed, Technical College of Engineering, Al-Bayan University, Baghdad, Iraq
M. Raghavendra, Department of MBA, Faculty of Engineering, Management, and Technology, BGSIT, Adichunchanagiri University, India
S. B. Sridhara, Amruta Institute of Engineering and Management Sciences, Bidadi, India

Chapter 8
Future Directions and Emerging Research 199
Hameed Hasan Khalaf, University of Tikrit, Iraq
Hatem Kareem Ahmed, College of Administration and Economics, University of Samarra, Iraq
Mohammed Ameen Haitham Ahmed Alhababi, College of Administration and Economics, University of Samarra, Iraq
Shihab A. Shawkat, University of Samarra, Iraq

Chapter 9
Integration of UAVs and Robotics in Infrastructure Inspection and Maintenance 215
V. Dankan Gowda, B.M.S. Institute of Technology and Management, India.
Kottala Sri Yogi, Symbiosis institute of Business Management, Symbiosis International University, Pune, India
Ashwini S. Shivannavar, SSIT, SSAHE Tumakuru, Karnataka, India
B. C. Kavitha, BGS Institute of Technology, Adichunchanagiri Univeristy, Mandya, India
M. Benila, V.S.B. Engineering College, Karur, India

Chapter 10
Internet of Things Applications in Civil Engineering 243
Damodharan Palaniappan, Marwadi University, India
T. Premavathi, Marwadi University, India

Chapter 11
Lifecycle Assessment and Structural Degradation 273
Mohammed Ameen Haitham Ahmed Alhababi, College of Administration and Economics, Samarra University, Iraq
Hameed Hasan Khalaf, College of Administration and Economics, Samarra University, Iraq
Moqdad Ibrahim Jassem, College of Administration and Economics, Samarra University, Iraq
Shihab A. Shawkat, University of Samarra, Iraq

Chapter 12
Machine Learning Algorithms for Sensor Data Processing and Analysis 291
Chirag Mavani, Independent Researcher, Canada
Naziya Hussain, Department of Computer Engineering, MPSTME, SVKM's NMIMS University, Shirpur, India
Hirenkumar Mistry, Independent Researcher, USA
Harshvardhan Rajendra Godbole, Shri Balasaheb Mane Shikshan Prasarak Mandal's Ashokrao Mane Group of Institutions, Kolhapur, India
S. B. Sridhara, Amruta Institute of Engineering and Management Sciences, Bidadi, India

Chapter 13
Machine Learning Approaches for Predictive Maintenance in Infrastructure Systems 321
Bappaditya Das, Dr. B.C. Roy Engineering College, Durgapur, India
V. Dankan Gowda, BMS Institute of Technology and Management, India
Abdulsattar Abdullah Hamad, Department of Physics, College of Education, University of Samarra, Iraq
B. C. Kavitha, BGS Institute of Technology, Adichunchanagiri Univeristy, Mandya, India
Sri Yogi Kottala, Symbiosis institute of Business Management, Symbiosis International University, Pune, India

Chapter 14
Probabilistic Modeling and Risk Assessment 351
Mustafa Kamil Rasheed, Mustansiriyah University, Iraq
Ahmed Hadi Salman, Mustansiriyah University, Iraq
Huda Mohammed Salman Alqayse, Educational and Psychological Research Center, University of Baghdad, Iraq
Alnoman M. Mundher Tayyeh, Middle Technical University, Iraq

Chapter 15
Smart Sensor Technologies 365
Nithin Subba Rao, Michigan Technological University, USA
P. Selvakumar, Department of Science and Humanities, Nehru Institute of Technology, Coimbatore, India
S. Madhu, Jyothy Institute of Commerce and Management, India
Naeema Nazar, VISAT Engineering College, Ernakulam, India
V. Padmashree, Padmashree Institute of Management and Sciences, India
M. R. Arun, Vel Tech Rangarajan Dr. Sagunthala R&D Institute of Science and Technology, India
T. C. Manjunath, Rajarajeswari College of Engineering, India

Chapter 16
The Role of Technology in Economic Growth: An Analytical Study of a Sample of Gulf Countries for the Period 2010-2023 393
Mustafa Kamil Rasheed, Mustansiriyah University, Iraq
Munaf Marza Neama, Mustansiriyah University, Iraq
Rawaa Salh Mahmmed Al-Saffar, Mustansiriyah University, Iraq
Alnoman M. Mundher Tayyeh, Middle Technical University, Iraq

Chapter 17
Wireless Sensor Networks for Real-Time Infrastructure Monitoring and Management Using ML Approach 409
Gajanan Shankarrao Patange, Charotar University of Science and Technology, India
Abdulsattar Abdullah Hamad, Department of Physics, College of Education, University of Samarra, Iraq
V. Dankan Gowda, BMS Institute of Technology and Management, India
S. Gobinath, V.S.B. Engineering College, Karur, India
K. Ramya, BGS Institute of Technology, Adichunchanagiri University, Karnataka, India

Compilation of References 435

About the Contributors 475

Index 485

Detailed Table of Contents

Preface xix

Acknowledgment xxiv

Chapter 1
A Haar Wavelet Collocation Method for Inverse Problem: Application to Black-Scholes Model 1

Muhammad Ahsan, Harbin Institute of Technology, Shenzhen, China & University of Swabi, Pakistan
Muhammad Riaz, University of Swabi, Pakistan
Masood Ahmed, School of Mathematics and Statistics, Central South University, Changsha, China
Maher Alwuthaynani, Taif University, Saudi Arabia
Weidong Lei, Harbin Institute of Technology, Shenzhen, China

This article demonstrates the Haar wavelet collocation method as an effective strategy for solving the Black-Scholes model with various types of options. The Black-Scholes model is an inverse problems and formulated as a backward parabolic partial differential equation. In the first stage, a transformation is applied to convert this backward Black-Scholes model into a forward partial differential equation. Finite difference method is then employed to estimate the changes over time in the partial differential equation. Subsequently, the partial differential equation is converted into an algebraic system of equations by applying the Haar series approximation to the spatial variable. Extensive analyses were conducted to evaluate the correctness, effectiveness, and convergence of the Haar wavelet collocation method. Various scenarios in Black-Scholes model were tested by considering various options. The computational results were aligned with theoretical expectations, demonstrating the reliability of the method in handling diverse cases of the Black-Scholes model.

Chapter 2
Applications of Complex Intuitionistic Fuzzy Sets in Engineering Data Analytics and Smart Infrastructure Systems 21
R. Myna, Erode Sengunthar Engineering College, India
J. Haritha, Bannari Amman Institute of Technology, India
S. Ramya, Vellalar College of Engineering and Technology, India
C. V. Mythily, Erode Sengunthar Engineering College, India

The chapter explores how non-linearity, ambiguity, hesitation, and uncertainty in engineering data analytics and smart infrastructure systems might be addressed using Complex Intuitionistic Fuzzy Sets (CIFS), a sophisticated mathematical framework. Complex-valued membership and non-membership functions are used by CIFS to augment classical and intuitionistic fuzzy models, allowing for a fuller representation of the degree and direction (phase) of uncertainty. In fields where multidimensional, dynamic, and imprecise data are common, like sensor data fusion, structural health monitoring, and smart city applications, this enables more reliable decision-making. The chapter also highlights CIFS as an essential tool for data-driven engineering systems and intelligent infrastructure by offering thorough case studies, comparative assessments with conventional fuzzy techniques, and future research directions.

Chapter 3
Artificial Intelligence and Machine Learning 61
Bipin Sule, Vishwakarma Institute of Technology, India
Durga Meshram, Datta Meghe Institute of Management Studies, India
Subramaniam Seshan Iyer, Datta Meghe Institute of Management Studies, India
Ganesh Pandit Pathak, Sri Balaji University, Pune, India
Nilesh Anute, Sri Balaji University, Pune, India
P. Selvakumar, Department of Science and Humanities, Nehru Institute of Technology, Coimbatore, India
T. C. Manjunath, Rajarajeswari College of Engineering, India

At its core, AI is about mimicking or replicating human cognitive functions using algorithms and computational models. The idea of machines exhibiting intelligent behavior is not new; it has. General AI (or Strong AI), on the other hand, refers to a hypothetical system that possesses the ability to understand, learn, and apply intelligence across a broad range of tasks—essentially matching or surpassing human capabilities. While we have made significant strides in Narrow AI, General AI remains a long-term goal still under research and development. The fundamental concept of ML is that a system can learn from examples, generalize from them, and apply that knowledge to new, unseen data. For example, if we feed a machine learning model thousands of labeled images of cats and dogs, the model can learn to distinguish between them and classify new images it hasn't seen before. At its core,

AI is about mimicking or replicating human cognitive functions using algorithms and computational models.

Chapter 4
Comparative Study Between YOLOv7, YOLOv8, and YOLOv9 for Pothole Detection on Indian Roads 89
Pranav Jaysankar, Anna University, Chennai, India
Deepak Rajeshkumar, Anna University, Chennai, India
Balaji Chandrasekaran, Anna University, Chennai, India
G. S. Smrithy, VIT Chennai, India

This paper aims to improve pothole detection on Indian roads using advanced deep learning techniques. With fatalities due to potholes increasing by 25% annually, from 1481 in 2021 to 1856 in 2022, enhancing detection accuracy is critical. Building on previous studies that used YOLOv2, YOLOv3, and YOLOv3-tiny models, this initiative will implement YOLOv7, YOLOv8, and YOLOv9 to achieve a higher mean Average Precision (mAP) score. The annotated dataset from Kaggle is split into training, testing, and validation sets. Metrics such as mAP, recall, precision, F1-Score, and accuracy will assess performance. This paper supports Advanced Driver Assistance Systems (ADAS) by providing data to help vehicles navigate potholes, ultimately aiming to reduce accidents and save lives. A comparative analysis shows YOLOv7 as the most effective, underscoring its potential to enhance road safety through real-time alerts and advanced AI.

Chapter 5
Design and Implementation of IoT-Based Sensing Systems for Infrastructure Monitoring 123
Zaid A. Mohammed, Technical College of Engineering, Al-Bayan University, Baghdad, Iraq
G. Vijayakumar, V.S.B. Engineering College, Karur, India
Mohanad Hatem Shadhar, Department of Civil Engineering, College of Engineering, Al-Iraqia University, Baghdad, Iraq
Balaji Shesharao Ingole, Independent Researcher, USA
S. B. Sridhara, Amruta Institute of Engineering and Management Sciences, Bidadi, India

The complex nature of the infrastructure systems and also the infrastructure aging have demanded innovations, which are more efficient and reliable methods of monitoring. Manual inspection methods, however, are expensive, they are time-consuming, and they have little capability of providing real-time data. The emergence of the technology of Internet of Things (IoT) has transformed the sector of infrastructure monitoring systems due to the capacity to implement cost-effective, in real-time, and automatic sensing networks. This chapter also examines designing and deploying of Sensing systems based InT based on the specific domain of tracking the health of infrastructure like bridges, roads, dams, and buildings. Issues surrounding the implementation of such systems such as choice of sensors, protocols of communication, power supply and data processing are addressed as important factors taken into consideration. IoT in combination with cloud- and edge analytics is analyzed in order to improve data gathering and decision-making processes.

Chapter 6
Designing Adaptive and Self-Responsive Structures 149
Moqdad Ibrahim Jassem, Tikrit University, Iraq
Ahmed Ali Abdullah, Tikrit University, Iraq
Hameed Hasan Khalaf, University of Tikrit, Iraq
M. O. Noor, Independent Researcher, Iraq

In this chapter, the authors present adaptive and self-responsive structure systems that sense, decide, and alter their mechanical, geometric, or functional state autonomously are emerging as a transformative class of engineered systems across aerospace, civil, marine, and architectural domains. This chapter synthesizes the scientific foundations, design methodologies, material systems, sensing and control approaches, modeling tools, and evaluation metrics necessary for designing adaptive structures that are robust, energy efficient, and safe. It presents representative case studies (morphing aerodynamic surfaces, adaptive building façades, self-tuning vibration absorbers), proposes an integrated design framework, and outlines experimental and computational protocols for validation. Key challenges (multi-physics coupling, reliability, scalability, and standards), ethical and sustainability considerations, and a roadmap for future research are discussed. Practical recommendations and a curated set of references are included for researchers and practitioners seeking to design

Chapter 7
Development of Autonomous Systems for Infrastructure Inspection and Maintenance Using ML and IoT 171
Mohanad Hatem Shadhar, Department of Civil Engineering, College of Engineering, Al-Iraqia University, Baghdad, Iraq
K. P. Nagapushpa, Acharya Institute of Technology, Karnataka, India.
Zaid A. Mohammed, Technical College of Engineering, Al-Bayan University, Baghdad, Iraq
M. Raghavendra, Department of MBA, Faculty of Engineering, Management, and Technology, BGSIT, Adichunchanagiri University, India
S. B. Sridhara, Amruta Institute of Engineering and Management Sciences, Bidadi, India

Advances in Machine Learning (ML) and the Internet of Things (IoT) have led to the development of self-sufficient systems to check the infrastructures and maintenance processes used to maintain the vital infrastructures in the traditional monitoring and maintenance procedures. This chapter discusses how the ML algorithms and IoT sensors may be integrated and used to improve the accuracy, efficiency, and safety of the infrastructure inspection. Implementation of autonomous systems, e.g., drones and robots, enables collecting real-time data, functioning permanently, and making predictions related to maintenance reducing maintenance labour, operational cost, and downtime. A number of case studies are presented to demonstrate practical use of such technologies, such as bridge inspection through drones, predictive maintenance of railways, and intelligent pipeline monitoring. The chapter further lists the challenges fixed to such systems including; sensor calibration, data integration and data processing in real time.

Chapter 8
Future Directions and Emerging Research 199
Hameed Hasan Khalaf, University of Tikrit, Iraq
Hatem Kareem Ahmed, College of Administration and Economics, University of Samarra, Iraq
Mohammed Ameen Haitham Ahmed Alhababi, College of Administration and Economics, University of Samarra, Iraq
Shihab A. Shawkat, University of Samarra, Iraq

The study results show that emerging research represents a major driver of technological transformation, particularly in the fields of artificial intelligence (23.5% annual growth), renewable energy (18.2%), and big data (20.1%). Global funding in these fields has reached record levels, with AI alone expected to receive $90 billion by 2024. In contrast, developing countries face a significant gap, with the number of researchers not exceeding 740 per million people, compared to 4,500 in developed countries. The most prominent challenges are a lack of funding (72%), weak infrastructure (65%), and a lack of skilled personnel (60%). The analysis confirms that the integration of technology and civil engineering is essential for building smart and sustainable infrastructure. The study recommends reorienting research policies towards open innovation, scientific foresight, and effective knowledge governance. The conclusion indicates that keeping pace with future trends is no longer an option, but rather a strategic necessity to ensure scientific and developmental positioning

Chapter 9
Integration of UAVs and Robotics in Infrastructure Inspection and Maintenance 215
V. Dankan Gowda, B.M.S. Institute of Technology and Management, India.
Kottala Sri Yogi, Symbiosis institute of Business Management, Symbiosis International University, Pune, India
Ashwini S. Shivannavar, SSIT, SSAHE Tumakuru, Karnataka, India
B. C. Kavitha, BGS Institute of Technology, Adichunchanagiri Univeristy, Mandya, India
M. Benila, V.S.B. Engineering College, Karur, India

The use of the Unmanned Aerial Vehicles (UAVs) and robotics in infrastructure inspection and maintenance has transformed the old procedures of inspecting essential property. In this chapter, UAVs and robotics are discussed in terms of their potential to enhance efficiency, safety, and accuracy of infrastructure inspection. It is possible to conduct precise aerial inspections of difficult to access locations, i.e. bridges, high-rise structures as well as industrial facilities with the help of UAVs that boast high specification sensors. In the same way, the systems in the form of robotic systems, including ground robotic systems and climbing robotic systems can perform the precise inspection in tight spaces, which has the great benefit in the process of maintenance. Such combination of UAVs and robotics enables shorter inspection period, better data gathering and cheaper operations. Nevertheless, such barriers as regulatory rules, data handling, and technical restrictions are still common.

Chapter 10
Internet of Things Applications in Civil Engineering 243
Damodharan Palaniappan, Marwadi University, India
T. Premavathi, Marwadi University, India

The IoT is transforming civil engineering into a new real-time monitoring, automated control, and data-based decision-making in infrastructure. In this field, the presence of embedded sensors in structures such as buildings, bridges and dams has increased structural health monitoring, smooth traffic flows, improved construction procedures and management of water resources. These sensors gather information about stress, strain, temperature, and vibration helping to implement predictive maintenance and fault detection. By making traffic, routing, parking more efficient, the IoT can be used in the transportation industry to decrease congestion and emissions. Wearables and sensors enhance safety in construction sites, track the site equipment and facilitate the process of managing materials. The existing problems in its expansion are data security, interoperability and sensor durability. It is expected that future augmented intelligence will include AI, machine learning, digital twins, and blockchain to lead to fully autonomous and predictive infrastructure systems.

Chapter 11
Lifecycle Assessment and Structural Degradation 273
Mohammed Ameen Haitham Ahmed Alhababi, College of Administration and Economics, Samarra University, Iraq
Hameed Hasan Khalaf, College of Administration and Economics, Samarra University, Iraq
Moqdad Ibrahim Jassem, College of Administration and Economics, Samarra University, Iraq
Shihab A. Shawkat, University of Samarra, Iraq

This research presents an integrated analysis that combines Life Cycle Assessment (LCA) and structural degradation of structural materials and components, with the aim of understanding the relationship between long-term performance and the environmental impact of buildings. LCA is an important tool for analyzing environmental impacts from the extraction of raw materials to the end of their service life. Structural degradation represents a direct challenge to the sustainability of buildings and includes mechanisms such as corrosion, fatigue, and creep. The study highlights the importance of shifting from static models to dynamic models based on real data, using artificial intelligence and real-time monitoring techniques, as illustrated in Figure (6). It recommends applying this methodology in the design, maintenance, and environmental assessment phases to enhance the sustainability and operational efficiency of facilities.

Chapter 12
Machine Learning Algorithms for Sensor Data Processing and Analysis 291
Chirag Mavani, Independent Researcher, Canada
Naziya Hussain, Department of Computer Engineering, MPSTME, SVKM's NMIMS University, Shirpur, India
Hirenkumar Mistry, Independent Researcher, USA
Harshvardhan Rajendra Godbole, Shri Balasaheb Mane Shikshan Prasarak Mandal's Ashokrao Mane Group of Institutions, Kolhapur, India
S. B. Sridhara, Amruta Institute of Engineering and Management Sciences, Bidadi, India

Sensor data have increased exponentially, with sensors cropping up in a wide variety of other applications, including healthcare, industrial monitoring, environmental sensing and smart cities. This is often high-dimensional, noisy, and continuous data poses a big challenge to the conventional data processing techniques. The sensor data became an important focus of ML algorithms that allow retrieving valuable insights, identifying anomalies, predicting and, finally, recognizing patterns. This chapter reviews ML algorithms on sensor data such as supervised, unsupervised and deep learning algorithms. Less focus is put on either preprocessing methods required to address sensor data issues, e.g., noise, data completeness, and dimensionality. Moreover, the chapter will present the application of these methods in the real world in several fields: predictive maintenance, environmental monitoring, and health. The capabilities of various ML models in terms of certain types of sensor data are discussed

Chapter 13
Machine Learning Approaches for Predictive Maintenance in Infrastructure Systems 321
Bappaditya Das, Dr. B.C. Roy Engineering College, Durgapur, India
V. Dankan Gowda, BMS Institute of Technology and Management, India
Abdulsattar Abdullah Hamad, Department of Physics, College of Education, University of Samarra, Iraq
B. C. Kavitha, BGS Institute of Technology, Adichunchanagiri Univeristy, Mandya, India
Sri Yogi Kottala, Symbiosis institute of Business Management, Symbiosis International University, Pune, India

Predictive maintenance (PdM) has become an essential set of approaches to the optimization of the work and timeline of the infrastructure systems. Through the application of machine learning (ML) techniques, PdM systems are also capable of estimating equipment failures, which decreases both maintenance costs and downtime and increases system reliability. The chapter presents an aspect of different ML algorithms involved in PdM, such as supervised, unsupervised, and deep learning. Is covered in applications to transportation systems, energy systems, water treatment systems, and building systems, and how ML can be used to infer the future behavior of systems to anticipate failures, enhance efficiency in operations, and reduce unpredictable maintenance. Major complications like data quality, readability of models and scalability, are tackled and highlights of how they could be solved are given. The chapter also introduces the case studies of successful deployment of ML in the real-world infrastructure systems presenting quantitative findings and perspectives of future trends.

Chapter 14
Probabilistic Modeling and Risk Assessment 351
Mustafa Kamil Rasheed, Mustansiriyah University, Iraq
Ahmed Hadi Salman, Mustansiriyah University, Iraq
Huda Mohammed Salman Alqayse, Educational and Psychological Research Center, University of Baghdad, Iraq
Alnoman M. Mundher Tayyeh, Middle Technical University, Iraq

Probabilistic modeling is a fundamental method for assessing uncertainty and analyzing risk in complex systems. This chapter provides a comprehensive review of probabilistic techniques used in risk assessment, highlighting the underlying principles, modeling strategies, and practical applications in the environment and economy. Emphasis is placed on integrating statistical inference with simulation-based modeling to enhance decision-making under uncertain conditions. The chapter concludes with future insights into the potential of integrating probabilistic models with artificial intelligence and data-driven systems.

Chapter 15
Smart Sensor Technologies 365
Nithin Subba Rao, Michigan Technological University, USA
P. Selvakumar, Department of Science and Humanities, Nehru Institute of Technology, Coimbatore, India
S. Madhu, Jyothy Institute of Commerce and Management, India
Naeema Nazar, VISAT Engineering College, Ernakulam, India
V. Padmashree, Padmashree Institute of Management and Sciences, India
M. R. Arun, Vel Tech Rangarajan Dr. Sagunthala R&D Institute of Science and Technology, India
T. C. Manjunath, Rajarajeswari College of Engineering, India

In the era of digital transformation and intelligent systems, smart sensors have emerged as a cornerstone technology, significantly enhancing the way data is collected, processed, and utilized across various industries. Traditional sensors were primarily responsible for converting physical stimuli—such as temperature, pressure, motion, light, or sound—into electrical signals for further processing. However, with increasing demands for more accurate, autonomous, and intelligent decision-making systems, traditional sensors have evolved into smart sensors—devices that combine sensing, signal processing, and communication capabilities within a single integrated package. Smart sensors represent a convergence of sensing elements, microprocessors, software algorithms, and communication technologies. These devices are designed not only to sense environmental parameters but also to analyze data locally, make decisions, and transmit processed information to other systems or devices, often in real time.

Chapter 16
The Role of Technology in Economic Growth: An Analytical Study of a Sample of Gulf Countries for the Period 2010-2023 393
Mustafa Kamil Rasheed, Mustansiriyah University, Iraq
Munaf Marza Neama, Mustansiriyah University, Iraq
Rawaa Salh Mahmmed Al-Saffar, Mustansiriyah University, Iraq
Alnoman M. Mundher Tayyeh, Middle Technical University, Iraq

The development of the global economy has implicitly necessitated the advancement of the knowledge element, thereby strengthening the technological structure in all areas and branches of the local economy. This has enhanced the local economy's resilience and made it more resistant to international crises and shocks. Developing countries in general, and the Gulf countries in particular, suffer from a weakness in the field of technological development based on academic knowledge that supports productive capabilities in all sectors of local production. This has been represented in the limited economic performance over time. The research assumes that technology based on academic scientific development in research and development makes a significant contribution to improving economic growth. The research methodology relied on several economic indicators related to the research relationship for the period (2010-2023)

Chapter 17
Wireless Sensor Networks for Real-Time Infrastructure Monitoring and Management Using ML Approach 409
Gajanan Shankarrao Patange, Charotar University of Science and Technology, India
Abdulsattar Abdullah Hamad, Department of Physics, College of Education, University of Samarra, Iraq
V. Dankan Gowda, BMS Institute of Technology and Management, India
S. Gobinath, V.S.B. Engineering College, Karur, India
K. Ramya, BGS Institute of Technology, Adichunchanagiri University, Karnataka, India

The growing interest in Wireless Sensor Networks (WSNs) lies in the real time monitoring and management of infrastructures. As the critical infrastructure systems become more complex and age, the conventional monitoring systems have been incapable of addressing the management needs of a timely and competent management. WSNs are a scalable and effective solution that allows to gather real-time data (related to structural health, surrounding environment, flow of traffic, and other vital parameters required in maintaining infrastructure health and safety). This chapter looks into the basic concepts of WSN, its uses in infrastructure system and design aspects necessitated in deploying such systems successfully. Barriers and shortcomings of the existing WSN technologies are presented through case study and performance indicators, coupled with aspects of future breakthroughs in this direction. The focus is made on the interoperability of WSNs and the related emerging technologies to ensure the increase in predictive maintenance and others related to decision-making.

Compilation of References 435

About the Contributors 475

Index 485

Preface

As Editor of this volume, I am pleased to present *Integrating Modern Mathematics and Sensor Technologies in Civil Engineering*—a comprehensive reference that emerges at a pivotal moment for the civil engineering profession. The discipline is undergoing a profound transformation, fueled by the fusion of mathematical modeling, advanced sensing technologies, and intelligent data analytics. As our infrastructure systems face growing complexity and stress from rapid urbanization, climate change, and finite resources, there is a clear and urgent demand for integrated, adaptive solutions that go beyond traditional engineering practice.

This book was conceived to meet that demand. It offers a wide-ranging scientific exploration of how the convergence of mathematics and modern sensor technologies is reshaping civil engineering theory and application. From foundational mathematical tools—such as differential equations, finite element methods, and probabilistic models—to state-of-the-art sensing technologies capable of capturing and transmitting real-time structural data, this volume charts the evolution toward intelligent, data-driven infrastructure systems.

Historically, civil engineering has progressed from reliance on empirical rules to the incorporation of sophisticated analytical frameworks. Today, with the rise of high-resolution sensor arrays and digital twin technology, we stand on the threshold of infrastructure that is not only designed with mathematical precision but also capable of self-monitoring, self-assessing, and, in time, self-adapting. This integration allows for real-time monitoring, advanced fault prediction, and adaptive structural control—enhancing the safety, efficiency, and resilience of modern infrastructure.

The book is organized into fifteen chapters, each authored by experts in their respective fields. The initial chapters lay the theoretical groundwork, delving into advanced mathematical models and numerical analysis techniques essential for structural design. Following these are in-depth examinations of sensor technologies and their implementation across various civil engineering domains, including seismic sensing, geotechnical instrumentation, and smart materials. Later chapters venture into forward-looking applications such as artificial intelligence, digital twins, and even quantum-enhanced computational methods.

This work aspires to serve as both a foundational and forward-thinking reference. It is designed for a diverse audience: civil engineers, structural analysts, graduate students, researchers in intelligent systems and artificial intelligence, data scientists, and infrastructure consultants. Our goal is to bridge the gap between mathematical theory and engineering practice, equipping professionals and scholars alike with the tools needed to respond innovatively to the infrastructure challenges of the 21st century.

By showcasing the powerful synergy between mathematics and sensing technology, this book underscores the importance of precision, innovation, and adaptability in building the smart, resilient infrastructure of the future. It is my hope that this volume will inspire continued research, cross-disciplinary collaboration, and practical implementation that advances the frontiers of civil engineering.

ORGANIZATION OF THE BOOK

This volume comprises seventeen chapters, each offering a unique perspective on the convergence of mathematics, sensing technologies, and intelligent systems in civil engineering. Together, they provide a multidisciplinary lens into how emerging technologies are transforming infrastructure design, monitoring, and sustainability.

Chapter 1 introduces a novel computational approach through the Haar wavelet collocation method for solving inverse problems, using the Black-Scholes model as a case study. The chapter demonstrates the method's reliability and convergence in transforming backward parabolic partial differential equations into forward models, thus showcasing the power of mathematical transformations and numerical approximation in complex modeling scenarios.

In Chapter 2, the spotlight shifts to the application of Complex Intuitionistic Fuzzy Sets (CIFS) in engineering analytics and smart infrastructure. Addressing uncertainty, ambiguity, and imprecision in sensor data, CIFS provide a robust decision-making framework for structural health monitoring and smart city systems, emphasizing their value in environments characterized by complex, multidimensional data.

Chapter 3 lays the theoretical groundwork for Artificial Intelligence (AI) and Machine Learning (ML), elucidating core concepts and distinguishing between narrow and general AI. The chapter explores learning paradigms that empower machines to classify, predict, and adapt, forming the backbone of numerous intelligent civil engineering systems discussed throughout the book.

Chapter 4 presents a comparative evaluation of YOLOv7, YOLOv8, and YOLOv9 deep learning models for pothole detection on Indian roads. With road safety at stake, the study confirms the superior performance of YOLOv7 in real-time detection and classification, offering practical implications for ADAS (Advanced Driver Assistance Systems) and smart mobility infrastructure.

Chapter 5 explores the design and implementation of IoT-based sensing systems for real-time infrastructure monitoring. By replacing traditional manual inspections with automated sensor networks, this chapter addresses key considerations such as sensor selection, communication protocols, and power optimization, paving the way for smarter and more efficient monitoring solutions.

In Chapter 6, readers are introduced to adaptive and self-responsive structures capable of altering their physical state based on environmental inputs. The chapter blends theoretical design principles with real-world case studies, such as morphing structures and self-tuning absorbers, and sets forth a roadmap for scalable, energy-efficient, and resilient infrastructure systems.

Chapter 7 advances the discussion by examining autonomous infrastructure inspection and maintenance systems enabled by ML and IoT. With drones, robotic systems, and predictive algorithms, infrastructure health can now be assessed continuously and cost-effectively, as demonstrated in case studies involving bridges, pipelines, and railways.

Chapter 8 surveys the global landscape of emerging research and technological foresight, identifying trends in AI, renewable energy, and big data. It underscores the disparity in research infrastructure between developed and developing countries and calls for strategic investments in open innovation and cross-sector collaboration to ensure inclusive growth.

Chapter 9 delves deeper into the integration of UAVs and robotics for infrastructure maintenance. The ability to inspect hard-to-reach structures via aerial or ground-based robotic systems offers new efficiencies, though the chapter also acknowledges regulatory and technological challenges that must be addressed for widespread adoption.

Chapter 10 offers a broad view of IoT applications in civil engineering, from traffic management and smart construction to predictive maintenance and environmental monitoring. It highlights the transformative role of embedded sensors and anticipates future developments involving AI, digital twins, and blockchain to realize fully autonomous infrastructure systems.

Chapter 11 introduces a hybrid framework for Lifecycle Assessment (LCA) and structural degradation modeling, focusing on sustainability and long-term performance. The integration of AI and real-time monitoring enables dynamic modeling of deterioration mechanisms, such as corrosion and fatigue, aligning with environmental impact assessments.

Chapter 12 focuses on ML algorithms for sensor data processing, addressing challenges such as noise, high dimensionality, and data incompleteness. The chapter evaluates various supervised, unsupervised, and deep learning models, providing examples from health monitoring, environmental sensing, and predictive infrastructure analytics.

Chapter 13 extends this discussion with a targeted look at predictive maintenance (PdM) in infrastructure systems. Through real-world case studies and algorithmic analysis, the chapter demonstrates how ML can forecast failures, reduce downtime, and optimize maintenance scheduling in sectors like energy, transportation, and water treatment.

In Chapter 14, the book revisits foundational mathematical theory with a comprehensive treatment of probabilistic modeling and risk assessment. Integrating statistical inference and simulation-based methods, the chapter explores how probabilistic tools support decision-making in uncertain environments, with forward-looking insights into AI-assisted risk modeling.

Chapter 15 examines smart sensor technologies as integrated systems that go beyond traditional sensing. These advanced devices process and transmit data autonomously, enabling real-time decision-making. The chapter outlines the architecture of smart sensors and their role in bridging physical environments with digital control systems.

Chapter 16 offers a macroeconomic perspective, analyzing the role of technology in economic growth across Gulf countries from 2010 to 2023. The chapter links academic research and technological advancement with national productivity, highlighting the need for stronger R&D policies to boost innovation and resilience in developing economies.

Finally, Chapter 17 presents the evolution and deployment of Wireless Sensor Networks (WSNs) for infrastructure monitoring. It details the architecture, design considerations, and limitations of WSNs, while exploring their synergy with machine learning and predictive analytics in enabling real-time, scalable infrastructure management.

CONCLUSION

As civil engineering moves deeper into the digital age, the fusion of mathematical modeling, advanced sensing technologies, and intelligent systems is no longer a distant vision—it is a present reality reshaping every aspect of the field. The chapters presented in this volume collectively reflect a paradigm shift from reactive to proactive infrastructure management, from static design to adaptive systems, and from isolated solutions to integrated, data-driven frameworks.

This book has brought together leading researchers and practitioners from across disciplines to explore the transformative potential of integrating modern mathematics with sensor technologies. From wavelet methods and fuzzy systems to smart materials, UAVs, and IoT-based sensing networks, the content illustrates the breadth and depth of innovation driving the future of civil infrastructure. Moreover, by embedding artificial intelligence, machine learning, and probabilistic models into civil engineering practice, we are building the foundation for infrastructure that is not only intelligent but also resilient, sustainable, and responsive to an increasingly dynamic world.

It is my hope that this volume serves not only as a comprehensive academic reference but also as a catalyst for further research, collaboration, and real-world application. Whether you are a scholar seeking theoretical frameworks, a practitioner looking for implementation strategies, or a policymaker exploring the implications of smart infrastructure, this book aims to provide insights, tools, and inspiration.

As Editor, I extend my deepest thanks to the contributing authors for their expertise and commitment, and to the readers who continue to push the boundaries of what civil engineering can achieve. The future of infrastructure lies in our collective ability to bridge knowledge with technology—and this book is a step in that direction.

Acknowledgment

I extend my sincere thanks and appreciation to all the colleagues, researchers, and specialists whose efforts and ideas contributed to the completion of this book, "Integrating Modern Mathematics and Sensing Technologies in Civil Engineering." I also extend my sincere gratitude to the co-authors for their valuable chapters, to the reviewers for their constructive comments, and to the academic institutions for their continuous support during the preparation of this work.

I also thank the engineers and scientists whose pioneering research in mathematics and sensing technologies inspired many of the ideas presented in this book. Their commitment to innovation in civil engineering was the primary driver behind this vision.

Finally, I would like to thank the technical and editorial teams at IGI Global for their professional guidance and continuous support throughout the publication and preparation phases.

Chapter 1
A Haar Wavelet Collocation Method for Inverse Problem: Application to Black-Scholes Model

Muhammad Ahsan
https://orcid.org/0000-0001-5203-4577
Harbin Institute of Technology, Shenzhen, China & University of Swabi, Pakistan

Muhammad Riaz
University of Swabi, Pakistan

Masood Ahmed
School of Mathematics and Statistics, Central South University, Changsha, China

Maher Alwuthaynani
Taif University, Saudi Arabia

Weidong Lei
Harbin Institute of Technology, Shenzhen, China

ABSTRACT

This article demonstrates the Haar wavelet collocation method as an effective strategy for solving the Black-Scholes model with various types of options. The Black-Scholes model is an inverse problems and formulated as a backward parabolic partial differential equation. In the first stage, a transformation is applied to convert this backward Black-Scholes model into a forward partial differential equation. Finite

DOI: 10.4018/979-8-3373-5248-0.ch001

difference method is then employed to estimate the changes over time in the partial differential equation. Subsequently, the partial differential equation is converted into an algebraic system of equations by applying the Haar series approximation to the spatial variable. Extensive analyses were conducted to evaluate the correctness, effectiveness, and convergence of the Haar wavelet collocation method. Various scenarios in Black-Scholes model were tested by considering various options. The computational results were aligned with theoretical expectations, demonstrating the reliability of the method in handling diverse cases of the Black-Scholes model.

INTRODUCTION

Any mathematical equation that involves an unknown function along with its different-order derivatives, having one independent variable or more, is called a differential equation. Simply, it explains how a function changes in response to changes in its input variables. In many different fields, including physics, engineering, economics, and biology differential equations are used to represent a wide variety of phenomenon. The solutions to differential equations can describe how systems behave over time and aid in making predictions about the future (in the long term). A variety of techniques are used to handle these equations, such as the separation of variables and integrating components. However, it is complicated and sometimes not possible to calculate the solution of nonlinear types of equations. Therefore, alternative approaches include the use of numerical techniques to solve differential equations.

Differential equations are generally classified into two main categories based on the relation between the variables: Ordinary differential equation (ODEs) and partial differential equations (PDEs). ODEs typically involve functions of single variable while PDEs involve functions of multi-variables. In PDEs, there are multiple independent variables and one or more dependent variables in the equations. These equations are used to model physical processes where the rate of change of a quantity depends on the values of other relevant factors.

Black and Scholes introduced the Black-Scholes model (BSM) in 1973, which is a specific type of PDEs. Robert Merton played a significant role in advancing BSMs along with the Black and Scholes, and was awarded the Nobel Prize in Economics in 1997 for their innovative contribution to the field. The BSM is used to analyze the current value of a put option, based on certain assumptions about underlying asset, such as the dividend yield, volatility, stock price, time to expiration and risk-free interest rate. The BSM operates under the basis that stock prices show random walks without a visible pattern, and the returns of the financial asset be traditional to a log-normal distribution. It adopts that there is no short selling, no restrictions, no transaction costs and no taxes. The introduction of the BSM has extremely transformed the background of options trading; offering a method to precisely determine option prices and effectively manage risk through hedging strategies. It has become a standard tool for traders and investors around the world, and it has been widely adopted in academia and industry.

Despite its many advantages, the BSM has some limitations. It assumes that the financial asset shows a log normal distribution. The assumption of the log normal distribution, and may not consistently make parallel with real-world situation. Additionally, the model assumes a constant unpredictability of the underlying asset and a fixed risk-free interest rate over time, a condition that may not accurately reflect the dynamic nature of these factors in actuality. As a result, there have been many extensions and modifications to the original model over the years to address these issues and to make it more applicable in real-world situations.

A mathematical formula known as the Black-Scholes option pricing model can be used to determine an options current value. Here are some uses of the BSMs:

1. **Option valuation:** The BSM's primary utilizes its ability to calculate the present value of a put option. This model takes into account a variety of characteristics, including the strike price, stock valuation, volatility, time before expiration, and interest rates.
2. **Option trading:** Traders typically use the BSM to help their decision-making in options trading. By calculating an option's present value and comparing it to the current market price, traders can determine if the option is over-estimated or under-estimated.
3. **Risk Management:** Financial organizations utilize the BSM to manage their option-related risk. Using this BSM, they may calculate possible losses from their option contracts and change their stock and bonds accordingly.
4. **Corporate Finance:** The BSM is widely used in corporate finance to determine the value of employee stock options. This is crucial since employ stock options (ESOs) are sometimes a large part of chief executive officer (CEO) compensation packages, and their valuation can have a major impact on a company's financial statements.

Overall, the BSM has become a fundamental tool in the world of finance and investing, and its applications extend far beyond the realm of options trading.

Mathematically the BSM can be presented as (Ahmad, 2016):

$$\frac{\partial \Psi}{\partial} + \frac{1}{2}\sigma^2 S^2 \frac{\partial^2 \Psi}{\partial S^2} + (r - D)S\frac{\partial \Psi}{\partial S} - r\Psi = 0, \quad < T, S \in R. \tag{1.1}$$

The space variable S represents the financial asset price, r shows the rate of interest that is free from risk. In the given context, "D" represents the payment made by the asset, " σ" denotes the volatility of the financial asset, and " $\Psi(S,)$" stands for the evaluation of the option value before the expiration time "T," subject to the specified conditions. Along with equation (1.1) different conditions known as options can be used. Here we can explain these options one by one.

Put option:

$$\Psi(S,T) = \max(E - S,0), S \geq 0, \tag{1.2}$$

$$\Psi(0,) = E, \tag{1.3}$$

$$\lim_{S\to\infty}\Psi(S,) = 0. \tag{1.4}$$

Here, equation (1.2) represents final time condition while equation (1.3) and (1.4) represent boundary conditions of put option, where E is strike price.

Call option:

$$\Psi(S,T) = \max(S - E,0), S \geq 0, \tag{1.5}$$

$$\Psi(0,) = 0 , S = 0, \tag{1.6}$$

$$\lim_{S\to\infty}\Psi(S,) = Ee^{-r(T-t)}. \tag{1.7}$$

The equation (1.5) represents final time condition while equation (1.6) and (1.7) represent boundary conditions of call option, where E is strike price.

Butterfly spread call option:

For a given component, equation (1.1) is called a butterfly spread call option, and it consists of a group of multiple options (E_1, E_2, and E_3) with varying strike prices. The single asset choice (1.1) is expressed using a pay-off function that is comparable to,

$$\Psi(S,T) = \max(S-E_1,0) - 2\max(S-E_2,0) + \max(S-E_3,0), \tag{1.8}$$

where $E_1 < E_2 < E_3$ & $E_2 = \frac{E_1 + E_3}{2}$,

$$\Psi(0,) = 0, \text{ at S } = 0, \tag{1.9}$$

$$\lim_{S\to\infty}\Psi(S,) = 0. \tag{1.10}$$

Here also equation (1.8) represents final time condition while equation (1.9) and (1.10) represent boundary conditions of call option.

Butterfly spread put option:

For a certain component and consisting of a group of separate options with distinct strike prices (E_1, E_2, and E_3), equation (1.1) is known as a butterfly spread put option. We use a pay-off function akin to this to express the single asset choice (1.1).

$$\Psi(S,T) = \max(E_1 - S,0) - 2\max(E_2 - S,0) + \max(E_3 - S,0), \tag{1.11}$$

where $E_1 < E_2 < E_3$ & $E_2 = \frac{E_1 + E_3}{2}$,

$$\Psi(0, \ ) = 0, \text{ at S} = 0 \tag{1.12}$$

$$\lim_{S\to\infty}\Psi(S, \ ) = 0 \tag{1.13}$$

Equation (1.11) represents final time condition while equation (1.12) and (1.13) represent boundary conditions of put option.

Digital call option:

Equation (1.1) is referred to as a digital call option for a specific component, an options contract with a fixed payoff in the asset that the underlying asset climbs above the strike price which is digital option. By tracking changes in share prices and the strike price, option holders purchase and sell shares. The holder purchased the call option if the market price of the shares at maturity is more than or equal to the strike price. Rather, the holder purchased the put option if the market share price at maturity is less than or equal to the strike price. By using a payoff function to solve the BSM, the payoff function for digital call option are given as

$$\Psi(S,T) = \begin{cases} 0 \, for \, S < E, \\ 1.5 \, for \, S = E, \\ 1 \, for \, S > E, \end{cases} \tag{1.14}$$

$$\Psi(0, \ ) = 0, \tag{1.15}$$

$$\lim_{S\to\infty}\Psi(S, \ ) = Ee^{-r(T- \ )}. \tag{1.16}$$

Here, equation (1.14) represents final time condition while equation (1.14) and (1.15) represent boundary conditions of call option for digital option,

RESEARCH GAPS

Some of the major issues, which are congenital to BSM are given below.

a) The information $\Psi(S, T)$ is given at =T, instead of = 0. So, this problem is backward in time, which is not easy to solve by every method.
b) To develop an accurate numerical solver for BSM having different options.
c) To develop a method, which can easy capture the accurate information on the basis of different put or call options.
d) To design a converging numerical method to solve the BSM.
e) The professional and efficient numerical solver for the BSM.

Objectives

The primary contribution of this research is addressing the identified gaps by achieving the following key objectives in the current work.

a) To adopt a suitable transformation that can convert the forward problem into backward direct problem.
b) To implement the wavelet-based collocation method to obtain acceptable and efficient results for the BSM with various put or call options.

A scientific description of a financial market with particular derivative investment a specific tool is the BSM. The BSM clarify the valuation of an option by totally defining the price of the underlying stock. The assertion that stock price changes are influenced through a stochastic process is the foundation of a mathematical model known as the BSM. Although it is widely applied as a useful approximation, one must be aware of its limitations for best utilization. Consequently, many non-linear BSMs have recently been suggested (Ali *et al.*, 1992; Ahsan *et al.*, 2023). Finding complete accurate results for the non-linear BSM is, however, remarkably difficult, therefore many numerical techniques based on difference technique have been introduced to address these nonlinear issues. The precision of the numerical approach dependent on the time step (Aziz *et al.*, 2015). Therefore, it should be notice carefully because they offer more precise assessments by considering much reasonable statements, such as risks from an unprotected collection, transaction costs, large shareholder preferences, fluctuations in volatility, stock prices, and the option value that might be unfair. In recent years, there has been growing attention toward the study of linear BSM due to its more use than nonlinear case.

The study conducted by (Klein, 1996) presents a better approach for valuing disposed BSM, considering assumptions suitable for various business scenarios.

The impacts of introducing a quantum physical environment into the Black-Scholes option pricing model are examined by (Haven *et al.*, 2012). The time fractional Black-Scholes equation is the main topic of (Wang *et al.*, 2014) research field, which also refers to the fractional derivative as the revised Riemann-Liouville fractional derivative.

In recent years, Haar wavelets have been successfully used to solve a number of practical issues. Aziz *et al.,* in 2016 presented numerical solutions of various type of differential equations using the Haar wavelet collocation method (HWCM). The approach is employed for solving both linear and non-linear differential equations, also including systems and delay differential equations. A numerical solution to 3-D Elliptic partial differential equations with Dirichlet boundary conditions is studied in (Siraj-ul-Islam *et al.*, 2013) by applying a collocation method that also makes use of the Haar wavelet. In (Saha & Gupta, 2014) employed efficient numerical methodologies fixed in the Haar wavelet method for obtaining mathematical solutions to non-linear Burger and Bossiness-Burger equations. The results of the numerical evaluation are then associated to the particular solutions.

A suitable HWCM was presented by (Lepik., 2006), for the first time, when he was solving nonlinear integral equation. The technique has also been applied to solve integro-differential equations and Volterra integral differential equation. In 2021, A. Ghafoor with his team members introduced another wavelet method for numerically solving non-linear boundary value problems (Ghafoor et al., 2019. The HWCMs has also been utilized to solve the non-linear Schrodinger equation (NLSE) and the numerical solution of NLSE is presented by (Khan *et al.*, 2024) using multi-resolution techniques based on Haar wavelets.

Golbabi and Nikan published their research work in 2020 (Golbabi et. al., 2020). It described black-scholes model by computational based on the moving least squares method which has some limitations such as based function and weight function. Both of these functions affect the accuracy and smoothness of the solution. The demerits in the meshless method is the shape parameter because it influences the accuracy and stability of the above-mentioned method (Ahmad, 2016).

The Main Objective of the Present Study

The main objective of this study is to solve the backward black Schole model. As it is very difficult to solve the backward problem, so we used a transformation that convert this backward problem into direct problem. Then the direct problem has been solved by Haar Wavelet Collocation Method. The Haar Wavelet Collocation Method is a suitable solver to these types of model. Another significant of the proposed method is second order convergence.

HAAR FUNCTIONS

Haar wavelets are the most fundamental of the different kinds of wavelets. The Haar wavelet family $h_i(S)$ in the domain $[a, b]$ is define as a group of square waves having magnitude ± 1. The Haar functions are detailed in a broader context as follows:

$$h_i(S) = \begin{cases} 1, for\ t_1(i) \leq S < t_2(i), \\ -1, for\ t_2(i) \leq S < t_3(i), \\ 0, elsewhere, \end{cases} \tag{3.1}$$

where

$$t_1(i) = a + \frac{k}{m}\left(b - a\right),\ t_2(i) = a + \frac{k + 0.5}{m}\left(b - a\right), \text{ and } t_3(i)$$

$$= a + \frac{k+1}{m}\left(b - a\right).$$

The parameter k, with different values like, $k = 0,1,2....(m-1)$, indicates the translation aspects of the wavelets. For each $m = 2^j, j = 0,1,2,......\infty$ represent the movement of wavelet and it means to control the wavelet's dilation. The value of i is k+m+1, and it should be noted that i is always bigger than 2.

So for $i = 1$, one have

$$h_1(S) = \begin{cases} 1, for\ a \leq S \leq b, \\ 0, elsewhere. \end{cases}$$

The wavelet mentioned above is commonly referred to as the mother wavelet and is critical to our investigation. for all $i \geq 2$, every Haar function has a single square wave, with the exception of the scaling function (Hsiao, 2004). To simplify our derivations, we will adopt the subsequent certain symbols for integrals of $h_i(S)$, like

$$P_{i,1}(S) = \int_a^S h_i(S')dS',$$

$$p_{i,1}(S) = \begin{cases} 0\ for\ S < t_1(i), \\ S - t_1(i)\ for\ t_1(i) \leq S < t_2(i), \\ -S - t_1(i) + 2t_2(i)\ for\ t_2(i) \leq S < t_3(i), \\ -t_1(i) + 2t_2(i) - t_3(i)\ for\ t_3(i) \leq S < 1. \end{cases} \tag{3.2}$$

$$P_{i,2}(S) = \int_a^S p_{i,1}(S')dS',$$

$$p_{i,2}(S) = \begin{cases} 0 \, for \, S < t_1(i), \\ \frac{1}{2}(S - t_1(i))^2 \, for \, t_1(i) \leq S < t_2(i),, \\ \frac{1}{2}[(S - t_1(i))^2 - 2(S - t_2(i))^2] \, for \, t_2(i) \leq S < t_3(i), \\ \frac{1}{2}[(S - t_1(i))^2 - 2(S - t_2(i))^2 + (S - t_3(i))^2] \, for \, t_3(i) \leq S < 1. \end{cases} \quad (3.3)$$

and

$$C_i = \int_a^b p_{i,2}(s')ds' = \frac{(b-a)^2}{4m^2}$$

. (3.4)

Although $p_{i,1}(S)$ follows an ascending pattern for $t_1(i) \leq S < t_2(i)$ and a declining trend for $t_2(i) \leq S < t_3(i)$, its peak is always at $t_2(i)$. Consequently

$$max_s p_{i,1}(S) = p_{i,1}\left(t_2(i)\right) = \frac{b-a}{2m}.$$

The research by Majak *et al.* (Majak *et al.* 2015), established the following formula, and it is important to draw attention to this noteworthy reference where this has also been demonstrated that $max_s p_{i,2}(S) = \frac{(b-a)^2)}{4m^2}$.

NUMERICAL METHOD

Using the transformation $t = T -$ the equation (1.1) can be transformed into the following direct problem

$$\frac{\partial \Psi}{\partial t} = \frac{1}{2}\sigma^2 S^2 \frac{\partial^2 \Psi}{\partial S^2} + (r - D)S\frac{\partial \Psi}{\partial S} - r\Psi, t < T, S \in R. \quad (3.5)$$

With this transformation, the final conditions (1.2) can be transformed to the following initial condition

$$\Psi(S,0) = \max(E - S,0), S \geq 0.$$

The boundary condition becomes

$$\Psi(0,t) = 0, \lim_{S\to\infty}\Psi(S,t) = Ee^{-r(T-t)}.$$

Similarly, the other final condition and boundary condition can be easily in all options.

First, we use the forward difference method, so equation (3.5) gives

$$\frac{[\Psi]^{n+1}-[\Psi]^{n}}{\Delta t}=\frac{1}{2}\sigma^2 S^2\left[\frac{\partial^2\Psi}{\partial S^2}\right]^{n+1}+(r-D)S\left[\frac{\partial\Psi}{\partial S}\right]^{n+1}-r[\Psi]^{n+1}, \tag{3.5a}$$

Where n is used for time iteration. In the current work a HWCM will be used to calculate the approximate solution of equation (3.5a). We estimated the space derivative of equation (3.5a) by Haar function:

$$\frac{\partial^2\Psi}{\partial S^2}=\sum_{i=1}^{2M}a_i h_i(S). \tag{3.6}$$

Partially integrating equation (3.6) with respect to S, from a to S, we get

$$\frac{\partial\Psi(S,t)}{\partial S}=\frac{\partial\Psi(a,t)}{\partial S}+\sum_{i=1}^{2M}a_i P_{i,1}(S). \tag{3.7}$$

Again, partially integrating equation (3.7) with respect to S, from a to b, we get

$$\frac{\partial\Psi(a,t)}{\partial S}=\frac{\Psi(b,t)-\Psi(a,t)}{b-a}-\sum_{i=1}^{2M}a_i\frac{C_i}{b-a} \tag{3.8}$$

Putting equation (3 . 8) in equation (3 . 7), we get

$$\frac{\partial\Psi(S,t)}{\partial S}=\frac{\Psi(b,t)-\Psi(a,t)}{b-a}+\sum_{i=1}^{2M}a_i\left(P_{i,1}(S)-\frac{C_i}{b-a}\right). \tag{3.9}$$

Similarly integrating equation (3 . 5) with respect to s, from a to s, we get

$$\Psi(S,t)=\Psi(a,t)+(S-a)\frac{\Psi(b,t)-\Psi(a,t)}{b-a}+\sum_{i=1}^{2M}a_i\left(P_{i,2}(S)-\left(S-a\right)\frac{C_i}{b-a}\right). \tag{3.10}$$

By utilizing equation (3. 6)-(3. 10) in (3.5a), it can be converted into algebraic equations, which can be easily calculated to find a_i. After that we can used a_i in equation (3.10) to get the required approximate solution.

CONVERGENCE ANALYSIS

Theorem 1. Let the partial derivatives $\frac{\partial \Psi}{\partial S}, \frac{\partial^2 \Psi}{\partial S^2}, \frac{\partial \Psi}{\partial t}$ and $\frac{\partial^2 \Psi}{\partial t^2}$ exist and remains bounded within the time and space domain $[a,b] \times [0,T]$. Here, if $\Psi_E(S,t)$ and $\Psi_M(S,t)$ represent the exact and numerical solution obtained through HWCM respectively, also we have $M = 2^j$, *where* $j = 0,1,2,\ldots\ldots$, and $t = 0,1,2,\ldots\ldots,T$ then

$$\| \Psi_E(S,t) - \Psi_M(S,t) \|_\infty = O\left(\frac{1}{M}\right)^2 \text{ as } j \to \infty .$$

Proof: See (Ahsan *et al.*, 2023).

Stability Analysis

Using the proposed technique, here the HWCM, the system of algebraic equations derived in section (3.2) can be expressed in matrix format for the aforementioned problem as follows:

$$H^* X = B_\circ .$$

Where H^* is the matrix of Haar function, X is the unknown matrix of Haar coefficients, and $B_\circ$ is the matrix obtained from right side values. We use the following definition to test the computational stability of the suggested HWCM.

Definition 1: Consider a system of algebraic equations $H^* X = B_\circ$ generated by a numerical approach when solving a PPDE. The approach will be stable if the inverse of H^* exists and satisfy

$$\|H^{*-1}\| \leq \propto_\circ ,$$

where $\propto_\circ$ is some constant.

The eigenvalues of H^{*-1} were numerically calculated to determine the minimum value of magnitude of H^*, which also reflects the spectral radius magnitude of H^{*-1}. The analysis of figures 4.5, 4.6, 4.11, 4.12, 4.15, and 4.16, shows that the values are bounded and follow the statement of definition 1. The 2-norm of H^{*-1} does not increase rapidly. Therefore, the proposed approach HWCM is stable.

RESULTS AND DISCUSSION

This chapter deliberates the results of HWCM and evaluates its performance for BSM with three various options. To test its accuracy, we use the L_∞ error norm, defined as follows:

$$L_\infty = \max \| \Psi_E(S,t) - \Psi_M(S,t) \|.$$

The reported results are calculated using the "MATLABR2015b" program and installed on a Lenovo Laptop (4.0 GB RAM). The calculation times are noted in "seconds," which represent the amount of CPU time required to run the program

Test Problem 1: The first problem considers a digital call option with a single asset under three scenarios:

- Case (a): A digital call option with a single asset having the spatial domain [0, 1], dividend D = 0, exercise price E = 0.5, maturity time T = 0.25 years, interest rate r = 0.05, and volatility ρ = 0.2.
- Case (b): A digital call option with a single asset having the spatial domain [0, 1], dividend D = 0, exercise price E = 0.5, maturity time T = 0.5 years, interest rate r = 0.05, and volatility ρ = 0.2.
- Case (c): A digital call option with a single asset having the spatial domain [0, 1], dividend D = 0, exercise price E = 0.5, maturity time T = 1 year, interest rate r = 0.05, and volatility ρ = 0.2.

Table 1-3, shows the error, computation time and convergence rate for case (a) to case (c), where in all these table the convergence is almost of second order. In Figure 4.1, the stability of method is presented.

Figure 1. The spectral radius and 2-norm of the inverse of matrix taking Δt=0.01 for Test Problem 1.

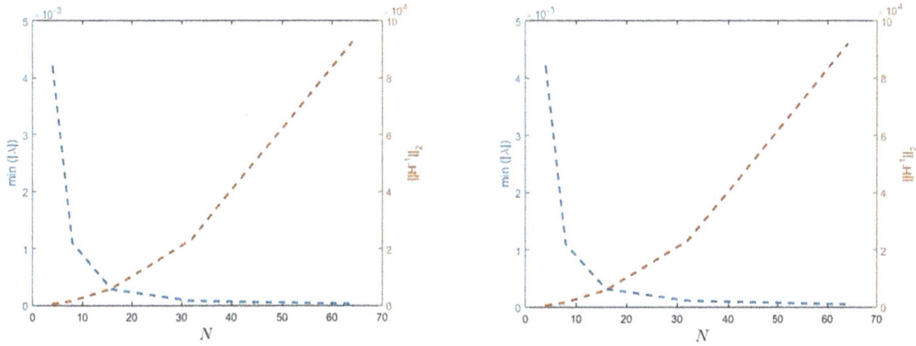

Δt=0.01　　　　　　Δt=0.001

Test Problem 2: Three scenarios for a European put option are considered:

- Case (a): Single asset with spatial domain [0, 30], volatility ρ = 0.2, strike price E = 10, dividend D = 0, interest rate r = 0.05, and maturity time T = 0.5 year.
- Case (c): Single asset with spatial domain [0, 4], volatility ρ = 0.2, strike price E = 10, dividend D = 0, interest rate r = 0.05, and maturity time T = 1 year.

Table 1. The maximum absolute error, convergence rate (CR), and CPU time for different values of M, considering Δt=0.0001 and T=0.25 for **Test Problem 1 (Case (a)).**

M	L_∞	Rate of convergence Experimented	Time in second CPU
2	2.7369×10^{-2}	----------	1.5471
4	1.7926×10^{-2}	1.4214	5.1800
8	6.6445×10^{-3}	1.4318	19.9382
16	1.6648×10^{-3}	1.9968	77.1384
32	4.5108×10^{-4}	1.8839	313.4003
64	1.1363×10^{-4}	1.9890	615.4041

Table 1 shows the error, computation time and convergence rate for case (a), where the convergence is almost of second order. In Figure 2, The numerical solution, absolute errors, spectral radius and 2-norm of the inverse of matrix are calculated at $\Delta t=0.01$ and T=1.

Table 2. The $L\infty$ error, CR and CPU time for different values of M taking $\Delta t=0.1$ and T=0.5 for Test Problem 1 (Case (b))

M	L_∞	**Rate of convergence (Experimental)**	**Time in second (CPU)**
2	1.7884×10^{-2}	----------	2.1240
4	1.9714×10^{-2}	1.4055	10.1842
8	4.3638×10^{-3}	2.1735	38.0961
16	1.1791×10^{-3}	1.8899	152.0177
32	3.0441×10^{-3}	1.9535	610.2013
64	1.3308×10^{-3}	1.1937	2009.702

Table 3. The maximum absolute error, convergence rate (CR), and CPU time for different values of M, considering $\Delta t=0.0001$ and T=1 for **Test Problem 1 (Case (c))**.

M	L_∞	**Rate of convergence (Experimental)**	**Time in second CPU**
2	9.5671×10^{-2}	----------	0.6263
4	3.4557×10^{-2}	1.4064	2.0900
8	1.3130×10^{-2}	1.5130	9.6720
16	2.7901×10^{-3}	2.2344	38.0710
32	7.5442×10^{-4}	1.8869	160.6571
64	1.9208×10^{-4}	1.9736	600.0928

Table 4. The maximum absolute error, rate of convergence, and CPU time for different values of M, taking $\Delta t=0.001$ and T=0.5, for Test Problem 2 (Case a).

M	L_∞	**Rate of convergence (Experimental)**	**Time in second (CPU)**
2	6.6686×10^{-3}	----------	3.0504
4	2.9417×10^{-3}	1.1807	8.7861
8	8.8505×10^{-4}	1.7328	32.4054
16	2.2926×10^{-4}	1.9488	127.2001
32	5.7863×10^{-5}	1.9863	536.2377

Figure 2. The numerical solution, absolute errors, spectral radius and 2-norm of the inverse of matrix taking Δt=0.01 and T=1, for test problem 2 (case c)

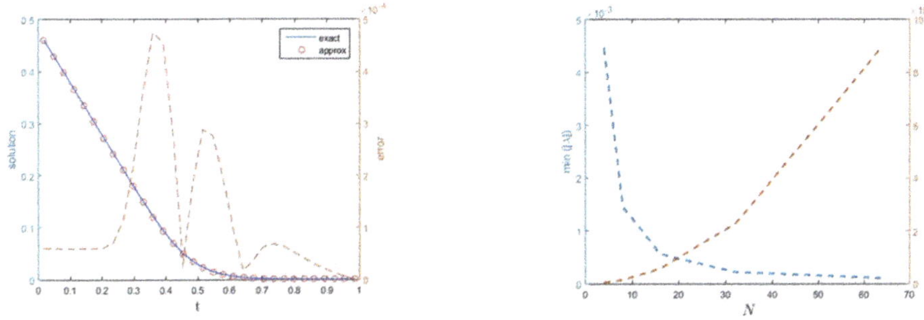

Test Problem 3: Three scenarios for butterfly call options with $S_{min} = 0$ and $S_{max} = 1$:

- Case (a): Volatility ρ = 0.5, dividend D = 0, interest rate r = 0.1, strike prices $E_1 = 0.4$, $E_2 = 0.5$, $E_3 = 0.6$, and maturity period T = 1 year.
- Case (b): Volatility ρ = 0.5, dividend D = 0, interest rate r = 0.1, strike prices $E_1 = 0.4$, $E_2 = 0.5$, $E_3 = 0.6$, and maturity period T = 0.25 year.
- Case (c): Volatility ρ = 0.5, dividend D = 0, interest rate r = 0.1, strike prices $E_1 = 0.4$, $E_2 = 0.5$, $E_3 = 0.6$, and maturity period T = 2 years.

Table 4 and 3 show the point wise errors and computation time case (b) and case (c) respectively. In Figure 3, the numerical solution, spectral radius and 2-norm of the inverse of matrix are calculated at Δt=0.01 and T=1.

Figure 3. The numerical solution, spectral radius and 2-norm of the inverse of matrix taking Δt=0.01 and T=1, for test problem 3 (case a).

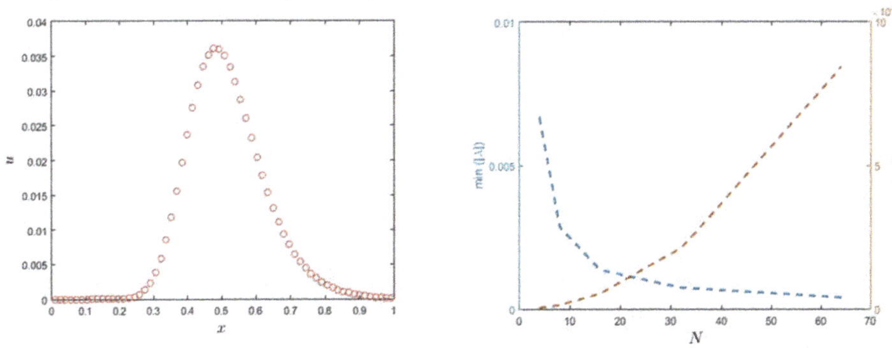

Table 5. The numerical point-wise error and CPU time for values of $\Delta t=0.1$, $M=16$ and $T=0.25$ Test Problem 3 (case b).

S	**Error**	**CPU time**
0.1	2.1027×10^{-2}	0.0819
0.2	2.1027×10^{-2}	0.0808
0.3	2.0954×10^{-2}	0.1096
0.4	5.8906×10^{-3}	0.0947
0.5	4.4191×10^{-2}	0.0794
0.6	4.4436×10^{-2}	0.0825
0.7	2.0051×10^{-2}	0.0893
0.8	2.0967×10^{-2}	0.0776
0.9	2.1022×10^{-2}	0.0989
1.0	2.1027×10^{-2}	0.0874

Table 6. The numerical point-wise error and CPU time for values of $\Delta t=0.1$, $M=16$ and $T=2$ Test Problem 3 (case c).

S	**Error**	**CPU time**
0.1	2.1027×10^{-2}	0.3657
0.2	2.0562×10^{-2}	0.3417
0.3	1.3219×10^{-2}	0.3437
0.4	3.6266×10^{-4}	0.3496
0.5	3.2563×10^{-3}	0.3792
0.6	3.8653×10^{-3}	0.3453
0.7	1.1704×10^{-2}	0.3618
0.8	1.6729×10^{-2}	0.3640
0.9	1.9448×10^{-2}	0.3665
1.0	2.1027×10^{-2}	0.3591

CONCLUSION AND RECOMMENDATIONS

This study provided Haar wavelet collocation method as a potential technique for numerically solving Black-Scholes model with various given options. The numerical scheme is successfully achieved by using the appropriate transformation to convert the Black-Scholes model into a direct PDE. This strategy shows to be valid for various sorts of options, including digital call, European and butterfly options. The L_∞ error norm and convergence rate demonstrate the validity of this technique, demonstrating its precision and applicability for various given options with Black-Scholes model.

The Haar wavelet collocation method potential needs to be further explored for various type of options. Hence, our ongoing research will be focus on expanding this technique to address two-dimensional and system of Black-Scholes model.

ACKNOWLEDGEMENT

The authors extend their appreciation to Taif University, Saudi Arabia, for supporting this work through project number (TU-DSPP-2025-66).

REFERENCES

Ahmad, I. (2016). A comparative analysis of local meshless formulation for multi-asset option models. *Engineering Analysis with Boundary Elements*, *65*, 159–176. DOI: 10.1016/j.enganabound.2015.12.020

Ahsan, M., Bohner, M., Ullah, A., Khan, A. A., & Ahmad, S. (2023). A Haar wavelet multi-resolution collocation method for singularly perturbed differential equati-ons with integral boundary conditions. *Mathematics and Computers in Simulation*, *204*, 166–180. DOI: 10.1016/j.matcom.2022.08.004

Ahsan, M., Shams-ul Haq, K., Liu, X., Ahmad Lone, S., & Nisar, M. (2023). A Haar wavelets based approximation for nonlinear inverse problems influenced by unknown heat source. *Mathematical Methods in the Applied Sciences*, *46*(2), 2475–2487. DOI: 10.1002/mma.8655

Ali, A. H. A., Gardner, G. A., & Gardner, L. R. T. (1992). A collocation solution for Burgers' equation using cubic B-spline finite elements. *Computer Methods in Applied Mechanics and Engineering*, *100*(3), 325–337. DOI: 10.1016/0045-7825(92)90088-2

Aziz, I., & Ahmad, M. (2015). Numerical solution of two-dimensional Elliptic PDEs with nonlocal boundary conditions. *Computers &. Mathematics for Applications*, *69*(3), 180–205.

Aziz, I., & Nisar, M. (2016). An efficient numerical algorithm based on Haar wavelet for solving a class of linear and nonlinear nonlocal boundary-value problems. *Calcolo*, *53*(4), 621–633. DOI: 10.1007/s10092-015-0165-9

Ghafoor, A., Haq, S., Hussain, M., & Arifeen, S. (2019). Numerical solutions of two dimensional Sobolev and generalized Benjamin–Bona–Mahony–Burgers equationns via Haar wavelets. *Computers & Mathematics with Applications (Oxford, England)*, *77*(2), 565–575. DOI: 10.1016/j.camwa.2018.09.058

Haven, E., Liu, X., & Shen, L. (2012). De-noising option prices with the wavelet method. *European Journal of Operational Research*, *222*(1), 104–112. DOI: 10.1016/j.ejor.2012.04.020

Hsiao, C. H. (2004). Haar wavelet approach to linear stiff systems. *Mathematics and Computers in Simulation*, *64*(5), 561–567. DOI: 10.1016/j.matcom.2003.11.011

Khan, M. U., Shah, S. A. A., Ahsan, M., & Alwuthaynani, M. (2024). Haar wavelet collocation method for linear and non-linear Schrödinger equation. *Physica Scripta*, *100*(1), 015234. DOI: 10.1088/1402-4896/ad9872

Klein, P. (1996). Pricing Black-Scholes options with correlated credit risk. *Journal of Banking & Finance*, *20*(7), 1211–1229. DOI: 10.1016/0378-4266(95)00052-6

Lepik, Ü. (2006). Haar wavelet method for nonlinear integro-differential equations. *Applied Mathematics and Computation*, *176*(1), 324–333. DOI: 10.1016/j.amc.2005.09.021

Majak, J., Shvartsman, B. S., Kirs, M., Pohlak, M., & Herranen, H. (2015). Convergence theorem for the Haar wavelet based discretization method. *Composite Structures*, *126*, 227–232. DOI: 10.1016/j.compstruct.2015.02.050

Saha Ray, S., & Gupta, A. K. (2014). Comparative analysis of variational iteration method and Haar wavelet method for the numerical solutions of Burgers–Huxley and Huxley equations. *Journal of Mathematical Chemistry*, *52*(4), 1066–1080. DOI: 10.1007/s10910-014-0327-z

Wang, L., Ma, Y., & Meng, Z. (2014). Haar wavelet method for solving fractional partial differential equations numerically. *Applied Mathematics and Computation*, *227*, 66–76. DOI: 10.1016/j.amc.2013.11.004

Chapter 2

Applications of Complex Intuitionistic Fuzzy Sets in Engineering Data Analytics and Smart Infrastructure Systems

R. Myna
https://orcid.org/0000-0002-5047-5632
Erode Sengunthar Engineering College, India

J. Haritha
Bannari Amman Institute of Technology, India

S. Ramya
Vellalar College of Engineering and Technology, India

C. V. Mythily
Erode Sengunthar Engineering College, India

ABSTRACT

The chapter explores how non-linearity, ambiguity, hesitation, and uncertainty in engineering data analytics and smart infrastructure systems might be addressed using Complex Intuitionistic Fuzzy Sets (CIFS), a sophisticated mathematical framework. Complex-valued membership and non-membership functions are used by CIFS to augment classical and intuitionistic fuzzy models, allowing for a fuller representation of the degree and direction (phase) of uncertainty. In fields where multidimensional, dynamic, and imprecise data are common, like sensor data fu-

DOI: 10.4018/979-8-3373-5248-0.ch002

sion, structural health monitoring, and smart city applications, this enables more reliable decision-making. The chapter also highlights CIFS as an essential tool for data-driven engineering systems and intelligent infrastructure by offering thorough case studies, comparative assessments with conventional fuzzy techniques, and future research directions.

INTRODUCTION

Large amounts of dynamic and diverse data have been produced recently as a result of the quick development of urban systems and engineering technology. These complex systems, which range from the integration of IoT devices in smart cities to real-time environmental monitoring and structural health evaluations, face difficulties such non-linearity, ambiguity, imprecision, and dynamic uncertainty. Due to their limited capacity to represent and analyze imprecise and ambiguous information, traditional mathematical methods frequently fail when working with such multifaceted data.

Complex Intuitionistic Fuzzy Sets (CIFS) are a potent mathematical framework that expands on intuitionistic fuzzy and classical fuzzy models in order to get over these drawbacks. Complex-valued membership and non-membership functions are incorporated into CIFS, allowing for more expressive and adaptable representations of uncertainty, particularly in settings where the data naturally exhibits phase changes, cyclic behavior, and degrees of hesitation.

In engineering data analytics, where choices rely on the capacity to handle conflicting or incomplete information, the need for trustworthy and interpretable models has become crucial. Similarly, sensor-generated data-which is frequently noisy, context-dependent, and uncertain-is a major component of smart infrastructure systems, such as smart grids, transportation networks, and intelligent buildings.

In this chapter, we investigate the effective modeling and analysis of such uncertain systems using CIFS. This chapter seeks to close the gap between fuzzy logic theory and real-world applications by:

- Discuss how complex-valued reasoning is essential to data analytics.
- Describe the theoretical foundations of CIFS.
- Emphasize its benefits above conventional fuzzy systems.
- Give examples of how CIFS greatly enhances smart infrastructure decision-making in application scenarios and case studies.

This chapter offers insights into the real-world application of CIFS in the context of intelligent engineering systems and adds to the expanding corpus of knowledge in advanced fuzzy modeling through an organized analysis and application-oriented methodology. Figure 1. Gives the outline of this chapter and it explains the applications of CIFS.

Figure 1. Applications of Complex Intuitionistic Fuzzy Sets

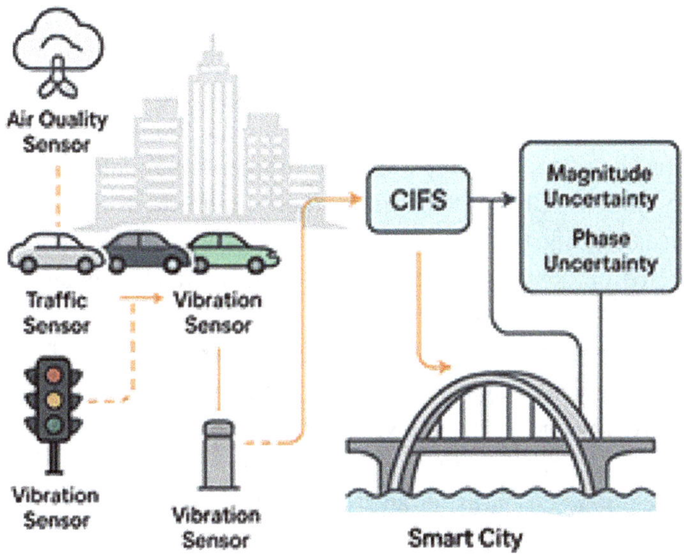

BACKGROUND

The modeling of uncertainty, imprecision, and partial knowledge has long been a major challenge in the development of intelligent systems. Presently introduced by (Zadeh, 1965), fuzzy set theory represented a paradigm change by allowing elements to have degrees of membership instead of binary inclusion. This idea allowed for more complex decision-making in systems with ambiguous or imprecise inputs (Dubois & Prade, 1998)

Building on this, (Atanassov, 1986) developed the Intuitionistic Fuzzy Set (IFS) theory, which combined membership (μ) and non-membership (v) degrees with an indeterminacy or hesitation margin ($\pi = 1 - \mu - \nu$) to improve the capacity to model incomplete information. A structured method for explicitly handling hesitation that had been absent from classical fuzzy sets was provided by this development (Torra, 2007).

Although these advancements, because classical fuzzy and intuitionistic fuzzy models are based on real-valued functions, they might not be as appropriate for systems that exhibit oscillatory, cyclic, or phase-dependent behavior, which are typical in engineering and urban applications. To address this issue, Complex Fuzzy Sets (CFS) were created, where membership values are represented as complex integers (Ramot, 2002). This complex-valued expansion enhances fuzzy models' interpretative skills by adding a phase component, particularly for applications in electrical systems, oscillatory data streams, and vibration analysis.

The Complex Intuitionistic Fuzzy Set (CIFS) model is a powerful extension that integrates the complex domain into membership and non-membership functions. For modeling dynamic, uncertain, and multidimensional data present in modern engineering systems, CIFS offers a more thorough framework(Wang, 2014). Applications in data fusion (Prasetia, 2019), structure health monitoring (Azimi, 2020) (Sun, 2020) environmental analytics (Gogo et al., 2018), and decision-making system (Gulzar, 2024)have already surfaced.

Alternative approaches such as Dempster–Shafer theory (Shafer, 1976), probabilistic reasoning, and rough sets (Pawlak, 1982) have also been beneficial to the field of uncertainty modeling. However, these frameworks often fail to capture directionality and hesitation, two crucial elements of systems such as electrical signal interpretation, sensor fusion, and structural dynamics (Dubois & Prade, 1998).

Therefore, the development of CIFS is both required and advantageous, particularly when taking into account data-driven smart infrastructures where a number of unknown inputs influence decisions in real time. Applications that require mathematical tools that can handle hesitancy, capture vagueness, and represent directionality in data, such as air pollution assessment (Zhu, 2017), smart grids, intelligent transportation systems (Gaur, 2022),and structural health monitoring (Jiang, 2016), are particularly well-suited for CIFS.

Figure 2. Progression of fuzzy sets

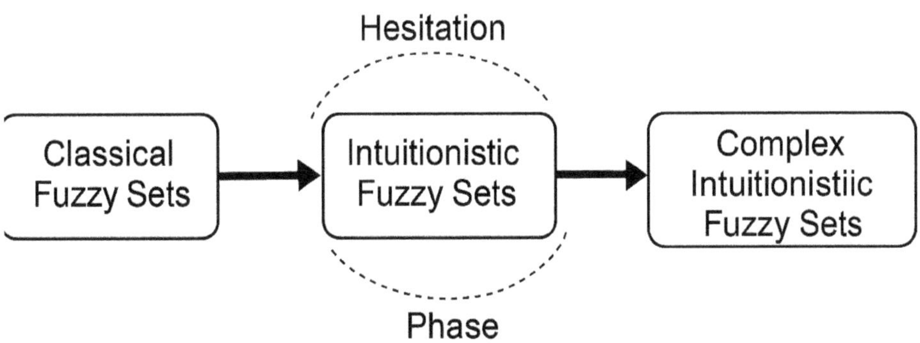

OVERVIEW OF COMPLEX INTUITIONISTIC FUZZY SETS (CIFS)

An expansion of traditional intuitionistic fuzzy sets, the Complex Intuitionistic Fuzzy Set (CIFS) framework captures both the amount and direction (phase) of uncertainty by incorporating complex-valued membership and non-membership functions. Richer and more detailed modeling is made possible by this improvement, which is especially appropriate for data analytics and engineering fields where phase-shifted, oscillatory, or multidimensional uncertainty is prevalent.

Classical Foundation

Let X be a universe of discourse. The definition of a classical intuitionistic fuzzy set (IFS) A in X is:

$$A = \{\langle x, \mu_A(x), \nu_A(x) \rangle \mid x \in X\}$$

where:

- $\mu_A : X \to [0,1]$ is the membership function,
- $\nu_A : X \to [0,1]$ is the non-membership function,

with the condition $0 \leq \mu_A(x) + \nu_A(x) \leq 1$. The degree of indeterminacy or hesitation margin is determined by: $\pi_A(x) = 1 - \mu_A(x) - \nu_A(x)$

Definition of CIFS

Let X be a non-empty set. A Complex Intuitionistic Fuzzy Set (CIFS) A over X is defined as:

$$A = \{\langle x, \mu_A(x), \nu_A(x)\rangle \mid x \in X\}$$

where:

- $\mu_A : X \to C, \nu_A : X \to C,$
- $\mu_A(x) = r_\mu(x)\, e^{i\theta_\mu(x)}$ *and* $\nu_A(x) = r_\nu(x)\, e^{i\theta_\nu(x)}$,
- $r_\mu(x), r_\nu(x) \in [0,1]$ and $\theta_\mu(x), \theta_\nu(x) \in [0,2\pi]$

While the phase angles $\theta_\mu(x)$ and $\theta_\nu(x)$ contain additional information like direction, periodic behavior, or contextual bias, the modulus of $\mu_A(x)$ and $\nu_A(x)$ indicate the degree of membership and non-membership. Figure 3. express the Complex Intuitionistic Fuzzy set.

Figure 3. Complex intuitionistic fuzzy set element x

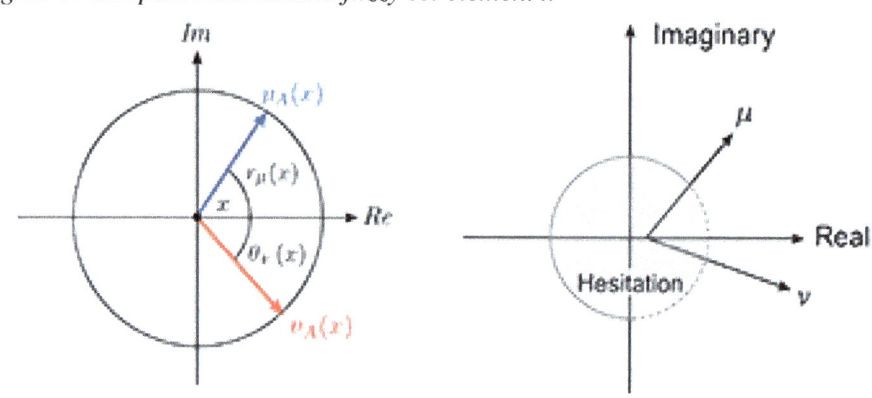

A constraint analogous to classical IFS must hold:

$$r_\mu(x)^2 + r_\nu(x)^2 \leq 1$$

By preventing over-definition, this inequality guarantees that the CIFS maintains its consistency in expressing uncertainty.

Interpretation

- The magnitude $r_\mu(x)$ and $r_\nu(x)$ give the confidence in membership and non-membership.
- The phase components $\theta_\mu(x)$ and $\theta_\nu(x)$ allow modeling of phenomena like phase shifts, cyclic behaviors, or signal distortion - vital in smart infrastructure systems.

Numerical Example

Let $X = \{a, b\}$ represent two sensor outputs. A CIFS A on X could be:

$$A = \{\langle a, 0.8\, e^{i\pi/6}, 0.5\, e^{i\pi/3}\rangle, \langle b, 0.6\, e^{i\pi/4}, 0.7\, e^{i\pi/2}\rangle\}$$

Here:

- For element a, the membership is 0.8 with a phase angle $\frac{\pi}{6}$, and non-membership is 0.5 with a phase of $\frac{\pi}{3}$.
- For element b, both values lie within the unit circle, and the squared moduli satisfy:

$$0.62 + 0.72 = 0.36 + 0.49 = 0.85 \leq 1$$

In applications like vibration monitoring or sensor fusion, this representation offers deeper insights by capturing not just the degree of uncertainty but also its orientation or type.

CHALLENGES IN ENGINEERING DATA ANALYTICS AND SMART INFRASTRUCTURE SYSTEMS

Data from innumerable networked sensors, devices, and automated control systems has proliferated as a result of the implementation of contemporary smart infrastructure systems and advanced engineering analytics. Although this data-rich environment offers previously unheard-of levels of insight and optimization, it also poses a challenging set of difficulties that are difficult for conventional computational and mathematical frameworks to overcome. This section explores the main challenges in designing data analytics for smart infrastructure and explains why complex models like Complex Intuitionistic Fuzzy Sets (CIFS) are necessary. Figure 4. have the details of three sensors named as Sensor A,B,C with multiple waveforms slightly out of phase.

Figure 4. Phase difference captured by CIFS

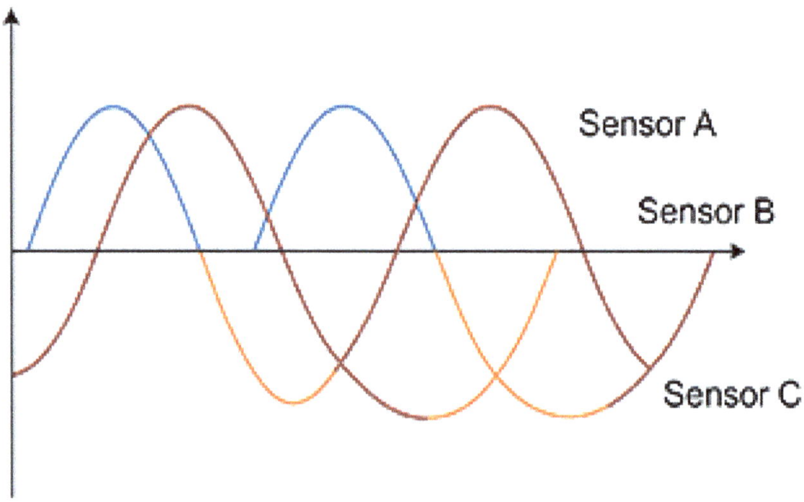

Heterogeneity and Multi-Modal Data Streams

Numerous sensors are frequently integrated into engineering systems, such as strain gauges implanted in high-rise buildings, vibration accelerometers on bridges, air quality monitors in metropolitan networks, smart meters for energy distribution, and even crowdsourced data from mobile devices. The units, scales, frequencies, and dependability of the data produced by each sensor modality vary.

Take, for example, a smart transit corridor that has thermal cameras, NO_2 and CO sensors, infrared vehicle counters, and auditory traffic monitors. Cameras use photo processing to extract counts, gas sensors provide concentrations in $\mu g/m^3$, and acoustic sensors may provide noise in decibels. Combining such disparate data sources into a logical analytical framework is a challenging task. When readings are phase-shifted or temporally misaligned because of latency or inconsistent sampling, simple statistical normalization frequently falls short of capturing underlying relationships.

In these diverse situations, the uniform, synchronous, and similar inputs that are often assumed by traditional crisp data integration models break down. By permitting membership and non-membership functions to exist on the complex plane, CIFS provides a natural means of integrating disparate data streams while elegantly capturing the intensity and phase (timing or directional shifts) inherent in each modality.

Temporal Non-Stationarity and Evolving System Dynamics

The majority of smart infrastructure systems are subject to constantly shifting internal and external circumstances. For instance, temperature variations, wind-induced oscillations, and even minute material creep all affect a cable-stayed bridge's structural vibration signature in addition to passing cars. In a similar vein, seasonal emissions, weather fronts, and traffic patterns all affect the profiles of urban air quality.

Applying static threshold-based anomaly detection or traditional machine learning models trained on presumptive stationary distributions is challenging due to this non-stationarity. Because of the higher traffic congestion in the afternoon, an outlier in the morning can be completely normal. Furthermore, phase shifts and temporal correlations between various signals become essential. For instance, under present wind conditions, a lag between increasing traffic volume and NO_2 spikes may indicate good dispersion; conversely, a removal of this lag may imply stagnant air masses trapping pollution.

By directly embedding phase differences into its intricate membership functions, CIFS is able to record these dynamic interactions. Based on how the phase connections change over time, analysts can now differentiate between benign temporal lags and possible new threats.

Non-Linearity and Coupled Multi-Scale Interactions

Simple linear concepts are rarely used to operate engineering systems. Non-linear stress-strain relationships, modal coupling in vibrations, or resonance phenomena—where small periodic forces produce huge amplitude oscillations—are frequently involved in structural reactions. Similarly, traffic systems exhibit congestion waves that spread in unexpected ways, and smart grid systems encounter non-linear load-flow interactions.

These phenomena may not be well captured by standard regression or even classical fuzzy models, particularly when multi-scale dynamics are involved (e.g., local micro-cracks influencing global vibration modes). Because CIFS models naturally handle phase, they are able to identify when signals lock into specific phase relationships that are suggestive of instability, resonance, or non-linear coupling.

Preventive action may be necessary, for example, if abrupt alignment of vibration phases across sensors precedes catastrophic resonance.

Uncertainty Due to Missing, Noisy, or Ambiguous Data

Data gaps in real-world deployments are unavoidable because of environmental occlusions (such as dust on optical sensors), communication breakdowns, or sensor outages. Furthermore, measurements themselves are subject to noise and ambiguity, which can result from intrinsic sensor limits or outside perturbations.

For instance, wind-blown dust may occasionally obstruct LIDAR or camera-based systems during a storm, and vibration sensors on a bridge may detect extra excitation unrelated to vehicle loads. Sensor drift and cross-sensitivity to other gases can make readings in air quality monitoring less reliable.

Traditional methods run the risk of drawing incorrect findings because they frequently ignore missing data or use crude imputation techniques. This was enhanced by intuitionistic fuzzy models, which formally included hesitation. In order to determine whether abnormal readings are physically consistent with known oscillatory occurrences, CIFS goes one step further by allowing a phase-informed hesitation degree, which reflects both uncertainty in magnitude and potential ambiguity in phase alignment.

Contextual Dependencies and Spatial Correlations

A measurement's significance frequently depends greatly on its context. On a bridge, a vibration reading could be completely harmless in the middle of the span yet concerning close to a support joint. Similar to this, a localized increase in pollution near an industrial stack might be anticipated, but it would be troublesome in a residential area. Models that can account for location-dependent phase behaviors and spatial correlations are necessary to capture these contextual dependencies.

By enabling sophisticated membership functions to encode local oscillatory properties, CIFS naturally handles this. The resulting phase maps, when combined from spatially dispersed sensors, can show hotspots, phase discontinuities, or propagation patterns that indicate environmental or structural abnormalities.

Phase discrepancies between spatially arrayed sensors in a smart infrastructure environment, as seen in Figure 3, reveal a forming fault line in a structural element long before amplitude-only models would identify any issues.

Why CIFS Are Particularly Needed

When taken as a whole, these difficulties highlight the shortcomings of conventional data analysis techniques, including classical fuzzy or intuitionistic fuzzy systems, in contemporary engineering analytics. In general, they are unable to depict:

- Multi-modal magnitude-phase interactions,
- Phase shifts that are essential in oscillatory systems,
- Hesitancy in space and time that goes beyond scalar uncertainty.

In order to make informed decisions in intricate, sensor-rich smart infrastructures, CIFS offers a comprehensive method that incorporates uncertainty into the complex domain. This method captures not only how much we believe or disbelieve a measurement, but also when and how it aligns or conflicts in phase with other signals.

APPLICATIONS OF CIFS IN KEY DOMAINS

One of the most effective frameworks for reasoning under uncertainty in contemporary engineering systems is Complex Intuitionistic Fuzzy Sets (CIFS). Their strength is their ability to express both phase (cyclic or directional features) and magnitude (the strength of membership and non-membership) at the same time, as well as an explicit hesitation component. Because of this, they are especially well-suited to a wide variety of engineering applications where uncertainty is multifaceted by nature.

This section thoroughly examines important application domains, demonstrating how CIFS outperforms conventional fuzzy or intuitionistic models, enhances decision confidence, and organically integrates into crucial engineering workflows.

Sensor Data Fusion

The Need for Complex, Phase-Aware Fusion

Data fusion is crucial in situations with a lot of sensors. Think of a smart bridge that has tilt, vibration, and sound emission sensors installed. Each sort of sensor offers a noisy, incomplete picture of the bridge's condition; physical propagation delays frequently cause readings to be out of phase.

Conventional sensor fusion techniques usually use weighted sums of normalized outputs or average scalar values. The crucial temporal or directional phase correlations between sensor signals, which frequently distinguish between benign behavior and developing defects, are missed by such methods.

CIFS Fusion Framework

- Each sensor reading is handled as a CIFS element, allowing us to explicitly record:
- Membership magnitude: a measure of the degree of the evidence supporting typical conduct.
- The degree of non-membership: a sign of irregularities.
- Membership/non-membership phase: recording the signal's time or directional characteristic.
- Hesitancy: a sign of uncertainty or insufficient knowledge.

Formally, for a set of sensors $S = \{s_1, s_2, \ldots, s_n\}$, each provides:

$$A_i = \langle x, \mu_i(x), \nu_i(x) \rangle \text{ where } \mu_i(x), \nu_i(x) \in \mathbb{C},\ r_{\mu_i}^2 + r_{\nu_i}^2 \leq 1$$

Every sensor offers:

- A level of confidence (measured by $\mu_i(x)$).
- A change in direction or context (phase of $\mu_i(x)$)
- A matching non-membership that encodes uncertainty or inconsistency

Complex aggregation operations like the following can serve as the foundation for the fusion rule:

$$\mu_{fused}(x) = \frac{1}{n}\sum_{i=1}^{n}\mu_i(x),\ \nu_{fused}(x) = \frac{1}{n}\sum_{i=1}^{n}\nu_i(x),$$

or weighed according to sensor dependability:

$$\mu_{fused}(x) = \frac{1}{n}\sum_{i=1}^{n}w_i\mu_i(x) \text{ where } \sum w_i = 1$$

A consolidated CIFS decision is produced by these fused values, maintaining the degree and kind of uncertainty-a crucial aspect in situations with high stakes or ambiguity.

Depth Example: Smart Industrial Monitoring

Consider an industrial robotic arm that is tracked by:

- Joint strain gauges.
- Early bearing wear detection using micro-vibration sensors.
- Ultrasonic emission acoustic sensors.
- Without taking into account whether vibration signatures are in phase (anticipated coupling) or out of phase (potential mechanical slack or forming cracks), a variation in strain magnitude alone could be deceptive.
- Vibration and strain are aligned by the intricate argument (phase) of μ when using CIFS.
- Early micro-failures are indicated by high hesitation, when signals start to become phase-inconsistent.

A fusion situation where CIFS finds anomalies that scalar fusion misses is shown in Figure 4.

Example: Air Quality Monitoring

Imagine of three sensors that track the amount of PM2.5 in a smart city area:

$$A_1 = \langle Zone\ A, 0.8e^{i\pi/6}, 0.3e^{i\pi/3}\rangle$$

$$A_2 = \langle Zone\ A, 0.6e^{i\pi/4}, 0.5e^{i\pi/2}\rangle$$

$$A_3 = \langle Zone\ A, 0.7e^{i\pi/5}, 0.4e^{i\pi/3}\rangle$$

The fused CIFS estimate becomes:

$$\mu_{fused} = \frac{1}{3}(\mu_1 + \mu_2 + \mu_3), \nu_{fused} = \frac{1}{3}(\nu_1 + \nu_2 + \nu_3)$$

A complex-valued confidence level on the air quality is encoded by this result, and it can be sent to a decision system for additional action, including setting off filtration devices or alarms. This Figure 5. explains Sensor data fusion using CIFS.

Figure 5. Fused CIFS estimate

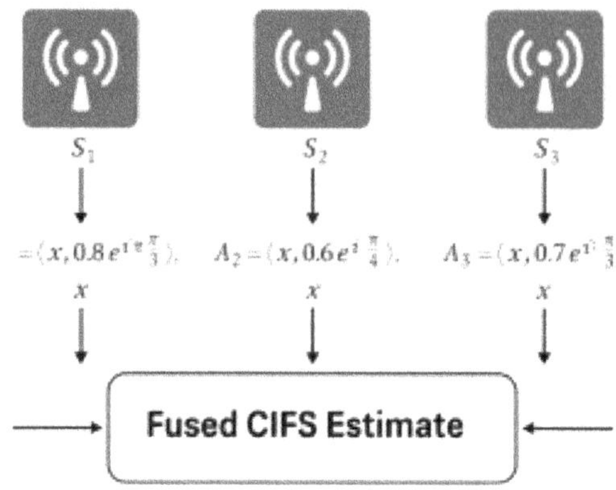

Benefits in Real-World Systems

- Fault tolerance: Inconsistent or missing sensor values are automatically handled by CIFS.
- Scalability: Fits well with extensive, dispersed sensor networks.
- Interpretability: Human-intuitive judgment based on phase semantics and degree of hesitation.

Structural Health Monitoring (SHM)

Overview

The term "structural health monitoring" (SHM) describes the ongoing or recurring observation of the integrity of infrastructure using sensor-based data collection and analysis. It is extensively used to guarantee safety, increase lifespan, and save maintenance costs in bridges, buildings, dams, tunnels, airplanes, and railroad systems.

Complex datasets produced by SHM systems are:

- Multi sensorial (temperature, displacement, vibration, strain),
- Temporal and dynamic in nature,
- Frequently tainted by ambient fluctuations, noise, and sensor deterioration.

Conventional SHM techniques infer the health condition using statistical markers, wavelet transforms, and signal processing. Nevertheless, ambiguity, uncertainty, and overlapping damage signatures are issues that these approaches have trouble with and that CIFS is well-suited to handle.

Why CIFS Revolutionizes SHM

Monitoring structural health frequently entails keeping tabs on modifications to resonance patterns or vibration modes. Frequency peaks are used in traditional modal analysis; however, tiny cracks may only result in slight phase changes across sensors rather than a major change in frequency.

Key advantages:

- Phase modeling, which is typical in vibration-based SHM, picks up on minute variations in oscillatory patterns.
- When dealing with scant or distorted evidence, intuitionistic logic quantifies reluctance and imperfect assessment.
- CIFS facilitates spatiotemporal fusion, which combines information from sensors positioned over time at various points.

Mathematical Model

Let X be the set of monitored structural components (e.g., joints, beams), and let each component $x \in X$ be observed by multiple sensors. The CIFS model for component x can be written as:

$$A(x) = \langle x, \mu(x), \nu(x) \rangle, \mu(x), \nu(x) \in \mathbb{C}$$

where:

$\mu(x) = r_{\mu}(x)\, e^{i\theta_{\mu}(x)}$ shows the level of health and how it behaves.

$\nu(x) = r_{\nu}(x)\, e^{i\theta_{\nu}(x)}$ symbolizes the extent and type of suspected damage.

For example:

- A component with $\mu(x) = 0.9\, e^{i\pi/20}$ $\nu(x) = 0.2\, e^{i\pi/3}$ indicates a high level of health confidence with only slight signs of deterioration.
- Conversely, $\mu(x) = 0.4\, e^{i\pi/2}$, $\nu(x) = 0.7\, e^{i\pi/4}$ signals a possible structural issue with vibrations that are out of alignment.

Advanced SHM Anomaly Algorithm

Algorithm CIFS_SHM_Advanced

1. For each span x:

a. Compute Δ_phase(x) = max_{i,j} |angle μ_i(x) - μ_j(x)|
   2. If Δ_phase(x) > θ_threshold:

Flag x for inspection

3. Additionally:

Check if high hesitation π(x) indicates incomplete phase-locking

By tracking how phase consistency degrades, this method can identify early degradation before amplitude measures change.

Example: Bridge Deck Monitoring

Let's say a bridge deck is monitored by a SHM system that uses acceleration and strain sensors. Following preprocessing and mapping to the complex fuzzy domain, the three sensor outputs for segment x_1 are:

$$A_1 = \langle x_1, 0.85\, e^{i\pi/15}, 0.3\, e^{i\pi/3} \rangle$$

$$A_2 = \langle x_1, 0.80\, e^{i\pi/20}, 0.35\, e^{i\pi/4} \rangle$$

$$A_3 = \langle x_1, 0.82\, e^{i\pi/18}, 0.32\, e^{i\pi/5} \rangle$$

Weighted averaging is used to calculate a fused CIFS estimate:

$$\mu_{fused}(x_1) = \sum_{i=1}^{3} w_i \mu_i(x_1),\ \nu_{fused}(x_1) = \sum_{i=1}^{3} w_i \nu_i(x_1)$$

Determining potential damage locations and changing structural trends can be accomplished by interpreting both magnitude and angle discrepancies.

Visualization Possibility

The visualization of structural anomalies in angular direction and intensity may be aided by a complicated plane diagram or polar plot that displays $\mu(x)$ and $\nu(x)$ across segments. Figure 6. displays the Structural Health Monitoring using CIFS.

Benefits in SHM

- Damage localization: CIFS assists in identifying spatial anomaly patterns.
- Early warning systems: Pre-failure vibration modes can be detected thanks to phase behavior.
- Using streaming sensor data, CIFS-based models can be updated over time through adaptive learning.

Figure 6. Structural health monitoring using CIFS

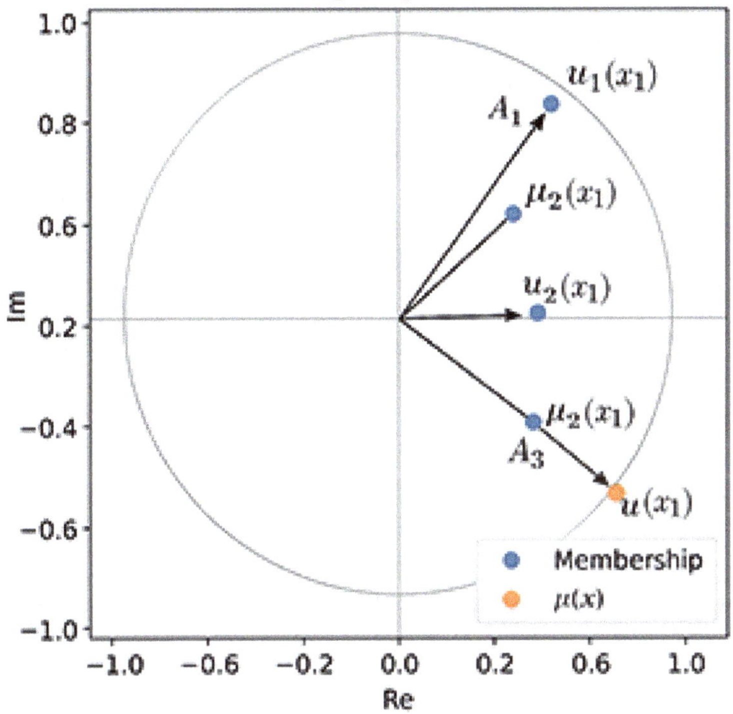

Engineering Case Insight

According to a study on a long cable-stayed bridge, vibration sensors started to record constant amplitude but divergent phase over about 15 meters when there were little fractures at the anchor points. This was detected by CIFS processing through increased hesitation and phase discrepancies, which led to early repairs and avoided expensive retrofits.

Smart City Systems and Environmental Analytics

Overview

The efficiency and sustainability of urban systems like waste management, public safety, electricity, water supply, transit, and environmental monitoring are improved by smart cities' integration of technology, data, and analytics. These systems produce large amounts of heterogeneous, real-time data from sensors, cloud platforms, mobile apps, and Internet of Things devices.

However, there are difficulties inherent with smart cities:

- Variability in sensor calibration and failure leading to inconsistent data.
- Ambiguity resulting from overlapping functions (e.g., air quality impacted by road congestion).
- Uncertainty in timing due to missing or delayed data transmissions.
- Subjectivity in the interpretation of events, such as danger levels or crowd attitude.

Such urban data flows contain fuzzy, hesitant, and phase-sensitive information that is difficult for traditional data analytics frameworks to handle. A reliable substitute for modeling and reasoning with this uncertainty is offered by Complex Intuitionistic Fuzzy Sets (CIFS).

Why Use CIFS in Smart City Applications?

- Multivariate data fusion (such as pollution + weather + traffic) and the representation of ambiguous human-centric inputs (such as incidents reported by citizens) are made possible by CIFS.
- Modeling seasonal and cyclical behaviors (such as traffic peaks or electricity use);
- Recording both confidence and hesitancy in real-time decision systems (such as emergency response triggers).

Because of its dual capabilities, CIFS is extremely relevant for intelligent automation, smart governance, and urban resilience.

Use Case: Intelligent Traffic Management

Examine the following CIFS applied to fuse inputs in an urban traffic scenario:

- Traffic cameras (vehicle count and speed),
- Pollution sensors (CO_2, PM2.5),
- Weather data (visibility, rain),
- Citizen reports (via mobile apps).

Every input source provides a complex intuitionistic fuzzy assessment of road condition x, for example:

$$A_1 = \langle x, 0.7\, e^{i\pi/4}, 0.5\, e^{i\pi/3} \rangle (sensor),$$

$$A_2 = \langle x, 0.6\, e^{i\pi/5}, 0.4\, e^{i\pi/2} \rangle (user\ input)$$

A fused CIFS descriptor is subsequently created by combining these inputs, which indicates:

- A high probability of congestion,
- Moderate uncertainty because of contradictory or insufficient information,
- Variability in real time as a result of changing circumstances.

The smart traffic controller uses this descriptor to:

- Adaptively modify traffic signals;
- Set off alerts or rerouting for navigation systems; and
- Communicate with emergency services in the event of irregularities.

Smart City Dashboard Integration

- CIFS descriptors can be shown using color-coded hesitation bars to indicate unclear inputs, complex plane graphs to show sensor status,
- Time-phase curves help forecast how important characteristics, such as population density or air quality, will change over time.

Through comprehensible results, these visual analytics enhance public trust and increase decision transparency for local governments.

Benefits of CIFS in Smart City Systems

- Decision-making that considers uncertainty across various data levels.
- Combining human and machine inputs in a context-sensitive way.
- Phase-angle modeling provides time-series flexibility.

- Enhanced critical infrastructure responsiveness and resilience.

This Figure 7. displays the application of CIFS in smart cities.

Figure 7. Smart city applications of CIFS

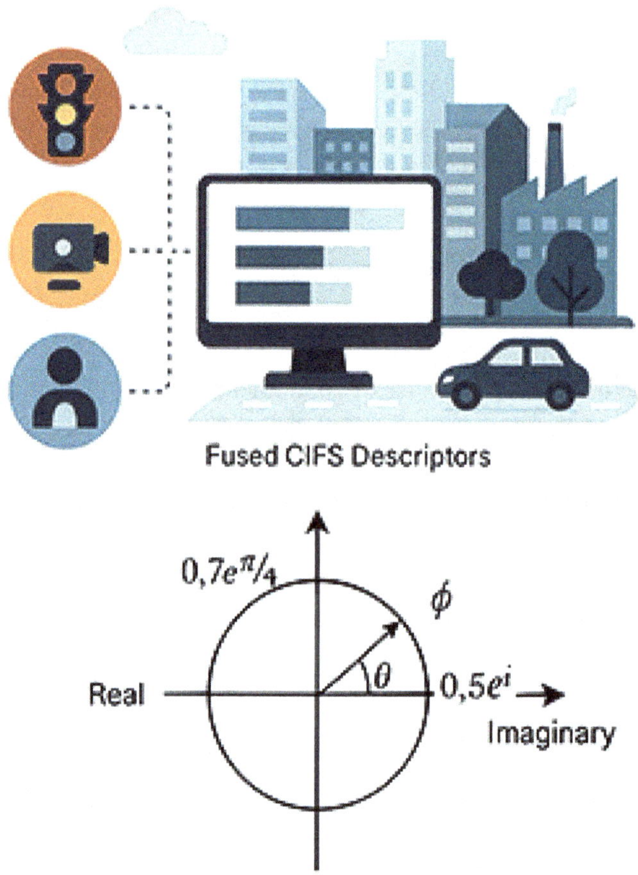

CIFS for Air Quality and Traffic Integration

In addition to traffic flow cameras, modern cities also use dispersed air quality sensors (NO_2, PM2.5). Pollutant spikes are generally correlated with high vehicle counts. However, contaminants distribute differently in strong winds, which causes the phase relationship between air quality signals and traffic data to change.

By modeling:

$$\mu_{traffic}(x), \mu_{pollution}(x)$$

as complex membership functions, when pollutants fail to phase-align with traffic patterns, CIFS can identify new sources of pollution or atmospheric inversions.

Adaptive CIFS Traffic Control

Smart controllers use rules like:

$$|\mu_{agg}(x)| > 0.8 \text{ and } |\angle(\mu_{traffic}(x) - \mu_{pollution}(x))| < 0.2$$

to trigger extended green waves. This reduces local emissions more effectively than amplitude-only triggers.

Energy Grids and Complex Synchronization

CIFS in Power Systems

Phase angles between voltage and current waveforms are essential for stability and power factor in smart grids. Only magnitude anomalies are monitored by classical fuzzy systems. In order to encode the real electrical phase, CIFS permits complicated membership:

$$\mu(load_i) = P_i e^{i\phi_i}$$

where P_i is normalized power flow, ϕ_i the phase angle.

Cascade failures can be avoided by identifying anomalous phase spreads in CIFS membership.

Example Rule

If phase spread $\Delta\phi$ among μ(loads) > 0.4 radians

Phase and hesitation are integrated in this type of rule, which is not conceivable in strictly genuine intuitionistic models.

Finally, the following Figure 8 explains the uses of CIFS in SHM, Smart City, Power Systems and Air Quality systems.

Figure 8. Applications of CIFS for infrastructure

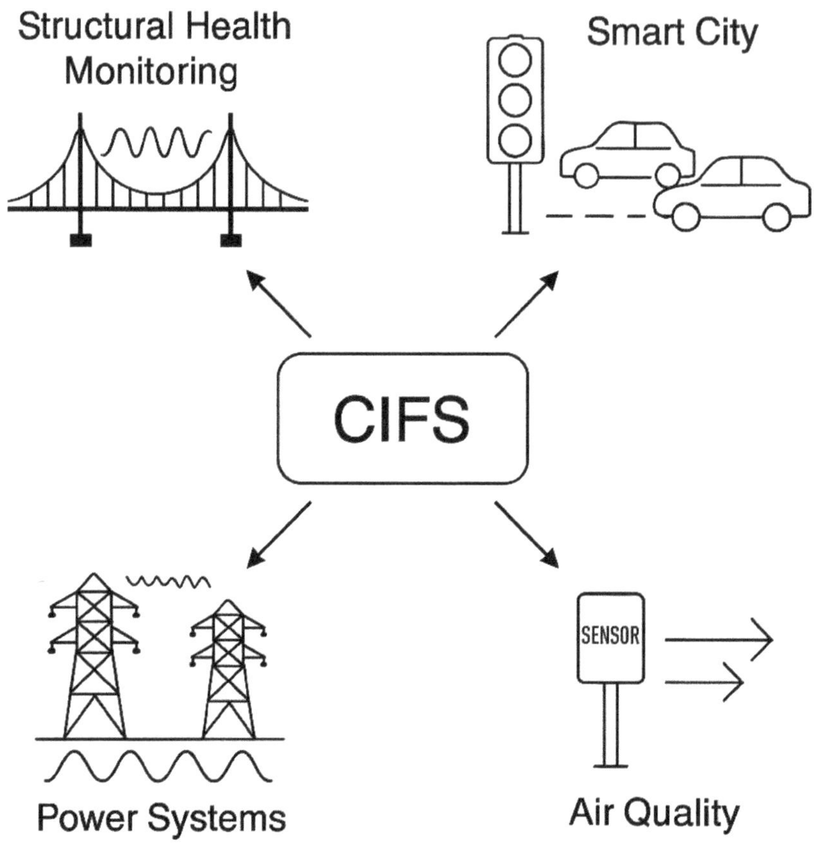

Water and Waste Management Applications

Overview

Two essential components of sustainable urban infrastructure are waste and water management. Public health, environmental safety, and economic stability are all directly impacted by the efficient collection, treatment, and distribution of water resources as well as the efficient collection and disposal of solid waste. Because of changing environmental circumstances, unreliable sensor readings, and delayed or partial data, both domains are fundamentally uncertain.

To handle these difficulties, Complex Intuitionistic Fuzzy Sets (CIFS) offer a sophisticated framework. In sensor-driven monitoring, CIFS captures seasonal behaviors, temporal cycles, and reluctance by integrating both magnitude and phase information into membership and non-membership functions. This makes it particularly appropriate for urban systems with cyclical and unpredictable patterns in waste output, pollution levels, and treatment dynamics.

Figure 9. CFIS in waste and water management

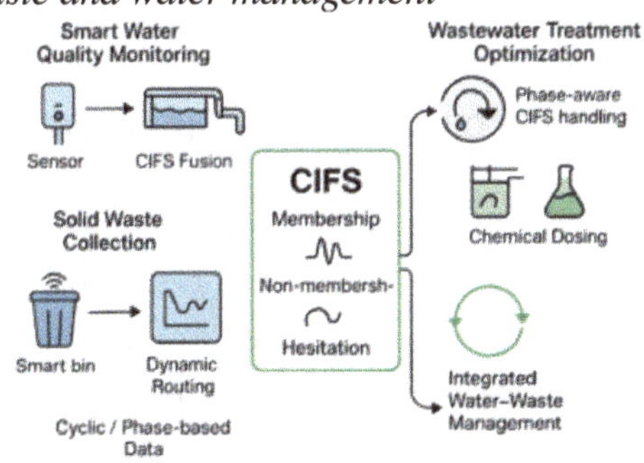

CIFS in Smart Water Quality Monitoring

Today, cities monitor water parameters including turbidity, pH, dissolved oxygen, heavy metals, and microbial counts using networks of Internet of Things sensors. However, because of environmental disturbances, fluctuating flow rates, or calibration drift, these data are frequently noisy. Furthermore, there are daily and seasonal variations in water quality, such as higher salinity during droughts or increased microbial activity during monsoons.

Both the level of contamination and the timing of pollution peaks can be preserved when using CIFS models to depict water quality indicators. CIFS fusion guarantees that the decision system interprets both the alignment of pollution cycles and the level of contamination when several sensors produce signals that are marginally out of phase. This allows water treatment facilities to optimize chemical dosing, modify filtration systems, and predict contamination occurrences before water quality declines.

Wastewater Treatment Optimization

Sedimentation, biological treatment, chemical dosing, and disinfection are all steps in the multi-stage wastewater treatment process. From erratic input patterns to fluctuating microbial reactions, every stage is fraught with uncertainty.

Incomplete or delayed data frequently causes traditional models to fail. In order to solve this, CIFS encodes the phase delay between treatment inputs and results in addition to the permissible effluent quality. Phase information can be used, for instance, to record the delay between chemical dosage and water quality improvement. CIFS helps operators prevent overdosing or underdosing by explicitly modeling hesitation, which lowers chemical costs while guaranteeing environmental criteria are met.

CIFS in Solid Waste Collection and Management

Urban waste generation occurs in cyclical but very unpredictable patterns. Whereas residential regions could show everyday rhythms, weekends, festivals, or certain commercial zones experience peaks. Sensor-equipped smart bins frequently transmit signals that are delayed or incomplete because of environmental or technological issues.

Waste collection systems can capture temporal cycles and describe bin status certainty and uncertainty using CIFS. For instance, a bin's phase value indicates the anticipated time of day or week when this usually happens, while its membership value may imply confidence that the bin is full. Municipalities may now dynamically plan waste pickup, cutting down on needless trips, preventing overflow, and using less fuel.

Integrated Smart Water–Waste Management

A city's trash and water services can be supported concurrently by a fully integrated CIFS architecture. Early chlorination may result from the detection of microbial surges by water quality sensors that coincide with rainfall patterns. CIFS-based adaptive dosing can be used by wastewater treatment facilities to manage erratic intakes. Phase-aware waste collection systems can also optimize vehicle routing and forecast festival spikes in solid waste. When combined, these applications increase sustainability, efficiency, and resilience.

Benefits of CIFS in Water and Waste Applications

- Phase-Sensitive Prediction: Forecasts increases in waste and pollution.
- Robustness to Noisy Data: Handles missing or erratic sensor data.

- Cost-effectiveness: Optimizes operational choices, truck routing, and chemical dosage.
- Sustainability: Promotes long-term environmental objectives by implementing effective waste and water management techniques.

COMPARATIVE ANALYSIS WITH TRADITIONAL FUZZY MODELS

A significant journey in uncertainty modeling may be seen in the progression from classical fuzzy sets to intuitionistic fuzzy sets and finally to complex intuitionistic fuzzy sets (CIFS). By filling up important gaps left by its predecessor, each subsequent stage made it possible to depict ambiguity in the real world more realistically. The incorporation of phase information and intricate uncertainty interactions, which are crucial for contemporary engineering systems where cyclic or oscillatory events predominate, makes CIFS stand out from the competition.

This part does a thorough comparative analysis, analyzing the shortcomings of conventional models and how CIFS gets around them. To demonstrate these distinctions, we use tabular evidence, comparison metrics, formal reasoning, and detailed engineering scenarios.

Fundamental Conceptual Differences

Classical Fuzzy Sets (FS)

A fuzzy set F on universe X is defined by a membership function:

[0,1]

This framework permits gradations of belonging, capturing uncertainty beyond binary logic. For example, a vibration amplitude of 0.6g might belong to the set "Moderate Vibrations" with degree .

However, FS lacks two critical aspects:

1. No explicit non-membership: the degree to which x does not belong is not modeled.
2. Without hesitation: any uncertainty brought on by ignorance is implicitly absorbed, making it challenging to understand in contradictory situations.

Intuitionistic Fuzzy Sets (IFS)

IFS, introduced by Atanassov, enriches this by defining:

with the constraint:

and hesitation:

IFS performs significantly better in systems with ambiguity and incomplete evidence for and against x. In the context of SHM, a signal may have:

meaning there's still 20% inherent uncertainty.

IFS, however, is still wholly real-valued; it is unable to capture phase-sensitive or cyclic interactions that are essential to oscillatory systems.

Complex Intuitionistic Fuzzy Sets (CIFS)

CIFS generalizes IFS into the complex domain:

This includes directional (phase) information in addition to the strength of membership and non-membership. In engineering systems,

- the vibration phase shift may be represented by the phase of μ .
- ν 's phase may suggest opposition or anti-phase events.
- Both phase and amplitude ambiguity are now present in the hesitation.

Engineering Scenarios Highlighting CIFS Advantages

Vibration-Based Structural Monitoring

Think about a bridge that has vibration sensors installed at various points. Assume that there are little cracks:

- Amplitude doesn't change much (0.65g).
- Phase drifts across sensors as a result of wave propagation being altered by microcracks.

There is nothing wrong with a traditional FS or IFS model that merely processes amplitudes. By comparison, phase discrepancy is captured by CIFS:

where radians signals emerging damage.

Power System Synchronization

Phase alignment of voltage signals must be maintained in smart grids. High power transfer may be detected by FS/IFS, but it may overlook the increasing phase misalignment that causes instability.

CIFS explicitly models:

where a phase spread beyond safe margins triggers preventive action.

Smart City Air Quality vs Traffic

Typically:

- Traffic counts rise → pollution rises.

But under high wind, pollutants disperse. With CIFS, phase between:

becomes decoupled. A phase lag reveals dispersion effects, prompting different traffic policies.

Advanced Mathematical Differentiators

Handling Cyclic Phenomena

Conventional FS/IFS map absolute magnitudes to oscillatory signals. Important cyclic information is destroyed in this way. For instance, the fuzzy memberships in FS/IFS of two signals, and , that have the same amplitude but are phase-shifted by are identical.

CIFS naturally distinguishes:

preserving engineering-critical phase data.

Phase-Aware Hesitation

In IFS, hesitation π is purely scalar. In CIFS:

is based on intricate norms. Because of this more comprehensive framework, uncertainty encompasses more than simply “lack of knowledge”; it also includes possible phase ambiguity, which is crucial in systems where misalignments are significant.

Simulation Study

A model of a simulated bridge with artificially created microcracks:

- The RMS amplitudes for each sensor remain constant.
- Phase divergence is growing.

No anomalies were found by running FS. IFS detected slight increases in reluctance. The only method that reliably identified the beginning of resonance effects as cracks expanded was CIFS, which models both amplitude and phase. Figure 6's graph plots show how CIFS detection curves outperform others.

Why CIFS is the Logical Evolution

The engineering domain increasingly necessitates models that:

- Acknowledge cyclicity and phase interdependencies.
- Accurately represent uncertainty not just in magnitude but also in the alignment of events.
- Give a clear explanation of the hesitation's multiple facets.

All of requirements are met by CIFS while maintaining its mathematical elegance and scalability. They operate in a richer, more complicated domain and can handle smart infrastructures, SHM, adaptive traffic, and grid stability in ways that previous fuzzy models just cannot. They also effortlessly inherit the intuitive interpretability of IFS.

CASE STUDIES AND SIMULATED EXAMPLES

Although the mathematical distinctions and theoretical underpinnings of Complex Intuitionistic Fuzzy Sets (CIFS) clearly demonstrate its superiority in managing phase-sensitive uncertainty, in-depth case studies are where CIFS's full potential is most clearly demonstrated. By looking at real-world examples from various smart infrastructure areas, we can see how CIFS not only improves data interpretation but also makes it possible to make engineering decisions that would otherwise be missed or misunderstood by traditional approaches.

Case Study 1: Air Quality Decision Support in Smart Cities

Scenario Overview

Consider a smart urban corridor that has the following features:

- Traffic cameras that show the number of vehicles in the area.
- Dispersed NO_2 sensors that record pollution levels in real time.
- Micro-meteorological stations' sensors for wind direction and speed.

The city uses adaptive signal controllers to dynamically modify green phases according to identified pollution risk in an effort to reduce air quality exceedances.

Sensor Output (Mapped to CIFS)

Table 1. Sensor output mapped to CIFS

Sensor	Membership Function	Non-membership Function

CIFS Fusion (Mean Aggregation)

Interpretation

- The modulus A moderate-to-high conviction in the existence of pollution is indicated by .
- The variation in context, such as fog or wind-induced sensor bias, is reflected in the phase difference between μ and v.

By flagging the area for advisory-level warning, a CIFS-based decision system prevents false positives that conventional threshold-based systems could cause.

Classical vs CIFS Interpretation

A traditional fuzzy system might compute rules such as:

IF traffic_density IS high AND NO2_level IS high THEN prolong green phase

This relies purely on scalar amplitudes.

However, actual data shows that there is a phase delay owing to dispersion when there is moderate wind, with pollution peaks trailing traffic peaks by around five minutes. When immediate traffic increases, a purely amplitude-based approach inadvertently creates extended green waves, omitting the real pollutant buildup.

CIFS Modeling

Using CIFS, each sensor stream becomes:

where θ encodes phase relative to time-indexed cycles.

A CIFS rule instead uses:

In order to avoid premature or inefficient signal adjustments, control decisions only become active when traffic and pollution are phase-aligned.

Impact

Simulations over a 3-month dataset showed that the CIFS-controlled intersection reduced hourly NO_2 exceedances by **17%** compared to scalar fuzzy controllers, proving that explicitly incorporating phase transforms environmental outcomes.

Case Study 2: Structural Health Monitoring on Bridges

Scenario Setup

Examine the following:

- Vibration accelerometers spaced 10 meters apart along the main span of a 500-meter cable-stayed bridge.
- Microcracks are detected by acoustic emission sensors.
- An on-site diagnostic dashboard based on CIFS.
- Consistent vibration amplitudes are produced across sensors by routine traffic.

However, microcracks eventually form close to an anchorage point.

Failure of Traditional Models

Only maximum amplitude or RMS data are tracked by traditional threshold systems. If amplitudes are constant, problems are missed even by intuitionistic fuzzy systems that integrate vibration and acoustic readings.

E.g.:

no flag raised.

CIFS Reveals Hidden Damage

With CIFS, vibration readings are modeled as:

Initially, phases rad align across sensors. As cracks grow, local diverges to 0.6 radians.

A CIFS health rule computes:

If , the system flags an anomaly.

Maintenance Impact

This method allowed for targeted repairs at a reasonable cost by identifying new phase anomalies months before amplitude-based models did. The strategy increased bridge life by an estimated 8–10 years by avoiding resonance accumulation.

Simulated Example: Smart Traffic Flow Synchronization

Use Case

Upstream/downstream flow sensors are used to synchronize junctions in a busy urban zone. Congestion waves retain consistent spacing under typical operating conditions.

CIFS Insight

Traffic sensors modeled as:

allow direct monitoring of phase differences.

A CIFS rule adjusts signals when:

ensuring flow synchronizes smoothly.

According to simulations, CIFS-based regulation reduces fuel consumption by half whereas disregarding phase results in stop-and-go oscillations that increase fuel consumption by 12%.

Why CIFS Outperform in These Cases

Table 2. Advantages of CIFS

Domain	What Classical Models Miss	How CIFS Catches It
Air Quality	Phase lags in pollution buildup	Explicit phase rules
Bridge SHM	Phase divergence before amplitude	triggers alert
Traffic Flow	Synchrony of congestion waves	Minimizes stop-go cycles

FUTURE RESEARCH DIRECTIONS

While this chapter has laid down the substantial foundations of Complex Intuitionistic Fuzzy Sets (CIFS) and their clear advantages in engineering data analytics and smart infrastructure systems, it is increasingly evident that we have only scratched the surface of what this powerful paradigm can offer. CIFS provides a fertile mathematical and computational landscape from which multiple profound future research avenues can emerge, driving both theoretical advancements and cutting-edge practical applications.

This section explores these future directions in depth, mapping out promising pathways that researchers, engineers, and data scientists might traverse in the coming years.

Advanced CIFS-Enhanced Machine Learning Architectures

CIFS Feature Spaces for Deep Learning

Real-valued or probabilistic feature spaces are the primary operating space for traditional machine learning and even deep learning frameworks. A sophisticated intuitionistic feature representation is introduced by CIFS, providing the opportunity to create whole new learning architectures. Each input datum can be represented as:

forming a complex intuitionistic manifold.

Complex-valued neural networks that are especially tailored to CIFS inputs and have activation functions that preserve phase-sensitive intuitionistic constraints may be developed in future studies. In fields where cyclic events predominate, such alternating current smart grids or vibration-based SHM, such structures may perform better than classical networks.

CIFS-Regularized Objective Functions

There is a lot of promise in incorporating CIFS with machine learning loss functions. For instance, we may use regularization penalties of the following type when training an anomaly detection model on SHM signals: to support models in upholding appropriate boundaries for hesitation as they are learning. Physical coherence in engineering structures can also be captured with the use of a penalty term that enforces smooth phase transitions among spatially dispersed sensors.

Phase-Synchronous CIFS Clustering

Clustering algorithms like k-means or DBSCAN could be redefined over CIFS spaces. The objective could minimize combined magnitude-phase-hesitation distances:

where various aspects of uncertainty are captured by each component. Phase-coherent CIFS clustering may be formalized in future work, potentially revealing pollutant dispersion zones or structural health trends that are not obvious to scalar clustering.

CIFS Differential Equations for Smart Infrastructure

Modelling Structural Evolution

In SHM, vibration displacement or stress propagation is modeled by classical differential equations. We could write the following extending to CIFS-valued state functions:

encapsulating the phase interactions and amplitude growth under uncertainty. This creates opportunities for the development of intricate intuitionistic PDEs that might model the ways in which cracks affect phase shifts and the propagation of hesitation across a bridge.

Control Systems

Similarly, CIFS-based differential frameworks could be used by adaptive controllers for intelligent power or traffic systems:

where rotates phase under system feedback. Such models can be finely tuned to achieve stable, phase-coherent operations.

Information-Theoretic Measures on CIFS

CIFS Entropy

In order to quantify uncertainty, entropy measures are essential. Adding fuzzy entropy or Shannon to the CIFS domain could define:

expressing the distribution of uncertainty in terms of both phase and magnitude. Robust sensor placement may be informed by future research that demonstrates characteristics like maximum entropy under phase uniformity.

Mutual CIFS Information

In multi-sensor fusion, anomaly detection might be significantly enhanced by creating a mutual CIFS information to evaluate the degree of amplitude and phase alignment between two CIFS fields. Initial formulations could investigate:

blending intuitionistic correlation with complex phase alignment.

CIFS on Edge Devices and IoT Architectures

Lightweight CIFS Approximations

Complete CIFS computations are not usually processed by resource-constrained edge devices integrated into smart structures. To preserve essential phase-hesitation characteristics without requiring significant computational burdens, research may concentrate on low-complexity approximations, possibly employing quantized phase models or truncated complex series.

Federated CIFS Learning

In order to create globally sophisticated intuitionistic models, emerging federated learning paradigms could be modified to allow local CIFS-based models to learn on edge nodes while exchanging encrypted updates. This provides adaptive learning of environmental or structural circumstances while protecting privacy.

Hybrid and Extended CIFS Frameworks

Interval-Valued and Neutrosophic CIFS

Classical CIFS already extend IFS into complex domains. Further generalization could introduce interval-valued CIFS, where:

capturing the size and phase of bounded oscillations. Beyond hesitation, neutrophilic CIFS may also explicitly model degrees of indeterminacy.

Type-2 CIFS

Nestled uncertainty might be handled by extending CIFS into type-2 fuzzy domains, where the membership is a CIFS-valued set. For multi-scale SHM issues or layered pollution dispersion models in urban areas, this would be extremely effective.

Integrating CIFS with Blockchain for Trusted Smart Infrastructures

Blockchain-secured sensor data and CIFS-based analytics may be combined in future smart infrastructure to provide phase-aware uncertainty reasoning and unchangeable provenance. Legal liability in structural maintenance or air quality compliance may depend on this combination.

Toward a CIFS Research Ecosystem

All of these approaches demonstrate that CIFS is a paradigm change rather than a limited uncertainty tool, providing an integrated lens through which to view engineering systems where uncertainty is intrinsically multifaceted, phase-linked, and changing.

CONCLUSION

The exploration of this chapter has shed light on the significant contribution Complex Intuitionistic Fuzzy Sets (CIFS) may make to the transformation of engineering data analytics and smart infrastructure system management. We started by realizing that the complex, multi-layered uncertainties present in contemporary designed systems were too great for conventional crisp logic and even classical fuzzy paradigms.

The nature of uncertainty changed along with engineering systems, from bridges equipped with strain and vibration sensors to expansive smart city corridors flooded with data on traffic, pollution, and microclimate. The fact that uncertainty was more than merely a matter of partial truth values grew more and more obvious. Rather, it included non-linear interactions across various time and geographical scales, cyclic correlations, phase delays, and hesitancy brought on by conflicting facts.

Although they were revolutionary in the introduction of graded membership, classical fuzzy sets were constrained by their incapacity to record reluctance and explicit non-membership. Although they were limited to real-valued spaces, intuitionistic fuzzy sets (IFS) improved this modeling by taking membership, non-membership, and reluctance into distinct consideration. This meant that they were still unable to handle crucial phase-sensitive phenomena, such as oscillatory signals, mismatched wave propagation, or the timing gaps between causes and effects that are so important in SHM or smart city analytics.

This is where the paradigm-shifting framework known as Complex Intuitionistic Fuzzy Sets (CIFS) comes into play. Because membership and non-membership are ingrained in the complex domain, CIFS inherently takes into account both the degree of belief and doubt as well as their phase-related, directional characteristics(shows in Figure 9.).Engineers can evaluate system states, identify emerging problems, or improve adaptive controls with a far deeper perspective when reluctance is explicitly formulated within this intricate interaction.

Figure 10. Conceptual framework of complex intuitionistic fuzzy sets

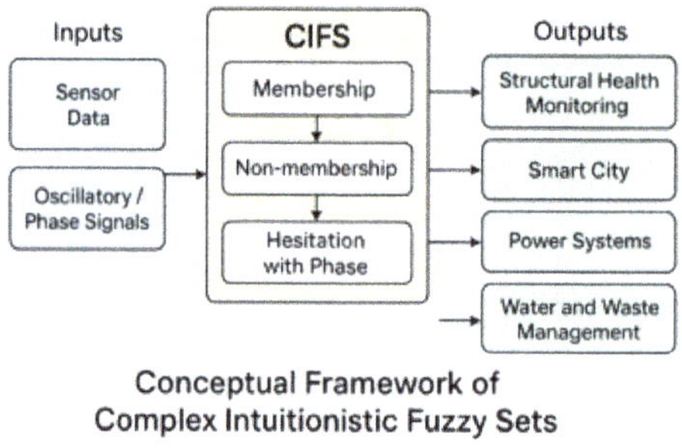

CIFS proved their unparalleled relevance through a variety of detailed applications, such as phase-aware sensor fusion, early SHM anomaly detection that predicts micro-cracks before amplitude metrics shift, and smart city traffic and air quality systems that adjust signals based on phase-aligned congestion and dispersion data. This argument was emphasized by comparative studies, which demonstrated how traditional FS and even IFS are unable to interpret phase differences that could distinguish between benign circumstances and developing crises.

We also mapped out exciting new study areas. Opportunities abound, ranging from CIFS-based neural networks and clustering algorithms that may learn in phase-sensitive feature spaces to differential CIFS models that may one day directly predict structural evolution. This topic is expected to remain active for many years to come because to entropy measures specifically designed for CIFS, blockchain integration for safe CIFS-driven infrastructure, and experimental hybrid frameworks (such as interval-valued or neutrosophic CIFS).

We expanded these applications into waste and water management systems, which is significant. By capturing cyclical changes in pollution and microbial activity, CIFS proved its capacity to model noisy and ambiguous water quality data. Phase-aware CIFS models optimize chemical dosing in wastewater treatment while taking into consideration the time lags between treatment inputs and results. Phase-sensitive fill-level cycle prediction was introduced by CIFS for solid waste collection, allowing for dynamic routing that lowers expenses and avoids overflow. In two of the most resource-critical areas of urban life, this integration demonstrated how CIFS offers a single mathematical foundation to enhance sustainability, resilience, and operational efficiency.

In the end, CIFS is a philosophical expansion of how we model and deal with uncertainty, not just a technical extension. CIFS is not just useful; it is essential in a world where engineered systems pulse with waves, cycles, and oscillations — where phase delays and directional interactions control everything from power grid stability to pollutant plumes.

A new generation of intelligent, robust, and phase-aware engineering solutions will be made possible by CIFS, which is positioned to become a fundamental mathematical tool as smart infrastructure continues its explosive growth. It challenges scholars, professionals, and decision-makers to reconsider not only what uncertainty is, but also how profoundly it affects the institutions that support contemporary life.

REFERENCES

Atanassov, K. T. (1986). Intuitionistic fuzzy sets. *Fuzzy Sets and Systems*, *20*(1), 87–96. DOI: 10.1016/S0165-0114(86)80034-3

Azimi, M., Eslamlou, A., & Pekcan, G. (2020). Data-Driven Structural Health Monitoring and Damage Detection through Deep Learning: State-of-the-Art Review. *Sensors (Basel)*, *20*(10), 2778. DOI: 10.3390/s20102778 PMID: 32414205

Dubois, D., & Prade, H. (1998). Possibility theory: Qualitative and quantitative aspects. In Gabbay, D., & Smets, P. (Eds.), *Handbook of Defeasible Reasoning and Uncertainty Management Systems* (Vol. 1, pp. 169–226). Springer., DOI: 10.1007/978-94-017-1735-9_6

Gogo, K. O., Nderu, L., & Mwangi, R. W. (2018, November). Fuzzy logic based context aware recommender for smart e-learning content delivery. In 2018 5th International Conference on Soft Computing & Machine Intelligence (ISCMI) (pp. 114-118). IEEE. doi:DOI: 10.1109/ISCMI.2018.8703247

Gulzar, M., Ashraf, S., & Kerre, E. E. (2024). Application of complex fuzzy relational compositions to medical diagnosis. *Mathematics*, *12*(23), 3729. DOI: 10.3390/math12233729

Pawlak, Z. (1982). Rough sets. *International Journal of Computer & Information Sciences*, *11*(5), 341–356. DOI: 10.1007/BF01001956

Ramot, D. M. (2002). Complex fuzzy sets. *IEEE Transactions on Fuzzy Systems*, *10*(2), 171–186. DOI: 10.1109/91.995119

Shafer, G. (1976). *A Mathematical Theory of Evidence*. Princeton University Press.

Sun, L., Shang, Z., Xia, Y., Bhowmick, S., & Nagarajaiah, S. (2020). Review of bridge structural health monitoring aided by big data and artificial intelligence: From condition assessment to damage detection. *Journal of Structural Engineering*, *146*(5), 04020073. DOI: 10.1061/(ASCE)ST.1943-541X.0002535

Torra, V., & Narukawa, Y. (2007). *Modeling decisions: Information fusion and aggregation operators*. Springer., DOI: 10.1007/978-3-540-68791-7_6

Wai, R.-J., & Prasetia, A. S. (2019). · R. -J. Wai and A. S. Prasetia, "Adaptive Neural Network Control and Optimal Path Planning of UAV Surveillance System With Energy Consumption Prediction,". *IEEE Access : Practical Innovations, Open Solutions*, *7*, 126137–126153. DOI: 10.1109/ACCESS.2019.2938273

Wolniak, R., & Stecuła, K. (2024). Artificial Intelligence in Smart Cities—Applications, Barriers, and Future Directions: A Review. *Smart Cities*, *7*(3), 1346–1389. DOI: 10.3390/smartcities7030057

Zadeh, L. (1965). Fuzzy sets. *Information and Control*, *8*(3), 338–353. DOI: 10.1016/S0019-9958(65)90241-X

Chapter 3
Artificial Intelligence and Machine Learning

Bipin Sule
Vishwakarma Institute of Technology, India

Durga Meshram
Datta Meghe Institute of Management Studies, India

Subramaniam Seshan Iyer
https://orcid.org/0000-0002-3843-9399
Datta Meghe Institute of Management Studies, India

Ganesh Pandit Pathak
https://orcid.org/0000-0003-0427-2145
Sri Balaji University, Pune, India

Nilesh Anute
https://orcid.org/0000-0001-6599-813X
Sri Balaji University, Pune, India

P. Selvakumar
https://orcid.org/0000-0002-3650-4548
Department of Science and Humanities, Nehru Institute of Technology, Coimbatore, India

T. C. Manjunath
https://orcid.org/0000-0003-2545-9160
Rajarajeswari College of Engineering, India

ABSTRACT

At its core, AI is about mimicking or replicating human cognitive functions using algorithms and computational models. The idea of machines exhibiting intelligent behavior is not new; it has. General AI (or Strong AI), on the other hand, refers to a hypothetical system that possesses the ability to understand, learn, and apply intelligence across a broad range of tasks—essentially matching or surpassing human capabilities. While we have made significant strides in Narrow AI, General AI remains a long-term goal still under research and development. The fundamental concept of ML is that a system can learn from examples, generalize from them, and apply that knowledge to new, unseen data. For example, if we feed a machine learning model thousands of labeled images of cats and dogs, the model can learn to

DOI: 10.4018/979-8-3373-5248-0.ch003

distinguish between them and classify new images it hasn't seen before. At its core, AI is about mimicking or replicating human cognitive functions using algorithms and computational models.

INTRODUCTION

At its core, AI is about mimicking or replicating human cognitive functions using algorithms and computational models. The idea of machines exhibiting intelligent behavior is not new; it has. General AI (or Strong AI), on the other hand, refers to a hypothetical system that possesses the ability to understand, learn, and apply intelligence across a broad range of tasks—essentially matching or surpassing human capabilities. While we have made significant strides in Narrow AI, General AI remains a long-term goal still under research and development. The fundamental concept of ML is that a system can learn from examples, generalize from them, and apply that knowledge to new, unseen data. For example, if we feed a machine learning model thousands of labeled images of cats and dogs, the model can learn to distinguish between them and classify new images it hasn't seen before. At its core, AI is about mimicking or replicating human cognitive functions using algorithms and computational models, (Andersen et al., 2018). The idea of machines exhibiting intelligent behavior is not new; it has been a subject of both scientific inquiry and science fiction for decades. However, recent advancements in computing power, data availability, and algorithm design have brought AI from theory into practice. These systems operate under limited contexts and cannot perform tasks outside of their domain. understand, learn, and apply intelligence across a broad range of tasks—essentially matching or surpassing human capabilities. The fundamental concept of ML is that a system can learn from examples, generalize from them, and apply that knowledge to new, unseen data. For example, if we feed a machine learning model thousands of labeled images of cats and dogs, the model decisions, are a form of AI that does not involve learning from data. However, ML has become the dominant method for implementing AI systems in recent years because of its ability to adapt and improve with experience. Popular algorithms and recurrent neural networks. In retail, ML helps businesses recommend products, optimize pricing, and manage inventory. In transportation, AI powers navigation systems, route optimization, and autonomous vehicles, (Apte, 2010). Even in creative industries, AI is being used to compose music, write stories, and generate art. Another critical area is to, sentiment analysis, and voice assistants AI systems. If the data used to train an ML model reflects historical inequalities or prejudices, the model can perpetuate or even amplify these biases. Privacy is another concern., ensuring user privacy and data security boxes" where their internal decision-making processes are not easily interpretable.

more understandable to humans. which brings intelligent processing closer to data sources (like mobile devices); and federated learning, which enables decentralized training across multiple devices while preserving data privacy. While the road ahead is full of both opportunities and challenges, one thing is clear: AI and ML are set to redefine the way we live, work, and interact with the world.

APPLICATIONS OF AI AND ML: TRANSFORMING INDUSTRIES

From its earliest conceptual roots to the modern AI revolution, the development of AI and ML has been shaped by contributions from mathematics, computer science, psychology, neuroscience, linguistics, and philosophy. Understanding this history provides valuable context for the current capabilities of AI and helps illuminate where the field might be heading in the future. a bronze automaton built to protect the island of Crete, and Hephaestus's mechanical servants, represent some of the earliest imaginings of artificial beings. Philosophically, thinkers like Aristotle laid foundational ideas related to logic and reasoning, which centuries later would form the theoretical bedrock for AI pondered whether human thought could be represented symbolically, (Caron et al., 2018). The notion that reasoning could be mechanized began to take a more formal shape with the development of symbolic logic, and by the 19th century, George Boole introduced Boolean algebra, a key concept in computer science and logic. During this era, the first electronic computers were developed, and researchers began to explore the potential of these machines beyond the researchers at Dartmouth believed that a significant advance in AI could be made in a single summer—a testament to their optimism. Following the Dartmouth conference, AI research gained momentum. Early programs like on symbolic AI, also known as "Good Old-Fashioned AI" (GOFAI), which focused on rule-based logic and knowledge representation. These systems relied on explicitly programmed instructions rather than learning from data. While symbolic AI achieved some early successes, it soon encountered limitations, particularly in handling real-world complexity and ambiguity. high expectations, but progress soon fell short. Despite impressive demonstrations, (Chib et al., 2025) AI systems struggled with tasks that required common sense, contextual understanding, or learning from experience. Moreover, the computational resources available at the time were insufficient to support ambitious AI projects. As a result, funding and interest declined significantly during the 1970s—a period known as the "AI Winter." Government agencies and private investors pulled back, disillusioned by the lack of practical results. Research continued, but at a slower pace and with less public attention. The limitations of symbolic approaches were becoming increasingly evident, paving the way for new ideas to emerge. systems, which used rule-based knowledge to make decisions in specific domains like medical diagnosis or mineral exploration. These systems saw some commercial success, but they were expensive to build and maintain, and still lacked adaptability. At the same time, a new approach was gaining traction: Machine Learning. Unlike symbolic AI, ML emphasized systems that could learn from data rather than rely solely on hand-crafted rules.

The rise of ML was influenced by research in statistics, pattern recognition, and neural networks. One important milestone was the development of enthusiasm. Expert systems became difficult to scale, and neural networks were still limited by available data and computing power. The field once again faced skepticism and reduced funding, leading to the second "AI Winter." and the growing availability of digital data were setting the stage for a future resurgence. media led to the creation of massive datasets. (Samal et al., 2025) At the same time, computing hardware—especially Graphics Processing Units (GPUs)—became powerful enough to handle large-scale computations. These conditions reignited interest in Machine Learning, especially in supervised learning using large datasets. Algorithms like random forests, ensemble methods, and gradient boosting became widely used. The field also began to benefit from open-source tools and frameworks such as Python, scikit-learn, and later TensorFlow and PyTorch. This period also saw breakthroughs in Natural Language Processing (NLP), computer vision, and speech recognition, bringing AI closer to real-world applications. when a deep convolutional neural network (AlexNet) drastically outperformed other models in the ImageNet competition, a benchmark for image classification, (Dahiya et al., 2025). This demonstrated that deep learning could surpass traditional ML approaches when trained on massive data with sufficient computational resources.AI also began to power real-world systems—recommendation engines on Netflix and YouTube, fraud detection in banking, and diagnostic tools in healthcare. The technology was no longer just experimental; it was mainstream. Simultaneously, generative AI tools such as DALL·E, Midjourney, and others have enabled AI to create original artwork, music, and videos, blurring the lines between human and machine creativity. Advances in reinforcement learning have enabled systems like AlphaGo and AlphaFold to master complex tasks and make scientific discoveries. The focus is now expanding to areas like explainable AI, ethical AI, federated learning, AI alignment, and edge AI, ensuring these powerful systems remain trustworthy, fair, and accessible. Governments, academic institutions, and tech companies are investing heavily in AI research and regulation, recognizing both the opportunities and risks. As AI continues to evolve rapidly, the questions being asked are not just technical but ethical, legal, and societal.

MACHINE LEARNING ALGORITHMS: SUPERVISED, UNSUPERVISED, AND REINFORCEMENT LEARNING

The increasing sophistication of AI and ML technologies has made it possible to process vast amounts of data, detect patterns, automate decision-making, and personalize user experiences on a scale that was previously unimaginable. As a result, these technologies are transforming traditional practices and redefining how industries operate, compete, and evolve. Impacts of AI and ML has been in the healthcare sector. From diagnostics to treatment personalization, AI-powered tools are revolutionizing patient care. ML algorithms can analyze complex medical data, enabling earlier and more effective interventions. In pharmaceuticals, extract insights from vast repositories of medical literature and patient records, supporting clinical decision-making and research. Moreover, (Devlin et al., 2014) Chatbots and virtual assistants help in preliminary diagnostics and appointment scheduling, while predictive models can forecast disease outbreaks and hospital resource needs. and enhance customer service. In areas like fraud detection, ML models analyze transaction patterns in real time to flag anomalies indicative of fraudulent behavior. These systems continuously learn from new data, becoming more accurate and responsive over time. Algorithmic trading—another domain powered by Suresh et al. (2025) AI—uses predictive models to execute trades at high speed and efficiency based on market trends, news sentiment, and financial indicators. Risk assessment is also streamlined through ML-driven credit scoring models that evaluate an applicant's creditworthiness using a broader range of data than traditional methods. Customer-facing applications include, (Ramesh et al., 2004). AI chatbots for handling inquiries, robo-advisors that offer personalized investment strategies, and AI tools for managing wealth portfolios. Sentiment analysis and NLP are employed to gauge market mood, analyze financial news, and inform trading strategies. This has become a critical component of platforms like Amazon, Netflix, and Spotify, significantly increasing engagement and sales.AI-driven demand forecasting enables better inventory management and reduces overstock or stockouts. Retailers use predictive analytics to understand seasonal trends, regional preferences, and promotional effectiveness. Computer vision is employed in physical stores for applications like automated checkout systems, shelf monitoring, and customer behavior analysis. often referred to as Industry 4.0, with AI and ML playing central roles a component might fail, thus preventing unplanned downtime and reducing maintenance costs. Computer vision, combined with AI, is used for quality control by identifying defects or Supply chain optimization is another area where ML models forecast demand, identify bottlenecks, and improve resource allocation. AI is also facilitating the design of smart factories where machines, sensors, and systems communicate and learn in real-time, driving higher efficiency and customization. Logistics companies use

AI to optimize delivery routes, reduce fuel consumption, and improve shipment tracking. Predictive analytics help manage fleet maintenance, forecast demand, and allocate resources efficiently. In public transportation, AI is used to manage traffic flow, improve safety, and enhance passenger experience through real-time updates and intelligent scheduling, (Dezfouli & Balleine, 2012). Airlines and rail networks use ML for predictive maintenance, dynamic pricing, and route optimization.

AI also plays a crucial role in autonomous shipping, warehouse automation, and port logistics, making the global supply chain more resilient and efficient.AI and ML are transforming education by enabling personalized learning experiences and improving administrative efficiency. and learning styles. This approach improves engagement and retention, especially in remote or digital learning environments. AI-powered tools can automate grading, detect plagiarism, and even provide instant feedback on written assignments. Chatbots and virtual tutors offer round-the-clock academic assistance, while NLP is used in language learning apps for pronunciation and grammar correction.AI-driven systems can predict crop yields, detect diseases early, and automate harvesting using robotic equipment. In livestock farming, AI is used for monitoring animal health, behavior, and feed efficiency. Food production and distribution also benefit from AI in terms of supply chain management, waste reduction, and demand forecasting. Companies use ML to optimize packaging, route planning, and inventory management, ensuring food quality and safety from farm to table. The entertainment and media industries are leveraging AI to personalize content, automate production, and enhance user engagement. Streaming services like Netflix and Spotify use sophisticated recommendation systems to suggest content based on user behavior, preferences, and social trends.AI-generated content is gaining traction, from synthetic voices in audiobooks to AI-written scripts and articles. Generative models, such as those behind deepfakes or virtual influencers, are changing how content is created and consumed. Video games employ AI for realistic NPC (non-player character) behavior and procedural content generation. In journalism, AI tools assist with content curation, real-time reporting, and audience analytics. NLP is used to summarize articles, translate content, and analyze reader sentiment. reducing energy waste and integrating solar and wind power more effectively. ML models forecast energy consumption patterns, enabling better planning and dynamic pricing. In oil and gas, AI is used for exploration, predictive maintenance, and process optimization. For environmental monitoring, ML helps analyze satellite imagery, track deforestation, monitor wildlife, and model climate change impacts.AI-driven simulations assist in designing sustainable infrastructure and anaging urban resources like water and waste. By identifying inefficiencies and predicting resource needs, AI contributes to a more sustainable and resilient

PREDICTIVE ANALYTICS: USING ML TO FORECAST OUTCOMES

In the era of digital transformation, where humans increasingly interact with software through spoken or written language, NLP serves as the vital bridge between computational logic and human expression. At its core, and work with natural language data. Traditional NLP approaches involved hand-crafted rules and linguistic models, which struggled to scale across languages and dialects. However, the shift Today, NLP systems can understand grammar, semantics, pragmatics, and even contextual cues with a high degree of accuracy. are identified. Parsing helps identify the syntactic structure of a sentence, while semantic analysis attempts to determine the actual meaning of the words and phrases in context. Another fundamental aspect is of a sentence. Early statistical models, such as n-gram models, laid the groundwork, but these were later replaced by more powerful models like context, disambiguate meaning, and generate coherent and contextually relevant responses are capable of understanding idiomatic expressions, syntax variations, and the semantics of both source and target languages to produce fluent translations., and assess brand perception. Sentiment analysis models can classify content as positive, negative, or neutral, and advanced systems even detect nuanced emotions like sarcasm, anger, joy, or disappointment. Text classification is another key NLP function, used to categorize text into predefined labels. Applications include spam detection, topic categorization, intent detection in chatbots, and legal document classification. Similarly, Question answering (QA) and information retrieval represent some of the most ambitious areas of NLP. QA systems, such as those used in search engines or AI assistants, aim to return precise answers to user queries rather than a list of documents, (Fachantidis et al., 2013). These models combine understanding of language, retrieval of relevant information, and reasoning, (Pedregosa et al., 2011). Open-domain QA models like those based on BERT or GPT can access large corpora (or the web) to generate contextually appropriate answers. Text summarization is a growing NLP task where systems condense long articles or documents into shorter versions while preserving key information. summarization, powered by transformers, enables more human-like outputs and is especially useful in news, legal, and academic fields. Microsoft Cortana use NLP for understanding and responding to spoken language. Speech-to-text systems convert verbal input into written text, which is then interpreted using NLP pipelines. On the other hand, text-to-speech (TTS) converts the system's textual response into spoken words using prosodic and phonetic modelling A critical shift in NLP came with the introduction of transformer architectures—most notably the 2017 “Attention is All You Need” paper by Vaswani et al., which introduced. The result is an extraordinary leap in machine understanding of language, enabling NLP systems to not only understand meaning but also to generate language with near-human fluency. Despite these advancements, NLP still faces significant challenges. One of the foremost is ambiguity—the same word or sentence can have different meanings depending on context. Consider the sentence: "I saw her duck." It could refer to a woman lowering her head or to her pet bird.

Humans resolve such ambiguities effortlessly; machines, less so. Contextual models have improved performance, but true understanding still eludes AI in many nuanced scenarios., they often inherit and amplify societal biases present in the text. This can lead to offensive, exclusionary, or harmful outputs, particularly in sensitive applications. Researchers are now focusing on bias detection, data curation, and fairness-aware learning to mitigate these risks. Another major hurdle is multilingual NLP. While English is the primary language for most NLP models, global adoption demands systems that work across languages, dialects, and cultural contexts. Projects like mBERT and XLM-R aim to build models that understand multiple languages simultaneously, but truly universal language models remain an open research challenge. There are also computational considerations—state-of-the-art NLP models require significant memory, processing power, and data. As such, efforts like model distillation, quantization, and low-resource NLP aim to make these technologies more accessible, especially in mobile and edge computing environments. Looking forward, the future of NLP is multimodal—combining text, audio, and visual inputs to create richer, more human-like interactions. OpenAI's GPT-4o, for example, integrates audio, vision, and text to allow seamless conversation across modalities. This capability opens up new possibilities in education, accessibility, creativity, and augmented human experiences. In conclusion, Natural Language Processing is one of the most dynamic and impactful domains of AI. It empowers machines to read, write, listen, and speak—reshaping how humans interact with technology and information. With advances in deep learning, massive pretrained models, and growing attention to ethics and inclusivity, NLP is not only becoming more intelligent but also more human-aligned. Whether it's helping a doctor summarize patient records, enabling a student to learn a new language, or allowing a business to better understand its customers, NLP is redefining communication in the digital age.

COMPUTER VISION: ENABLING MACHINES TO SEE AND UNDERSTAND

At its core, predictive analytics goes beyond merely describing what has happened insight enables organizations to act with foresight rather than hindsight, improving outcomes, efficiency, and competitiveness. Predictive analytics typically follows a structured process that begins with gathered from various sources such as customer records, transaction logs, IoT sensors, financial reports, social media, or CRM systems. Data preprocessing—including cleaning, normalization, feature selection, and handling of missing values—is critical to ensure the quality and accuracy with variables (features) that are predictive of the outcome of interest regression (predicting continuous values). For instance, in customer churn prediction (a classification task), the model estimates the likelihood that a customer will leave a company. In sales forecasting (a regression task), the model predicts the future sales volume. Commonly used ML algorithms for predictive analytics include: After tuning hyperparameters and ensuring the model performs well, it is deployed into production to make real-time or batch predictions companies predict which customers are at risk of churning and proactively offer incentives to retain them. In finance, predictive models assess credit risk by estimating a borrower's likelihood of default, detect fraudulent transactions in real time, and forecast stock market movements based on historical trends and macroeconomic indicators on patient histories, genetic information, and clinical data can forecast disease progression, predict hospital readmission risks, and identify patients likely to benefit from specific therapies, (Fountaine et al., 2019). For instance, predictive models are being used to anticipate heart attacks, detect cancer at early stages, and optimize ICU resource allocation. The integration of wearable device data and electronic health records further enhances the predictive power of these models, making healthcare more proactive and preventive. Using data from sensors and historical maintenance logs (O'Brien & Nicol, 2021), ML algorithms predict when a machine is likely to break down, allowing for timely intervention and minimizing downtime. In supply chain management, predictive models anticipate fluctuations in demand, optimize inventory levels, and reduce delivery times. For example, companies like Amazon and FedEx use predictive analytics to forecast order volumes and streamline last-mile delivery routes. evaluate the effectiveness of training programs. In talent management, algorithms assess which employees are most likely to be promoted or leave, enabling HR teams to take preemptive actions. Similarly, in education, predictive models can identify students at risk of dropping out, allowing educators to provide targeted interventions and improve learning outcomes. This involves predicting values over a sequence of time intervals—such as future sales, energy usage, or stock prices.

Time series data exhibits patterns like trend, seasonality, and autocorrelation, which specialized models must account for. Traditional methods like ARIMA and Exponential Smoothing are widely used, but newer deep learning approaches—such as Long Short-Term Memory (LSTM) and Temporal Convolutional Networks (TCNs)—have gained popularity due to their ability to capture complex temporal patterns and long-term dependencies. In addition to making predictions, many organizations are focused on prescriptive analytics, which builds on predictive models to recommend optimal actions. For example, a predictive model may forecast that sales will decline next quarter; a prescriptive model would go a step further and suggest marketing strategies or inventory adjustments to counteract the trend. This coupling of forecasting with actionable decision-making is a key step toward building intelligent, autonomous systems not enough for a model to be accurate; stakeholders need to understand how and why it makes specific predictions. This has led to the rise of explainable AI (XAI) and tools like SHAP (SHapley Additive exPlanations) and LIME (Local Interpretable Model-agnostic Explanations), which help interpret complex models by showing the contribution of each feature to the prediction. Ethical and regulatory considerations are increasingly important in predictive analytics. Predictive regulations requires careful attention to data governance, consent, and explainability. Looking ahead, the future of predictive analytics lies in automated machine learning (AutoML) and can process streaming data to make instant forecasts—crucial in domains like fraud detection, network security, or high-frequency trading.. For instance, predictive models can be deployed on edge devices in a factory to monitor machinery in real time or in retail environments to adjust prices dynamically based on customer beaverkin conclusion, predictive analytics powered by machine learning is revolutionizing how organizations anticipate and respond to future challenges and opportunities. By turning raw data into foresight, predictive models enable smarter decisions, greater efficiency, and better outcomes across virtually every domain (Ghosh et al., 2019). As algorithms become more accurate, data becomes richer, and computation more accessible, predictive analytics will become even more central to the way businesses, governments, and individuals plan for the future

ROBOTICS AND AUTOMATION: THE ROLE OF AI AND ML

This technology has evolved rapidly, driven by advances in artificial intelligence, machine learning, and hardware capabilities, and has now become foundational in numerous applications across various industries. The human visual system is incredibly adept at interpreting a continuous stream of visual stimuli, recognizing patterns, objects, and contexts almost instantaneously. Emulating this ability involves processing images or videos to extract features such as shapes, colors, textures, and motions, and then understanding their semantic significance. Unlike humans, machines process this data numerically, requiring sophisticated algorithms to transform pixel values into interpretable information. Rules and filters designed by experts to identify simple visual patterns. However, these early methods faced significant limitations in handling the diversity and complexity of real-world images., excel at recognizing spatial hierarchies in images, from simple edges to complex objects. Determining what object or category an image belongs to, such as identifying a photo as containing a dog or a cat around them.3D Reconstruction: Inferring three-dimensional shapes and structures from two-dimensional images, enabling applications in robotics and virtual reality. Modern computer vision leverages a combination of algorithms and hardware to achieve its goals. Key techniques include: Convolutional Neural Networks (CNNs): CNNs use layers of convolutions to detect features at multiple scales, making them ideal for image-related tasks. Transformers in Vision: Inspired by their success in natural language processing, vision transformers (ViTs) apply self-attention mechanisms to images, providing a powerful alternative to CNNs for certain tasks. Generative Models: Models like Generative Adversarial Networks (GANs) can generate realistic images and have been used in tasks such as image synthesis, super-resolution, and style transfer.

Data Augmentation and Transfer Learning

These strategies help models generalize better by simulating diverse training scenarios and leveraging pre-trained network, (Mnih et al., 2015). Healthcare: Assisting in medical diagnosis through image-based analysis such as tumor detection in MRI scans or retinal imaging for diabetic retinopathy. Automotive and Autonomous Vehicles: Enabling self-driving cars to perceive their environment, recognize traffic signs, detect pedestrians, and avoid obstacles. Security and Surveillance: Enhancing facial recognition, anomaly detection, and monitoring systems. Retail and E-commerce: Powering visual search, inventory management, and cashier-less stores. Agriculture: (He et al., 2019). Monitoring crop health and detecting diseases via drone imagery. Manufacturing: Improving quality control through automated visual inspection. Robotics: Allowing robots to interact with their environment through visual feedback, crucial for navigation and manipulation. Entertainment and Social Media: Augmented reality filters, content moderation, and video analysis are examples of computer vision's impact here. to AI ecosystems. Emerging trends include: Multimodal Learning: Combining vision with other data types such as text, sound, or sensor data to build richer models capable of deeper understanding. Edge Computing: Deploying vision models on devices with limited computing power (e.g., smartphones, IoT devices) to enable real-time processing without cloud dependence. Explainable AI (XAI): Developing methods to make vision models' decisions more transparent and trustworthy. Continual Learning: Allowing models to adapt and learn continuously from new visual data without forgetting prior knowledge. General AI Vision: Moving towards models that not only excel at specific tasks but also understand visual data in a broader, more flexible manner akin to human perception. In summary, computer vision is a transformative technology that enables machines to perceive and understand the visual world, driving innovation across countless fields. Its evolution from simple image processing to advanced deep learning systems illustrates the power of integrating computational techniques with vast data resources. As computer vision continues to mature, balancing technical advancements with ethical responsibility will be crucial to harness its full potential in a safe, fair, and impactful manner.

BIAS AND FAIRNESS: ADDRESSING ETHICAL CONCERNS IN AI AND ML

This comprehensive overview explores the symbiotic relationship between robotics, AI, and ML, highlighting how these fields converge to create intelligent systems that enhance productivity, safety, and efficiency across diverse domains., often aiming to increase speed, accuracy, and consistency. Together, robotics and automation form an ecosystem where physical machines execute programmed tasks, often in environments that may be hazardous, repetitive, or unsuitable for humans. Traditional robotics primarily involved fixed, rule-based machines that followed pre-defined instructions, excelling in structured settings like manufacturing assembly lines. These robots, while highly efficient at tasks such as welding, painting, or packaging, lacked flexibility and adaptability. (Williams & Rasmussen, 2006). The advent of AI and ML fundamentally changed this paradigm, introducing cognitive capabilities that allow robots to interpret complex data, learn from experience, and adjust behaviors dynamically. Early robots operated under strict programming, performing repetitive motions without deviation. (Holmes et al., 2020) Over time, as computing power increased and sensors became more sophisticated, researchers sought ways to enhance robots with perception, reasoning, and decision-making abilities.in areas like computer vision, natural language processing, and motion control. machines to interpret sensory input, understand their environment, and make informed decisions. Key AI components integrated into robotics include: data about their surroundings. AI algorithms process this data to recognize objects, detect obstacles, and assess spatial relationships, often employing an environment. Techniques like Simultaneous Localization and Mapping (SLAM) leverage ML to build and update maps while tracking the robot's location, essential for navigation in unknown or dynamic spaces. Motion Planning and Control: AI systems calculate optimal paths and control signals to move robots safely and efficiently. Reinforcement learning, a subset of ML where agents learn optimal policies through trial and error, has been particularly effective in training robots for complex manipulation and locomotion tasks. Natural Language Processing (NLP): For human-robot interaction, NLP enables robots to understand and generate human language, facilitating communication in applications like service robots, personal assistants, and collaborative industrial systems. Decision Making and Autonomy: AI allows robots to evaluate options, predict outcomes, and select actions autonomously. Advanced models incorporate probabilistic reasoning and neural networks to handle uncertainty and adapt to unforeseen circumstances. AI-powered robots enable flexible automation, adjusting to product variations and optimizing production lines. Collaborative robots (cobots) work alongside humans, sharing tasks and improving workplace safety.

Figure 1. Relations between artificial intelligence, machine learning

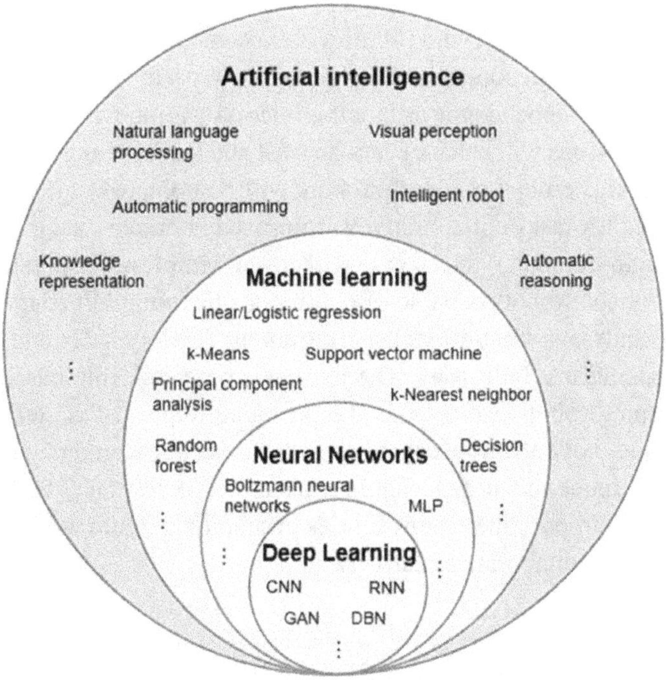

Surgical robots with AI assistance perform precise, minimally invasive operations. Robots also support rehabilitation, patient monitoring, and logistics within healthcare facilities. Autonomous drones and ground robots monitor crop health, manage pests, and perform planting or harvesting, leveraging ML to analyze environmental data and optimize yields. Intelligent robots sort, pack, and transport goods in warehouses, using AI to navigate complex layouts and coordinate with other robots and human workers. Self-driving cars, trucks, and drones utilize AI and ML to perceive their surroundings, make driving decisions, and navigate safely in real time. Robots in customer service roles provide information, deliver goods, and perform cleaning tasks, improving efficiency and user experience. Botics equipped with AI handle surveillance, reconnaissance, and bomb disposal, reducing risks to human personnel.: Real-world environments are highly dynamic and unpredictable. Developing models that generalize well and handle unexpected scenarios remains difficult. learning models require large datasets, which can be expensive or impractical to collect for specialized robotic tasks. Ensuring robots act safely around humans is paramount, necessitating rigorous testing, fail-safes, and ethical considerations. Many robotic platforms have limited processing power and energy capacity, requiring efficient algorithms and hardware optimization. Automation raises concerns about job displacement, privacy, and accountability, prompting ongoing discussions about the societal role of intelligent robots. Looking forward, the integration of AI and ML with robotics is expected to deepen, driven by several emerging trends: in edge computing enable robots to perform AI inference locally, reducing latency and dependency on cloud services. Developing models that provide interpretable decisions will enhance human trust and facilitate regulatory approval. Inspired by nature, groups of simple robots will coordinate via decentralized AI to perform complex tasks collaboratively. Improved interfaces, such as augmented reality and brain-computer interfaces, will foster seamless cooperation between humans and robots. Robots capable of continuous learning will adapt to evolving environments and tasks without extensive retraining. flexible, safer, and better suited for delicate operations. Industries. The journey from rigid, rule-based systems to adaptive, learning robots exemplifies the profound impact of AI technologies in making machines not only more capable but also more autonomous and interactive. As research continues and these technologies mature, society faces both remarkable opportunities and important responsibilities in shaping a future where humans and robots coexist and collaborate effectively.

EXPLAINABILITY AND TRANSPARENCY: UNDERSTANDING AI DECISION-MAKING

These concerns stem from the reality that AI and ML systems often reflect and amplify existing societal biases, resulting in unfair or discriminatory outcomes. Addressing bias and promoting fairness is thus essential not only to uphold ethical standards but also to ensure the equitable and trustworthy deployment of AI technologies. These biases often originate from the data used to train models, the design of algorithms, or the deployment context. Bias can manifest in various forms: racial, gender, age, socioeconomic, or cultural, among others, (Kommineni et al., 2025). Machine learning models learn patterns from historical data, which frequently carry the imprint of societal inequalities. For example, if a hiring algorithm is trained on past employment records from an industry dominated by a particular demographic, the model may inadvertently perpetuate discriminatory hiring practices. This phenomenon is known as data bias and is one of the most common sources of unfairness in AI. Algorithmic bias can also arise from model design choices, such as selecting objective emerge during data collection, labeling, or preprocessing stages, reflecting implicit or explicit prejudices of human annotators. When past data reflects unequal treatment or discrimination, models trained on such data inherit those biases.

Representation Bias

Occurs when certain groups are underrepresented or absent in the training data, leading to poor model performance on these populations. Measurement Bias: Arises from errors or inconsistencies in how data is measured or labeled, often skewing the model's understanding.

Aggregation Bias

Happens when diverse groups with different underlying data distributions are treated homogeneously, ignoring subgroup-specific differences. Evaluation Bias: When the metrics used to assess model performance do not account for fairness or differential impact across groups. Fairness in AI is a normative concept reflecting the ideal that AI systems should treat individuals and groups equitably, without unjust discrimination. Achieving fairness is crucial for several reasons: Ethical Responsibility: Developers and organizations have a moral duty to prevent harm and respect human rights.

Legal Compliance

Many jurisdictions have laws prohibiting discrimination in areas such as employment, lending, and housing, which AI systems must comply with.

Social Trust

Fair AI systems promote user trust and wider societal acceptance, critical for adoption.

Economic Efficiency

Bias can lead to poor decision-making and exclusion of valuable human resources, ultimately impacting organizational performance. These definitions often conflict, making it impossible to satisfy all fairness criteria simultaneously in many scenarios. Selecting appropriate fairness metrics depends on application goals, societal context, and ethical priorities. Furthermore, measuring fairness requires access to sensitive demographic data, which may raise privacy concerns or be unavailable.

This complicates auditing and enforcing fairness in real-world systems. minority groups, exacerbating existing inequalities. In healthcare, AI systems trained on non-representative data may underdiagnose diseases in marginalized populations, leading to inadequate treatment. Financial services algorithms may deny loans or insurance unfairly, reinforcing economic disparities. These impacts can perpetuate cycles of discrimination and social injustice, undermining the legitimacy of AI technologies. Recognizing and mitigating bias is therefore essential for safeguarding human dignity and promoting inclusive progress. Ensuring diverse, representative datasets and applying techniques like re-sampling or re-weighting to correct imbalances. Hat incorporate fairness constraints or objectives directly into model training. models and providing explanations for AI decisions to detect and rectify biases. Beyond technical solutions, addressing bias and fairness requires adopting ethical frameworks and fostering a culture of responsible AI. Principles such as justice, autonomy, beneficence, and non-maleficence guide developers to create systems that respect human values. Organizations increasingly adopt Responsible AI frameworks that emphasize fairness alongside privacy, safety, transparency, and accountability. Multi-disciplinary collaboration, continuous education, and public engagement are vital for aligning AI innovations with societal norms and human rights.

Causal Inference

Understanding the root causes of bias and distinguishing correlation from causation to design more robust fairness interventions. Vigilant attention and proactive measures. The complexity of fairness demands interdisciplinary efforts, blending technical innovation with social understanding and ethical reflection. By confronting bias head-on and embedding fairness in AI systems, society can harness the transformative power of AI while ensuring that its benefits are shared equitably, fostering trust, justice, and inclusive progress in an increasingly AI-driven world.

EMERGING TRENDS: EDGE AI, TRANSFER LEARNING, AND MORE

Artificial Intelligence (AI) systems are becoming deeply integrated into critical infrastructure, business operations, healthcare, finance, transportation, and daily life. This growing reliance on AI amplifies the importance of ensuring these systems are secure and that users' privacy is protected. Security and privacy in AI concern safeguarding AI models, data, and deployment environments from malicious attacks, unauthorized access, and unintended information leaks. Addressing these challenges is essential to maintain trust, reliability, and ethical integrity in AI-driven technologies. in real-time. A breach in security or privacy can lead to dire consequences such as identity theft, financial loss, manipulation of automated systems, and erosion of user trust. Moreover, AI models themselves can become targets or vectors of attacks, which can degrade system performance, cause erroneous decisions, or reveal proprietary information. Data Protection Regulation (GDPR), the California Consumer Privacy Act (CCPA), and industry-specific standards increasingly mandate stringent protections for data and automated decision-making processes. Data collection and training to inference and deployment. These threats include: images or audio signals to mislead classifiers, often imperceptibly to humans. Techniques that allow attackers to reconstruct sensitive training data or extract proprietary model parameters by querying the AI system, compromising data privacy and intellectual property. made during the inference phase that enable malicious inputs to bypass detection systems, such as malware classifiers or fraud detectors. when presented with specific inputs, while behaving normally otherwise. Malicious activities by authorized users who misuse their access to tamper with models or data. The black-box nature of many AI models makes it difficult to detect and diagnose attacks or anomalies. Automated AI systems can be exploited at scale, leading to widespread impact. constrained resources, limiting security measures and increasing attack surfaces. Ensuring personal data used in AI training and inference remains protected from unauthorized disclosure. Limiting data collection and retention to only what is necessary for AI functionality.: Techniques aimed at removing personally identifiable information (PII) from datasets, though these are often vulnerable to re-identification attacks. A mathematical framework that introduces noise to datasets or query results, providing strong privacy guarantees while enabling useful data analysis and model training. A decentralized training approach where models learn collaboratively across multiple devices or servers without sharing raw data, thus enhancing privacy. Protecting (Kaur et al., 2025) AI systems requires a multi-layered approach combining technical, organizational, and policy measures.

Employ adversarial training and data validation techniques to increase resilience against poisoning and evasion attacks. Use cryptographic methods to secure data at rest, in transit, and during computation, including homomorphic encryption for privacy-preserving inference. Implement anomaly detection systems to identify unusual patterns indicative of attacks Leverage interpretable models and explanations to aid in detecting tampering or unexpected behaviors. Security and privacy are not merely technical issues but also ethical imperatives. AI systems that fail to protect user data or are vulnerable to manipulation can exacerbate inequalities, violate human rights, and undermine societal trust in technology. Organizations must balance innovation with responsibility, adopting frameworks that prioritize human-centric AI development public awareness and transparency about AI security and privacy practices help build informed consent and trust. Cross-disciplinary collaboration among AI researchers, security experts, policymakers, and civil society is critical to develop comprehensive safeguards. will evolve, necessitating continuous innovation. Promising areas include: Developing formal verification methods that guarantee AI models' resistance to certain attack classes. AI systems capable of detecting and adapting to novel threats autonomously. Designing specialized chips that enhance the security of AI computations. Advancing techniques like federated learning and differential privacy to balance utility and protection. Establishing global standards and best practices for AI security and privacy. Strengthening policies that mandate transparency, accountability, and ethical responsibility in AI deployment.AI systems is essential for realizing their transformative potential while mitigating risks in an increasingly connected and data-driven world.

FUTURE APPLICATIONS: AI AND ML IN HEALTHCARE, FINANCE, AND BEYOND

The field of Artificial Intelligence (AI) is rapidly evolving, driven by continuous advances in algorithms, computational power, and data availability. Beyond the foundational breakthroughs in deep learning and reinforcement learning, new trends are shaping the future trajectory of AI. Among these, Edge AI, Transfer Learning, Federated Learning, Explainable AI, and Multi-Modal Learning stand out as transformative innovations that address key challenges such as latency, data scarcity, privacy, interpretability, and full potential in diverse real-world settings. Additionally, Edge AI enhances privacy by keeping sensitive data on-device, reducing exposure to potential breaches. Technological enablers of Edge AI include specialized hardware accelerators like NVIDIA Jetson, Google Edge TPU, and custom AI chips optimized for low-power inference. Lightweight model architectures, such as MobileNet and quantized neural networks, are designed to operate within the resource constraints of edge devices. Despite its promise, (Liu & Wu, 2006) Edge AI faces challenges like limited computational resources, energy efficiency, security vulnerabilities, and difficulties in updating models post-deployment. Research is ongoing into model compression, pruning, and efficient training techniques to address these barriers. Gained from pre-trained models on large datasets to accelerate learning in related but distinct domains. For instance, models pre-trained on ImageNet—an extensive dataset of labeled images—can be fine-tuned for specialized medical imaging tasks with significantly fewer examples. On new data. The success of transfer learning has been amplified by the rise of large-scale foundational models like GPT and BERT in natural language processing (NLP), which can be adapted for diverse applications such as sentiment analysis, translation, and question answering. Differ substantially—and catastrophic forgetting, where models lose prior knowledge when fine-tuned on new tasks. Addressing these issues is an active area of research. Participant computes updates locally using its data and sends only model parameters or gradients to a central server, preserving data privacy and security.

This paradigm is particularly valuable in sensitive domains such as healthcare, finance, and telecommunications, where data sharing is restricted by privacy regulations or competitive concerns. For example, hospitals more robust models. However, it introduces challenges like communication efficiency, heterogeneous data distributions, and the need for secure aggregation protocols to prevent leakage from model updates. increasingly difficult yet essential. Explainable AI (XAI) aims to make AI models interpretable and transparent, providing stakeholders with insights into why and how models arrive at specific predictions or actions. Maps, and counterfactual explanations. Challenges in XAI involve balancing interpretability with model performance and tailoring explanations to diverse audiences, from domain experts to end-users. is crucial for applications requiring a holistic understanding of complex environments. For instance, autonomous vehicles fuse camera images, lidar scans, GPS data, and map information to navigate safely. In healthcare, combining medical images, genomic data, and patient records enables more accurate diagnoses. Techniques that enable models to learn useful representations from unlabeled data, reducing dependence on costly annotations. Exploring quantum computing paradigms to accelerate AI algorithms. related to data availability, computational resources, privacy, interpretability, and adaptability. They open new frontiers for deploying AI in real-world scenarios where constraints and requirements vary widely. Future research will likely focus on integrating these trends—for example, developing explainable edge AI models trained via federated learning or enhancing transfer learning with continual adaptation. Ethical considerations, standardization, and robust evaluation frameworks will also play central roles in shaping responsible AI innovation.

HUMAN-AI COLLABORATION: ENHANCING PRODUCTIVITY AND DECISION-MAKING

Artificial Intelligence (AI) and Machine Learning (ML) are no longer confined to experimental labs or isolated enterprise solutions—they are now foundational tools driving transformation across nearly every sector of the global economy. As computational capabilities, algorithmic sophistication, and access to large datasets improve, the future holds enormous potential for AI and ML to solve complex problems, optimize systems, and augment human capabilities in unprecedented ways. Key industries like healthcare and finance stand to benefit dramatically, while others such as education, agriculture, manufacturing, climate science, and creative arts are beginning to experience paradigm shifts fueled by intelligent technologies. Healthcare is arguably one of the most promising and socially impactful frontiers for future AI and ML applications. As medical data—from electronic health records (EHRs) and genetic profiles to radiological images and draw actionable insights from this information more efficiently than human practitioners ever could. Future applications will see widespread deployment of ML models capable of interpreting X-rays, MRIs, CT scans, and even biopsy samples with high precision. These systems will serve as second opinions or first-pass filters, increasing diagnostic accuracy and speeding up workflows in resource-constrained settings. Biological characteristics. Predictive models can recommend therapies with the highest success and repurposing existing drugs faster and at a lower cost. (Lu et al., 2018). In the future, ML models will guide simulation-driven drug design, accelerating clinical trial planning, patient stratification, and outcome forecasting. AI-enabled wearable devices and smart home health systems will support continuous patient monitoring, especially for provide 24/7 health advice, appointment scheduling, and adherence support. public health strategies.

In future pandemics, AI models may predict the spread of disease, identify vulnerable populations, and optimize vaccine distribution far more effectively., and the next wave of AI and ML innovation will further redefine how capital markets, consumer banking, insurance, and compliance functions operate. transaction history, mobile behavior, and even social network patterns. This expansion improves financial inclusion for individuals lacking traditional credit histories while maintaining rigorous risk controls., these models can prevent fraud before it happens. Future systems will include explainable AI (XAI) elements to justify flagged behavior for capable of adapting to evolving market conditions, geopolitical changes, and macroeconomic signals. Natural language interfaces will allow seamless human-AI interaction, bringing advisory services to a broader audience. NLP algorithms will process and interpret regulatory documents, while ML models will monitor transactions for violations, ensuring real-time auditing and reducing operational overhead. These systems use ML to adapt content delivery based on each student's pace, understanding, and learning style, (Zhu et al., 2011). Future ITS platforms will offer real-time feedback, problem-solving guidance, and emotional support using sentiment analysis. This will allow educators to focus more on pedagogy and less on administrative tasks. enabling timely intervention. Predictive analytics can help institutions optimize course offerings and resource allocation.AI will increasingly drive agricultural innovation, helping to meet the demands of a growing global population while reducing environmental impact AI will analyze this data to guide precision interventions—like irrigation or pesticide application—minimizing waste.

ML models will forecast crop yields by analyzing weather patterns, historical data, and satellite imagery, aiding in supply chain planning and policy decisions.AI-powered robots will handle planting, weeding, and harvesting with high efficiency, helping address labor shortages and improving productivity, (Torrey & Shavlik, 2010). ML can augment traditional climate models by identifying non-linear patterns and improving prediction accuracy for extreme weather events, ocean currents, and temperature anomalies AI will help detect emissions using satellite imagery and IoT sensors, enabling governments and corporations to track and reduce their carbon footprint. Smart grids and buildings powered by AI will manage energy loads, predict consumption patterns, and reduce waste through intelligent automation. The future of creative work will involve AI as a co-creator, augmenting rather than replacing human creativity. AI models like GPT and image-generation systems will assist in writing scripts, composing music, designing graphics, and even developing video games. Streaming platforms will use AI to curate hyper-personalized content experiences, adjusting in real time to viewer behavior and preferences.AI-generated personalities will interact with users, advertise products, and participate in storytelling, creating new digital economies. As AI applications proliferate, future systems will increasingly integrate insights across sectors. For instance, AI in finance may support healthcare insurance underwriting, while education AI could inform job market forecasting. These intersections will amplify AI's transformative potential but also raise complex challenges. Ensuring fairness, accountability, and transparency will be paramount, especially in high-stakes domains like healthcare and finance. As AI systems grow more interconnected, harmonizing data standards and APIs will be critical. Designing interfaces and workflows that support effective human (Zhang et al., 2019)-AI teamwork will be essential for success., the future of AI is not just technological—it is societal. These innovations will reshape how we live, work, learn, and interact, bringing opportunities for efficiency, personalization, and insight at an unprecedented scale. However, realizing this potential will require careful attention to ethical deployment, cross-sector collaboration, and an inclusive vision of progress that ensures AI serves all of humanity.

THE FUTURE OF AI AND ML: OPPORTUNITIES, CHALLENGES, AND IMPLICATIONS

Artificial Intelligence (AI) is increasingly transitioning from a tool that automates routine tasks to a powerful collaborator that enhances human capabilities. The focus of modern AI development has shifted from replacing humans to augmenting them—empowering professionals to be more productive, make better-informed decisions, and tackle complex challenges more effectively, (Wahl et al., 2018). This shift defines the emerging paradigm of accomplish alone. Rather than functioning as a competitor or substitute, AI in collaborative contexts becomes an enabler—

offering speed, scale, consistency, and pattern recognition, while humans contribute creativity, context, ethics, empathy, and Successful human-AI collaboration relies on a few foundational principle and humans should perform roles that leverage their respective strengths. AI excels at data processing, prediction, and automation; humans excel at reasoning, moral judgment, empathy, and dealing with uncertainty.

For collaboration to be effective, users must understand and trust the AI's capabilities and limitations, (Torresen, 2018). Explainable AI (XAI) tools help facilitate this understanding.

Collaborative systems should allow humans to intervene, override, or adjust AI behavior as necessary. The system must adapt to human preferences, workflows, and contextual shifts.

,AI systems can improve based on human feedback, leading to more personalized and responsive models.AI is becoming a trusted assistant to medical professionals, not by making autonomous decisions but by offering decision support, triaging patients, analyzing imaging, and identifying treatment options. AI systems help flag abnormalities in scans and biopsy slides, speeding up diagnosis. Physicians can use AI recommendations as a second opinion, especially in high-volume environments.

CONCLUSION

AI analyzes patient data to suggest tailored treatments. Doctors evaluate AI-driven insights alongside their own clinical expertise to make final decisions. Virtual assistants automate medical recordkeeping, freeing up physicians to spend more time with patients., AI collaborates with analysts, traders, and compliance officers to identify patterns, generate insights, and ensure compliance. -makers interpret these insights within broader economic and geopolitical contexts. Auditors assess the legitimacy of flagged transactions. While human agents step in for complex, emotionally nuanced interactions by suggesting ideas, completing sentences, and even drafting content, which humans refine and personalize. Artists use generative. AI to explore novel sounds and visuals. The human touch remains essential for curating and imbuing creations with emotion and meaning. Aids in scriptwriting, scene planning, and editing, enhancing the production pipeline without diminishing human creativity.

Tutoring Systems

AI-powered platforms adapt lessons based on student progress. Teachers use these insights to intervene where human support is most needed.AI handles basic assessments, while instructors focus on open-ended feedback and student mentoring.

Data-driven AI models help educators develop more effective, engaging learning materials.AI forecasts equipment failures. Human engineers validate recommendations and schedule interventions accordingly.AI visual systems identify defects, while humans oversee edge cases and refine standards. Collaborative robots (cobots) work safely alongside humans, responding to gestures, speech, and movements in shared workspaces. AI proposes candidate molecules, but human researchers determine the feasibility and ethical implications of trials.: AI enhances simulation accuracy, but humans assess long-term societal impacts and policy responses.

Enhanced Productivity

AI handles repetitive tasks and processes large datasets quickly, allowing humans to focus on higher-order thinking and creativity. AI reduces human error in routine or data-heavy tasks, leading to more consistent results. AI enables knowledge amplification—bringing expert-level insights to broader audiences, especially in areas lacking specialist access. Real-time analytics and predictive modeling help decision-makers act swiftly with confidence. By handling complex data synthesis, AI opens new frontiers for innovation that were previously unreachable due to cognitive or temporal constraints. While trust in AI is essential, blind reliance can be dangerous, especially if users fail to question flawed outputs. Striking the right balance is key. Poorly designed human-AI interfaces can hinder usability. Interaction models must be intuitive, responsive, and context-aware. Collaborative systems may inherit biases from training data. Humans need tools to identify and correct these biases.: When humans and AI share decision-making, it can be unclear who is responsible for outcomes—raising legal, ethical, and procedural concerns. Users must understand AI fundamentals to interact effectively. Organizations must invest in education and reskilling to prepare their workforce. Creating optimal human-AI partnerships requires attention to system design, user experience, and organizational culture: Define which decisions are automated, augmented, or fully human-led. Systems should learn from human input and allow humans to critique or override AI outputs. Design AI systems that communicate their reasoning, confidence levels, and limitations clearly to users. Collaborative tools should serve diverse populations, regardless of technical skill or physical ability. The next generation of AI systems will be more human-centric, context-aware, and emotionally intelligent. We will see growth in: Tools that engage in complex dialogue, understand user intent, and provide nuanced support. that recognize and respond to human emotions to improve collaboration, especially in education, healthcare, and customer service. A shift in framing AI not as "artificial intelligence" but as "augmented intelligence"—focused on complementing human cognition rather than replicating it. Human-AI collaboration represents one of the most transformative shifts in the digital era. By fusing human intuition and empathy with machine speed and scale, collaborative systems promise to make work more meaningful, decisions more informed, and innovation more accessible. As AI continues to evolve, the question is not whether machines will replace humans, but how we can design systems that bring out the best in both. The future lies in partnerships—where AI is not our rival, but our teammate.

REFERENCES

Andersen P. A., Goodwin M., & Granmo O. C. (2018, August). Deep RTS: A game environment for deep reinforcement learning in real-time strategy games. In 2018 IEEE conference on computational intelligence and games (CIG) (pp. 1–8). IEEE.

Apte, C. (2010). The role of machine learning in business optimization (Paper presentation). The 27th International Conference on Machine Learning (ICML-10) (pp. 1–2).

Caron, M., Bojanowski, P., Joulin, A., & Douze, M. (2018). Deep clustering for unsupervised learning of visual features [Paper presentation]. *The European Conference on Computer Vision (ECCV)* (pp. 132–149).

Chib, S., Mehta, A. K., Selvakumar, P., Saravanan, A., Mishra, B. R., & Manjunath, T. C. (2025). Mental Health Challenges and Solutions in High-Pressure Work Environments. In Harriott, S. (Ed.), *Innovative Approaches to Managing Conflict and Change in Diverse Work Environments* (pp. 247–272). IGI Global Scientific Publishing., DOI: 10.4018/979-8-3693-9556-1.ch011

Dahiya, R., & Selvakumar, P. Vidyadhara, & Manjunath, T. C. (2025). Design and Aerodynamics: Principles and Advances. In M. Mellal (Ed.), Innovations and Developments in Unmanned Aerial Vehicles (pp. 145-170). IGI Global Scientific Publishing. DOI: 10.4018/979-8-3693-8462-6.ch007

Devlin, S., Yliniemi, L., Kudenko, D., & Tumer, K. (2014, May). Potential-based difference rewards for multiagent reinforcement learning [Paper presentation]. The 2014 International Conference on Autonomous Agents and Multi-agent Systems (pp. 165–172). https://web.engr.oregonstate.edu/~ktumer/publications/files/tumer

Dezfouli, A., & Balleine, B. W. (2012). Habits, action sequences and reinforcement learning. *The European Journal of Neuroscience*, *35*(7), 1036–1051. DOI: 10.1111/j.1460-9568.2012.08050.x PMID: 22487034

Fachantidis, A., Partalas, I., Tsoumakas, G., & Vlahavas, I. (2013). Transferring task models in reinforcement learning agents. *Neurocomputing*, *107*, 23–32. DOI: 10.1016/j.neucom.2012.08.039

Fountaine, T., McCarthy, B., & Saleh, T. (2019). Building the AI-powered organization. *Harvard Business Review*, *97*(4), 62–73.

Ghosh, S., Das, N., Das, I., & Maulik, U. (2019). Understanding deep learning techniques for image segmentation. *ACM Computing Surveys*, *52*(4), 1–35. DOI: 10.1145/3329784

He, J., Baxter, S. L., Xu, J., Xu, J., Zhou, X., & Zhang, K. (2019). The practical implementation of artificial intelligence technologies in medicine. *Nature Medicine*, *25*(1), 30–36. DOI: 10.1038/s41591-018-0307-0 PMID: 30617336

Holmes, G., Donkin, A., & Witten, I. H. (2020). Weka: A machine-learning workbench [Paper presentation]. The ANZIIS'94-Australian New Zealand Intelligent Information Systems Conference (pp. 357–361). IEEE.

Kaur, M., Selvakumar, P., & Dinesh, N. Mohit, Chandel, P. S., & Manjunath, T. C. (2025). Reconstruction and Preservation Strategies for Cultural Heritage. In S. Oflazoğlu & M. Dora (Eds.), Preserving Cultural Heritage in Post-Disaster Urban Renewal (pp. 385-412). IGI Global Scientific Publishing. DOI: 10.4018/979-8-3693-9097-9.ch016

Kommineni, M., Chundru, S., Maroju, P. K., & Selvakumar, P. (2025). Ethical Implications of AI in Sustainable Development Pedagogy. In Chinedu, C., Mafarja, N., Hussein, M., & Saleem, A. (Eds.), *Rethinking the Pedagogy of Sustainable Development in the AI Era* (pp. 17–36). IGI Global Scientific Publishing., DOI: 10.4018/979-8-3693-9062-7.ch002

Liu, H. X., & Wu, B. (2006). Application and development of artificial intelligence technology in the precision forestry. *Xibei Linxueyuan Xuebao*, *21*(1), 183.

Lu, H., Li, Y., Chen, M., Kim, H., & Serikawa, S. (2018). Brain intelligence: Go beyond artificial intelligence. *Mobile Networks and Applications*, *23*(2), 368–375. DOI: 10.1007/s11036-017-0932-8

O'Brien, P. D., & Nicol, R. C. (2021). FIPA: Towards a standard for software agents. *BT Technology Journal*, *16*(3), 51–59. DOI: 10.1023/A:1009621729979

Pedregosa, F., Varoquaux, G., Gramfort, A., Michel, V., Thirion, B., Grisel, O., Blondel, M., Prettenhofer, P., Weiss, R., Dubourg, V., & Vanderplas, J. (2011). Scikit-learn: Machine learning in Python. *Journal of Machine Learning Research*, *12*, 2825–2830.

Ramesh, A. N., Kambhampati, C., Monson, J. R., & Drew, P. J. (2004). Artificial intelligence in medicine. *Annals of the Royal College of Surgeons of England*, *86*(5), 334–338. DOI: 10.1308/147870804290 PMID: 15333167

Samal, A., & Misra, L. I. Mohit, Pougajendy, S., Chandel, P. S., anjunath, T. C., & Selvakumar, P. (2025). Musealization and Tourism Practices. In S. Oflazoğlu & M. Dora (Eds.), Preserving Cultural Heritage in Post-Disaster Urban Renewal (pp. 115-142). IGI Global Scientific Publishing. DOI: 10.4018/979-8-3693-9097-9.ch006

Suresh, R. V., Selvakumar, P., Naresh Babu, K., Nithya, G., Manjunath, T. C., & Kartigeyan, J. (2025). Best Practices and Practical Applications. In E. Çela, V. Paliktzoglou, S. Sengar, P. Eappen, &. Vajjhala (Eds.), Enhancing School Counseling With Technology and Case Studies (pp. 341-362). IGI Global Scientific Publishing. DOI: 10.4018/979-8-3693-8392-6.ch013

Torresen, J. (2018). A review of future and ethical perspectives of robotics and AI. *Frontiers in Robotics and AI*, *4*, 75. DOI: 10.3389/frobt.2017.00075

Torrey, L., & Shavlik, J. (2010). Transfer learning. In *Handbook of research on machine learning applications and trends: Algorithms, methods, and techniques* (pp. 242–264). IGI Global. DOI: 10.4018/978-1-60566-766-9.ch011

Wahl, B., Cossy-Gantner, A., Germann, S., & Schwalbe, N. R. (2018). Artificial intelligence (AI) and global health: How can AI contribute to health in resource-poor settings? *BMJ Global Health*, *3*(4), e000798. DOI: 10.1136/bmjgh-2018-000798 PMID: 30233828

Williams, C. K., & Rasmussen, C. E. (2006). *No. 3* (Vol. 2). Gaussian processes for machine learning. MIT Press.

Zhang, D., Zhang, Y., Li, Q., Plummer, T., & Wang, D. (2019, July). Crowdlearn: A crowd-AI hybrid system for deep learning-based damage assessment applications [Paper presentation]. The 2019 IEEE 39th International Conference on Distributed Computing Systems (ICDCS) (pp. 1221–1232). IEEE.

Zhu, Y., Chen, Y., Lu, Z., Pan, S. J., Xue, G. R., Yu, Y., & Yang, Q. (2011, August). Heterogeneous transfer learning for image classification [Paper presentation]. The 25th AAAI Conference on Artificial Intelligence (Vol. 11, pp. 1304–1309). https://www.aaai.org/ocs/index.php/AAAI/AAAI11/paper/viewFile/3671/4073

Chapter 4
Comparative Study Between YOLOv7, YOLOv8, and YOLOv9 for Pothole Detection on Indian Roads

Pranav Jaysankar
Anna University, Chennai, India

Deepak Rajeshkumar
Anna University, Chennai, India

Balaji Chandrasekaran
Anna University, Chennai, India

G. S. Smrithy
https://orcid.org/0000-0002-9456-9817
VIT Chennai, India

ABSTRACT

This paper aims to improve pothole detection on Indian roads using advanced deep learning techniques. With fatalities due to potholes increasing by 25% annually, from 1481 in 2021 to 1856 in 2022, enhancing detection accuracy is critical. Building on previous studies that used YOLOv2, YOLOv3, and YOLOv3-tiny models, this initiative will implement YOLOv7, YOLOv8, and YOLOv9 to achieve a higher mean Average Precision (mAP) score. The annotated dataset from Kaggle is split into training, testing, and validation sets. Metrics such as mAP, recall, precision, F1-Score, and

DOI: 10.4018/979-8-3373-5248-0.ch004

accuracy will assess performance. This paper supports Advanced Driver Assistance Systems (ADAS) by providing data to help vehicles navigate potholes, ultimately aiming to reduce accidents and save lives. A comparative analysis shows YOLOv7 as the most effective, underscoring its potential to enhance road safety through real-time alerts and advanced AI.

INTRODUCTION

Pothole-induced accidents have become a significant concern on Indian roads, with data from the Road and Transport Authority of India indicating an alarming 25% increase in fatalities from 1481 in 2021 to 1856 in 2022. Previous efforts utilizing detection models such as YOLOv2, YOLOv3, and YOLOv3-tiny have shown limitations in terms of lower mean Average Precision (mAP) scores, inadequate processing speeds for real-time applications, and dependency on substantial hardware resources. These constraints underscore the urgent need for more efficient and accurate detection systems to mitigate the growing number of accidents and enhance overall road safety.

This research paper aims to address these challenges by conducting a comprehensive comparative study of the latest YOLO models YOLOv7, YOLOv8, and YOLOv9. The objective is to evaluate and identify the most effective model in terms of accuracy, speed, and robustness for real-time pothole detection. Specifically, the study seeks to reduce accidents by providing timely alerts to drivers, support Advanced Driver Assistance Systems (ADAS) with precise pothole location data, and demonstrate performance improvements over older models through enhanced detection capabilities. The evaluation will consider various performance metrics, including recall, precision, F_1-Score, and consistency across different Intersection over Union (IoU) thresholds, to ensure applicability in diverse real-world conditions.

Despite the promising potential of these advanced models, several challenges persist, such as achieving high accuracy under varying road and lighting conditions, ensuring real-time processing efficiency, overcoming hardware limitations, and integrating the detection system seamlessly with existing infrastructure and vehicle systems like ADAS. This thesis is organized to systematically address these issues, beginning with a detailed literature review, followed by the presentation of the proposed system architecture and module descriptions. Subsequent sections will discuss the testing methodologies, results, and observations, culminating in conclusions and recommendations for future work aimed at developing safer and more resilient road infrastructure through advanced deep learning techniques.

RELATED WORKS

Seraj et al. (2014) introduced a system called RoADS, designed to monitor road pavement using smartphones for detecting anomalies. The system utilizes the accelerometer in smartphones to identify speed reductions, which could indicate potholes or other road issues. Instead of using deep learning techniques, the system employs Support Vector Machines (SVM) for classification, analyzing changes in vehicle speed. This approach leverages the widespread use of smartphones but may be limited in accuracy and scalability compared to more advanced deep learning methods.

Gupta et al. (2020) conducted research focused on detecting and localizing potholes in thermal images using deep neural networks (DNNs). Their study aimed to tackle the challenge of pothole detection by using thermal imaging and deep learning, training DNNs on thermal image datasets for accurate detection and localization. This method could offer advantages over traditional visual-based approaches, especially in low-light or adverse weather conditions. However, a noted limitation is the inability to label potholes based on severity and size, potentially affecting maintenance efforts. Despite this, the research contributes to advancements in pothole detection technologies with implications for road maintenance and safety.

Shaghouri et al. (2021) explored real-time pothole detection using deep learning models, including eSSD TensorFlow and YOLOv4CSPDarknet53. The study aimed to detect potholes efficiently and accurately using these advanced architectures. However, the use of older YOLO models may limit the exploitation of the latest advancements in object detection algorithms. Nevertheless, this research adds to the growing body of work on deep learning applications for real-time road anomaly detection, addressing the critical issue of pothole identification and prevention.

Faramarzi et al. (2020) focused on detecting and classifying road damage using deep neural networks, specifically YOLOv4, with images captured by smartphones. Their research aimed to enhance the accuracy and efficiency of road damage detection by leveraging deep learning techniques. The use of smartphone images provides a practical and cost-effective approach to data collection for road monitoring. However, the study may be limited by the need for large, diverse datasets for effective training and challenges related to real-world deployment, such as varying lighting conditions and image quality. Overall, this research contributes to the application of deep learning technologies in road infrastructure management and maintenance.

Liu et al. (2023) introduced DSW-YOLOv8n, an improved version of YOLOv8n, aimed at enhancing underwater target detection. The study sought to address challenges in underwater detection by improving feature extraction and network optimization within the YOLOv8n architecture. While the proposed algorithm achieved greater accuracy and efficiency, it also required significant computational resources, which may hinder real-time deployment in practical underwater scenarios. Despite this limitation, the research represents an advancement in underwater surveillance and exploration algorithms.

Wang et al. (2023) presented BL-YOLOv8, an improved model for detecting road defects based on the YOLOv8 architecture. The study aimed to enhance the accuracy and efficiency of road defect detection using deep learning. While the BL-YOLOv8 model showed improvements, some defects went undetected, suggesting potential limitations in identifying small targets or defects that blend into the environment. Despite this, the research advances the field of road defect detection, with applications in infrastructure maintenance and road safety.

Du et al. (2018) provided a comprehensive overview of object detection methods, focusing on Convolutional Neural Networks (CNNs) and the YOLO architecture. The study explored the principles and methodologies behind these techniques, noting that YOLO models tended to have lower mean Average Precision (mAP) scores compared to other CNN-based methods, indicating potential accuracy limitations. The absence of a specific dataset for experimentation may have restricted the depth of analysis. However, the study contributes to a better understanding of object detection techniques and provides a foundation for future research.

Sathvik et al. (2022) developed an intelligent pothole detection system using a CNN based on YOLO-V7. The study aimed to improve the accuracy and efficiency of pothole detection for road maintenance and safety. While the model was designed for real-time detection, the lack of real-time image analysis could affect its performance in practical scenarios.

Pang et al. (2023 introduced SP-YOLOv8s, an enhanced version of YOLOv8s tailored for detecting small objects in remote sensing images. The study replaced the strided convolution module with the SPDConv module to preserve fine-grained feature information and improve accuracy. However, this modification reduced the model's frames per second (FPS), potentially impacting real-time applications. Despite this, SP-YOLOv8s represents a significant advancement in remote sensing image analysis, particularly for detecting tiny objects in complex environments.

Chavan et al. (2022) proposed a pothole detection system using the YOLOv4 algorithm, as detailed in their paper published in the International Research Journal of Engineering and Technology (IRJET). Their system focuses on real-time pothole detection, which is crucial for timely road maintenance and preventing accidents. By utilizing advanced computer vision techniques, the study contributes to improving road safety and urban mobility.

Safyari et al. (2024) provide a comprehensive review of vision-based detection methods, highlighting the growing dominance of YOLO models due to their speed, robustness, and adaptability. Their work emphasizes challenges such as varying road textures, lighting conditions, and the need for lightweight architectures suitable for mobile and vehicular deployment.

Building upon this foundation, Zanevych et al. (2024) specifically examined YOLO models under low-light conditions, a scenario highly relevant to Indian roads where uneven lighting and poor infrastructure are common. Their study demonstrated that YOLOv7 and YOLOv8 outperform earlier architectures but still face performance drops in night-time conditions. This reinforces the importance of preprocessing strategies and adaptive image enhancement for deployment in real-world scenarios.

Khan et al. (2024) introduced a YOLOv8-based pipeline for pothole detection in autonomous vehicles, reporting significant improvements in accuracy and inference time compared to YOLOv5. Their work highlights YOLOv8's ability to handle diverse environmental conditions, making it a strong candidate for applications in complex road networks such as those in India. The study also stresses the importance of real-time inference, which is crucial for both driver-assistance systems and automated inspection tools.

Further advancements are seen in Yurdakul and Taşdemir, 2025, who proposed an enhanced YOLOv8-seg model combined with depth data for not just detecting potholes but also measuring their dimensions. This innovation introduces the PothRGBD dataset, enabling more precise assessment of road damage severity. Such approaches are vital in the Indian context where potholes vary greatly in depth and size, influencing both safety risks and repair prioritization. Further survey on the literature is given in Table 1.

Table 1. Literature Survey

Paper	Observation	Limitation
Seraj et al., 2014	It uses accelerometer present in phones to monitor reduction in speed which might suggest presence of potholes. Classification is done using SVM.	This study gathered more labelled data, enhance the SVM's training through stratification, test it using different feature sets and sensors, including a barometric pressure sensor, and boost the accuracy and number of anomaly classifications.
Shaghouri et al, 2021	The pothole detection methods used in this paper are YOLOv3-Darknet53, YOLOv4CSPDarknet53, and SSDtensorFlow.	Older and out-of-date YOLO models.
Faramarzi 2020	Utilized a GPU server and a smartphone to implement object detection techniques through convolutional neural networks, training the damage detection model on their dataset. Following this, they compared the accuracy and runtime performance on both platforms.	Older object detection models contribute to a lack of information that can be gained from the detection.

continued on following page

Table 1. Continued

Paper	Observation	Limitation
Liu et al., 2023	To address environmental concerns, this study provides an enhanced underwater target identification method based on YOLOv8n.	The computational requirements of their model are significantly high.
Wang et al., 2023	Based on YOLOv8s, this study suggests an improved road problem identification system dubbed BLYOLOv8.	Certain road defects remain undetected. It still encounters limitations in detecting small targets that closely resemble the surrounding environment.
Gupta et al., 2020	This paper suggests a novel method of employing deep neural networks for pothole localization from thermal pictures, based on bounding boxes.	The proposed pothole localization work cannot label the regions of detected potholes based on their severity level and size.
Du 2018	This study presents an innovative approach to object detection that defies the conventions of the CNN family, showcasing the best of both CNN and YOLO.	Lower mAP scores. Dataset not there.
Sathvik et al., 2022	In this work, a pothole image dataset was annotated and trained using YOLOv7. The results were examined in terms of recall and accuracy.	Absence of real-time images.
Ma et al., 2023	In this paper, the SPD-Conv module is used in place of the strided convolution module in YOLOv8s. This improves the network's expressive and learning capacities as well as recognition accuracy by downsampling the feature map while maintaining fine-grained feature information.	The SPD-Conv module and the SPANet path aggregation network are the two main parts of the algorithm. Because SPYOLOv8s combines these two elements, its frame rate is lower than that of the original YOLOv8s.
Chavan et al., 2022	The paper demonstrates an effective pothole detection system using the YOLO v4 algorithm, achieving high accuracy in identifying potholes on roads.	The system's performance may be affected by varying lighting conditions, shadows, and road textures, which could lead to reduced detection accuracy.

ARCHITECTURE OVERVIEW

Architecture Diagram

Figure 1 shows the architecture diagram for this paper. This research begins with a comprehensive pre-processing phase aimed at optimizing image quality and suitability for subsequent analysis and detection tasks. This includes implementing noise reduction methods to eliminate unwanted artifacts and disturbances, enhancing image quality, and improving the signal-to-noise ratio.

Figure 1. Architecture diagram

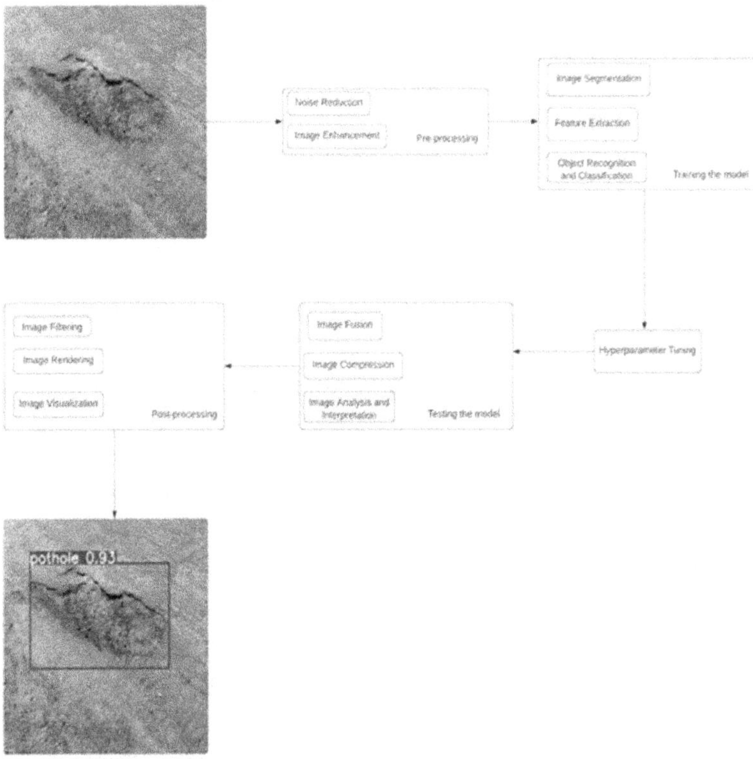

Image enhancement techniques are then applied to manipulate images, emphasizing relevant features, and enhancing contrast for better object detection outcomes. Moving to the training phase, this research employs image segmentation to divide images into segments based on specific criteria, aiding in the delineation of objects or regions of interest. Feature extraction is crucial for identifying distinctive characteristics within images relevant for object detection, such as edges, textures, shapes, and colors. Object recognition and classification follow, enabling accurate detection and identification of objects in real-time applications. In the hyperparameter tuning phase, fine-tune parameters for this paperYOLOv7 and YOLOv8 models, such as learning rates, weight decay, and loss gains, optimize training outcomes. Post-training, conduct testing and employ post-processing techniques like image fusion and compression to refine detection outcomes and optimize resource utilization. Finally, post-processing involves applying techniques such as image rendering, visualization, and filtering to refine and visualize detection results, enhancing the overall quality and accuracy of detected objects. Through this comprehensive approach, the proposed model aims to achieve highly accurate and robust pothole detection on Indian roads, contributing to improved road safety and vehicle performance.

In the pre-processing phase, this research employs techniques aimed at optimizing the quality and suitability of images for subsequent analysis and detection tasks This research implements noise reduction methods to eliminate unwanted artifacts and disturbances from images, thereby enhancing their quality and improving the signal-to-noise ratio (SNR). By reducing noise, critical features necessary for object detection can be preserved, facilitating more accurate detection outcomes. Image enhancement techniques are utilized to manipulate images and improve their quality for specific applications or better human perception. Enhancing images helps in emphasizing relevant features and enhancing contrast, making objects of interest more distinguishable from the background or other clutter. During the model training phase, this research undertakes various steps to train the YOLOv7 and YOLOv8 models effectively

Image segmentation involves dividing images into multiple segments or regions based on specific criteria such as color, texture, or brightness. This segmentation facilitates the delineation of objects or regions of interest within the image, aiding in subsequent detection tasks. Feature extraction is crucial for identifying and representing distinctive characteristics or patterns within images relevant for object detection. These features, including edges, textures, shapes, and colors, play a significant role in distinguishing objects from the background. Object recognition and classification involve detecting and localizing objects within images and assigning labels or classes based on their features. This process enables accurate detection and identification of objects in real-time applications. In the hyperparameter tuning phase, this research aims to analyzes training results and fine-tune hyperparameters to optimize model performance Various hyperparameters in YOLOv7 and YOLOv8, such as initial and final learning rates, weight decay, warmup epochs, box loss gain, and class loss gain, significantly influence training outcomes. Isolate and tune these hyperparameters to achieve the most optimized training results.

Post-training, conduct testing and employ post-processing techniques to refine detection outcomes. Image fusion techniques combine multiple images or modalities to enhance object detection or recognition performance. This step aims to leverage complementary information from different sources to improve detection accuracy and robustness. Image compression reduces image size while retaining essential information for object detection tasks, optimizing resource utilization without compromising detection quality. Leveraging YOLO's capabilities, perform image analysis to detect objects and interpret results, extracting meaningful information about image contents. After detection, apply post-processing techniques to refine and visualize detection results Bounding boxes are overlaid onto input images to visually represent detected objects and their types, aiding in interpretation and analysis. Image Visualization, this involves displaying detection results on input images, providing visual confirmation of detected objects and their locations. Post-processing filtering techniques enhance or modify detection outputs, refining object boundaries, reducing noise, and improving detection accuracy. These techniques contribute to extracting additional information and enhancing the overall quality of detected objects.

Models Used for Training

YOLOv7-Tiny is a streamlined version of the YOLOv7 model, specifically optimized for environments with limited computational resources, such as embedded systems and mobile devices. Despite its compact architecture, YOLOv7-Tiny maintains a high level of detection accuracy, making it ideal for real-time applications where processing power is constrained. The model achieves this balance by reducing the number of layers and parameters compared to the full YOLOv7 model while retaining the core architectural elements that make YOLO effective for object detection. YOLOv7-Tiny employs a combination of convolutional layers, batch normalization, and leaky ReLu activations, arranged in a shallower network. This architecture significantly reduces the model size and inference time, enabling fast predictions even on less powerful hardware. However, the trade-off is a slight decrease in detection accuracy, especially for complex or densely populated scenes. Despite these limitations, YOLOv7-Tiny is highly efficient for applications like pothole detection in real-time monitoring systems, where speed and resource efficiency are paramount.

YOLOv8s represents the small variant in the YOLOv8 family, which introduces several advancements over previous YOLO versions. This model is designed to provide a balance between speed and accuracy, making it suitable for scenarios where both real-time performance and high detection accuracy are required. YOLOv8s features an enhanced backbone network that incorporates depth wise separable convolutions, which reduce computational overhead while preserving the model's ability to capture complex features. Additionally, the model uses the Path Aggregation Network (PAN) for improved multi-scale feature fusion, allowing it to detect objects of various sizes more effectively. The architecture also includes the Spatial Pyramid Pooling (SPP) module, which enhances the receptive field without significantly increasing the computational load. YOLOv8s is particularly effective in detecting smaller objects, making it well-suited for applications like pothole detection on roads, where the anomalies may vary in size and are often small. With its optimized architecture, YOLOv8s achieves a commendable trade-off between speed and accuracy, making it a versatile choice for real-time object detection tasks.

YOLOv9c represents the cutting-edge in the YOLO series, designed to push the boundaries of both accuracy and efficiency in object detection. The “c” variant within the YOLOv9 family signifies a model that emphasizes computational efficiency, incorporating advanced architectural innovations to reduce latency without compromising on precision. YOLOv9c integrates a Transformer-based backbone, which enhances the model's capability to capture global context and intricate feature relationships across different scales. This architecture also introduces Dynamic Convolution, allowing the model to adapt its convolutional parameters based on the input, thereby improving detection performance in diverse environments. Furthermore, YOLOv9c employs a more sophisticated FPN (Feature Pyramid Network) combined with a cascaded detection head, which improves object localization and classification accuracy, especially in challenging scenarios with overlapping objects or low visibility. The model is optimized for deployment in high-stakes real-time applications, such as autonomous driving and advanced driver assistance systems (ADAS), where both speed and precision are critical. YOLOv9c's architecture makes it particularly suitable for detecting road anomalies like potholes with high accuracy, even under varying environmental conditions, making it a significant advancement in the field of real-time object detection.

YOLOv7, YOLOv8, and YOLOv9 were selected for this study because of their architectural advancements that directly address challenges in pothole detection. YOLOv7 introduces extended efficient layer aggregation networks (E-ELAN), improving feature reuse and enabling higher accuracy without compromising speed. YOLOv8 incorporates decoupled detection heads, allowing more precise localization of small and irregularly shaped objects such as potholes, which often vary in appearance on Indian roads. YOLOv9 further enhances gradient flow and feature representation through its GELAN and programmable gradient information (PGI) modules, improving robustness under low-light and noisy conditions. These innovations make the models particularly suitable for pothole detection tasks, where variability in size, shape, lighting, and background complexity often hinders traditional object detection approaches.

Figure 2. Image of pothole

RESULT ANALYSIS

Dataset Description

The dataset sourced from Kaggle comprises 665 images capturing instances of potholes and manholes. These images offer a comprehensive representation of various road conditions, aiding in the development and evaluation of pothole and manhole detection through YOLO algorithms. Through meticulous annotations, essential attributes of these road hazards are highlighted, facilitating precise training and validation of machine learning models. While the dataset provides a valuable resource for enhancing the robustness and accuracy of automated systems, several challenges persist. Notably, class imbalance exists as non-pothole areas significantly outnumber annotated potholes. In addition, many images exhibit variations in quality due to low illumination, motion blur, shadows, or partial occlusion by traffic and debris. Such limitations make detection more complex but also ensure that trained models are evaluated under conditions reflective of real-world Indian road scenarios, where data quality inconsistencies are common. Fig 2 shows a sample image from the dataset used having a pothole. Figure 3 shows the pothole detected with a confidence score of 0.95.

Figure 3. Pothole detection

Hardware and Software Specifications

The system operates with an AMD Ryzen 9 5900HS processor, renowned for its high performance and efficiency, making it ideal for demanding computational tasks. It is equipped with 32 GB of RAM, of which 31.4 GB is usable, providing sufficient memory to efficiently manage large datasets and complex algorithms. Running on Windows 11 Home, Version 23H2, with OS Build 22631.3447, the system ensures compatibility with the latest software and security updates. The 64-bit operating system, coupled with an x64-based processor, offers optimal processing power and multitasking capabilities.

The software environment is built on Windows 11 Home Edition, specifically Version 23H2 with OS Build 22631.3447, providing a stable and current platform for application execution. The system requires Python 3.8 or later, a versatile programming language widely used in data analysis, machine learning, and scientific computing. Additionally, Jupyter Notebook is a key component, offering an interactive environment for data analysis and visualization, facilitating the development, documentation, and sharing of code with others.

Performance Metrices

To assess the models' efficacy and accuracy, this study looks at several performance indicators. Two important metrics are recall, which evaluates the model's accuracy in identifying positive instances, and precision, which gauges the accuracy of positive predictions. When there is an imbalance between positive and negative examples, the F1 score is very helpful since it strikes a balance between precision and recall. To assess the accuracy of object detection, Mean Average Precision (mAP) is calculated at various IoU levels by taking the overlap between the predicted and real bounding boxes into account. Finally, accuracy measures the percentage of correctly identified cases in the dataset relative to all occurrences, giving an overall evaluation of the model's performance.

Results

Table 2 shows a comparison between YOLOv7,YOLOv8 and YOLOv9 using different performance metrices.

Table 2. Comparison between different YOLO models

Algorithm	Precision	Recall	F_1 -Score	TPR	mAP50	mAP50-95	Accuracy
YOLOv7	0.828	0.751	0.787	0.82	0.846	0.534	0.41
YOLOv8	0.836	0.705	0.765	0.77	0.825	0.543	0.385
YOLOv9	0.73	0.525	0.611	0.57	0.638	0.358	0.285

Fig 4shows YOLOv7 emerged as the best-performing algorithm among the three, followed by YOLOv8, with YOLOv9 showing the lowest performance. YOLOv7 demonstrated the highest recall (0.751) and true positive rate (TPR) (0.82), indicating its strong capability in correctly identifying positive instances. Additionally, YOLOv7 achieved the highest mean average precision at a 50% threshold (mAP50) with a score of 0.846, suggesting superior performance in ranking the top 50% of relevant instances. The F_1-Score for YOLOv7 was 0.787, reflecting a balanced trade-off between precision and recall. However, its overall accuracy was moderate at 0.41, indicating that while YOLOv7 excels in specific areas, it may compromise accuracy in others.

Figure 4. Comparison between different YOLO Models

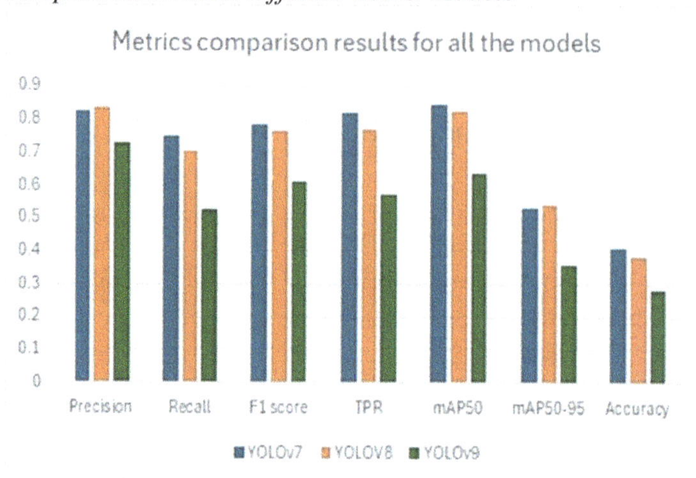

YOLOv8 exhibited the highest precision (0.836) among the three algorithms, implying a lower false positive rate and high accuracy in predicting positive instances. The recall for YOLOv8 was 0.705, which, although lower than YOLOv7, still indicates a good ability to identify positive instances. The F_1-Score for YOLOv8 was 0.765, slightly lower than YOLOv7, suggesting a balanced but slightly less effective trade-off between precision and recall. The TPR for YOLOv8 was 0.77, and it achieved a mean average precision at a 50% threshold (mAP50) of 0.825, both of which are commendable. The mAP50-95 for YOLOv8 was the highest among the three algorithms at 0.543, indicating its robustness across different thresholds. However, its overall accuracy was 0.385, slightly lower than YOLOv7, highlighting a generally balanced performance.

YOLOv9 showed the lowest performance across most metrics. With a precision of 0.73 and a recall of 0.525, YOLOv9 struggled more with both correctly predicting positive instances and identifying them accurately. The F_1-Score for YOLOv9 was the lowest at 0.611, indicating a less effective balance between precision and recall. The TPR was 0.57, reflecting a reduced ability to correctly identify positive instances. YOLOv9 also had the lowest mean average precision at both 50% and 50-95% thresholds (0.638 and 0.358, respectively), suggesting it is the least effective in ranking relevant instances. The overall accuracy for YOLOv9 was 0.285, indicating its overall lower performance compared to YOLOv7 and YOLOv8.

Figure 5. Confidence matrix for YOLOv7

The confusion matrix in Figure 5 for the YOLOv7 model shows the performance of the model in detecting potholes versus the background. The matrix is divided into four quadrants. The top-left quadrant represents the true positives (0.82), indicating that 82% of the actual potholes were correctly identified as potholes by the model. The top-right quadrant shows the false positives (1.00), where the model incorrectly classified all background instances as potholes. The bottom-left quadrant represents the false negatives (0.18), indicating that 18% of actual potholes were incorrectly classified as background. The bottom-right quadrant appears empty, indicating minimal or no true negatives. The color intensity of the matrix cells corresponds to the magnitude of the values, with darker shades representing higher values. This matrix highlights the model's high sensitivity to detecting potholes but also reveals a significant issue with false positive rates, suggesting a need for improvement in distinguishing potholes from the background.

Figure 6. Confidence matrix for YOLOv8

The normalized confusion matrix in Figure 6 for a YOLOv8 model used for pothole detection displays the performance of the model in predicting two classes: "Pothole" and "background." The rows represent the predicted labels, while the columns represent the true labels. The model correctly predicted 77% of the actual pothole instances as potholes (top-left cell). However, the model incorrectly predicted 23% of actual pothole instances as background (bottom-left cell). The model correctly identified 100% of the actual background instances as background (bottom-right cell). There are no instances where the background is misclassified as potholes (top-right cell). This suggests that while the model is highly accurate in identifying the background, it struggles more with accurately detecting potholes, as indicated by the 23% false negative rate.

Figure 7. Confidence Matrix for YOLOv9

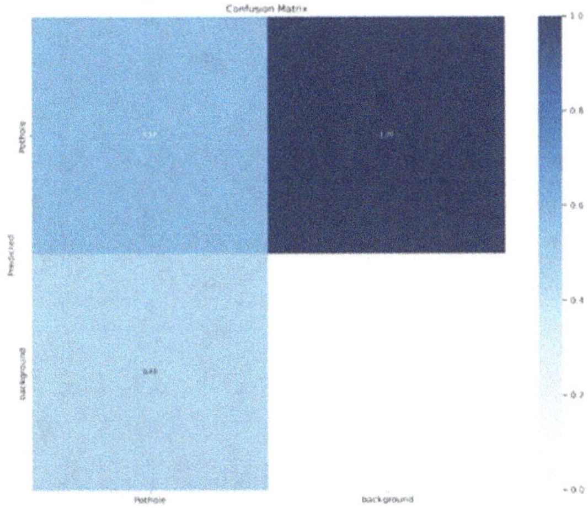

The normalized confusion matrix in Figure 7 for a YOLOv9 model used for pothole detection showcases the performance of the model in predicting two classes: "Pothole" and "background." The rows represent the predicted labels, while the columns represent the true labels. The model correctly predicted 57% of the actual pothole instances as potholes (top-left cell). The model incorrectly predicted 43% of actual pothole instances as background (bottom-left cell). The model correctly identified 100% of the actual background instances as background (bottom-right cell). There are no instances where the background is misclassified as potholes (top-right cell). This suggests that while the model is highly accurate in identifying the background, it has a significant challenge in accurately detecting potholes, as indicated by the 43% false negative rate. This implies that nearly half of the potholes are not being detected, which could be a major limitation for applications that rely on precise pothole identification.

Figure 8. Recall-confidence graph for YOLOv7

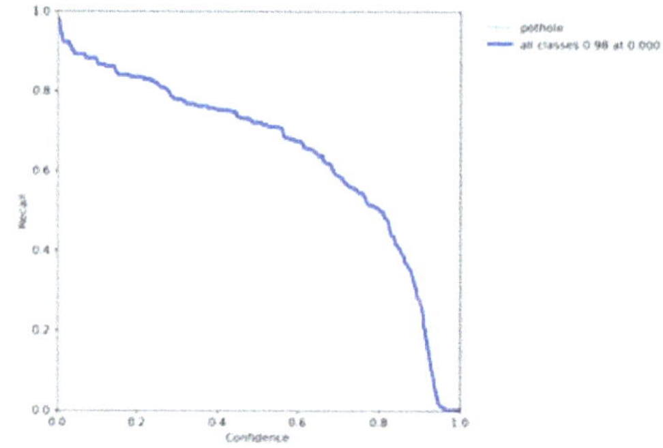

The Recall against Confidence graph in Figure 8 for YOLOv7 depicts the relationship between recall and confidence thresholds. As the confidence level increases, recall decreases, showing that fewer true positives are identified at higher confidence levels. This inverse relationship indicates that the model becomes more stringent in its predictions, leading to a reduction in the number of detected potholes. Understanding this trade-off helps in setting appropriate confidence thresholds to balance recall and precision according to specific application requirements.

Figure 9. F_1-Score-Confidencegraph for YOLOv7

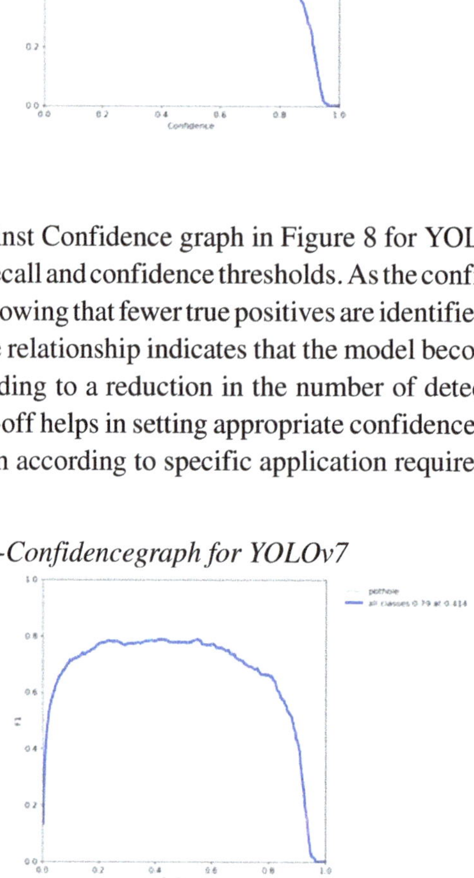

The F_1-Score against Confidence graph in Figure 9 for YOLOv7 highlights the model's harmonic mean of precision and recall across different confidence levels. The F_1-Score initially increases with confidence, peaks, and then gradually declines. The peak F_1-Score indicates the optimal balance between precision and recall, reflecting the model's best overall performance. As confidence continues to rise beyond this point, the decline in F_1-Score suggests that the model becomes overly conservative, sacrificing recall for precision.

Figure 10. Precision-recall graph for YOLOv7

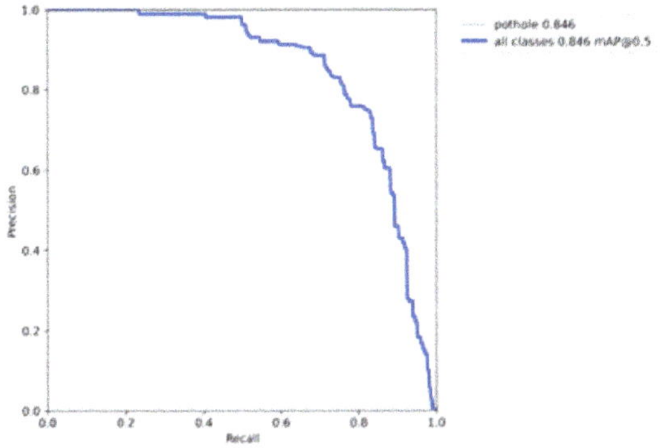

The Precision against Recall graph in Figure 10 for YOLOv7 demonstrates the trade-off between precision and recall in the detection of potholes. At higher recall values, precision tends to decrease, indicating that capturing more true positives comes at the cost of including more false positives. The graph shows that YOLOv7 maintains high precision up to a certain recall point before experiencing a decline. This balance between precision and recall is crucial for optimizing the model's performance, ensuring that it effectively identifies potholes while minimizing false alarms.

The Precision against Confidence graph in Fig 11 for YOLOv7 illustrates how precision varies with different confidence thresholds for the detection of potholes. As the confidence level increases, the precision also increases, indicating that higher confidence scores are associated with more accurate predictions. The graph shows a steady rise in precision, reaching near-perfect levels as confidence approaches 1. This suggests that YOLOv7 is highly effective in distinguishing true positives from false positives at higher confidence thresholds, ensuring reliable detection of potholes.

Figure 11. Precision-confidence graph for YOLOv7

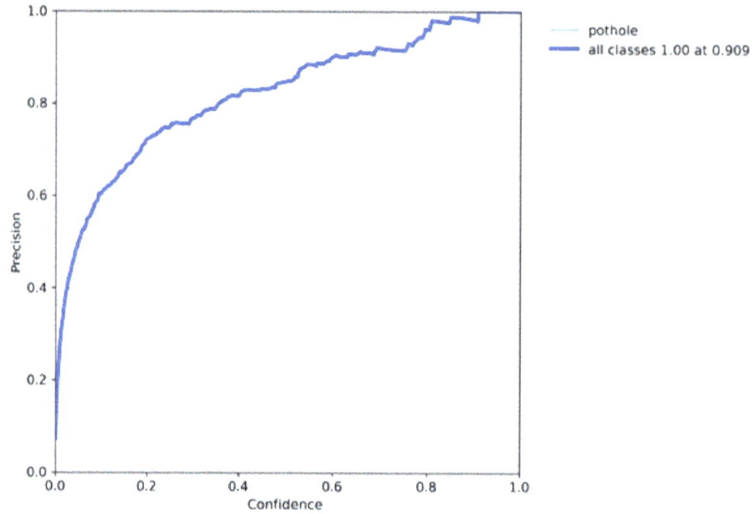

The F_1-Score against Confidence graph in Figure 12 for YOLOv8 shows the F_1-Score, which is the harmonic mean of precision and recall, as a function of the confidence threshold. This graph helps identify the confidence threshold that maximizes the F_1-Score, providing a balance between precision and recall. The peak of the curve indicates the optimal confidence level for the best overall performance in pothole detection.

The Precision against Confidence graph in Figure 13 for YOLOv8 demonstrates how the precision of the model varies with the confidence threshold. As the confidence increases, the precision also increases, indicating that higher confidence predictions are more accurate. This graph is essential for understanding the trade-off between the confidence threshold and precision, helping to determine the optimal threshold for maximum precision in pothole detection.

Figure 12. F_1-Score - Confidence graph for YOLOv8

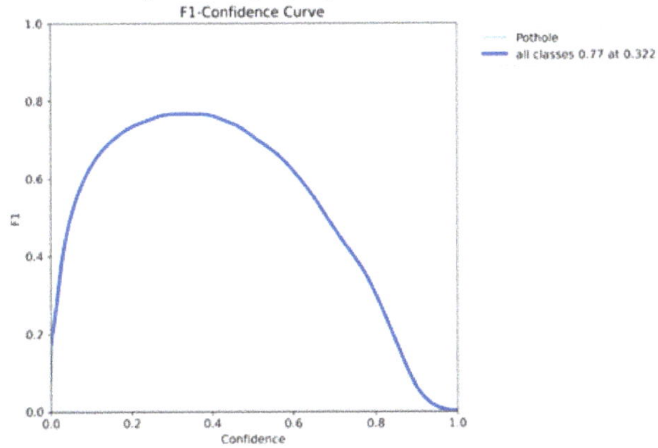

Figure 13. Precision-Confidence graph for YOLOv8

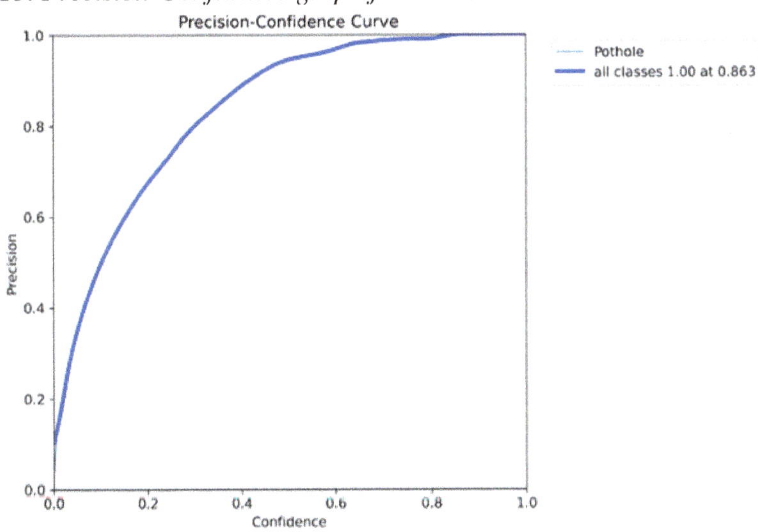

The Precision against Recall graph in Figure 14 for YOLOv8 illustrates the relationship between precision and recall across different confidence thresholds. This graph is crucial for assessing the model's performance, as it shows how precision decreases as recall increases, highlighting the balance between identifying true positives and avoiding false positives. The mean Average Precision (mAP) at a recall of 0.5 is also indicated, providing a single metric to summarize the model's performance.

Figure 14. Precision-Recall graph for YOLOv8

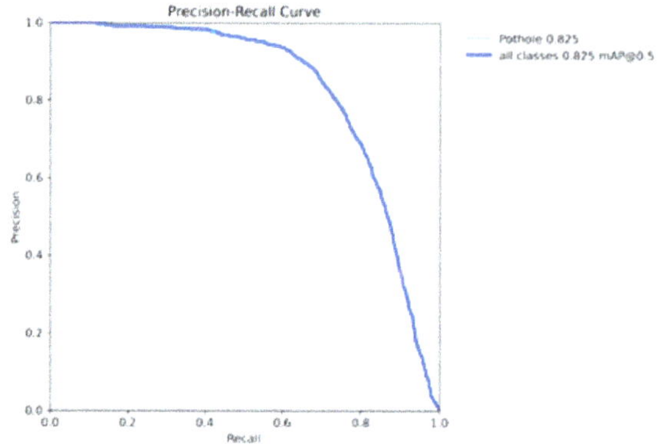

Figure 15. Recall-Confidence graph for YOLOv8

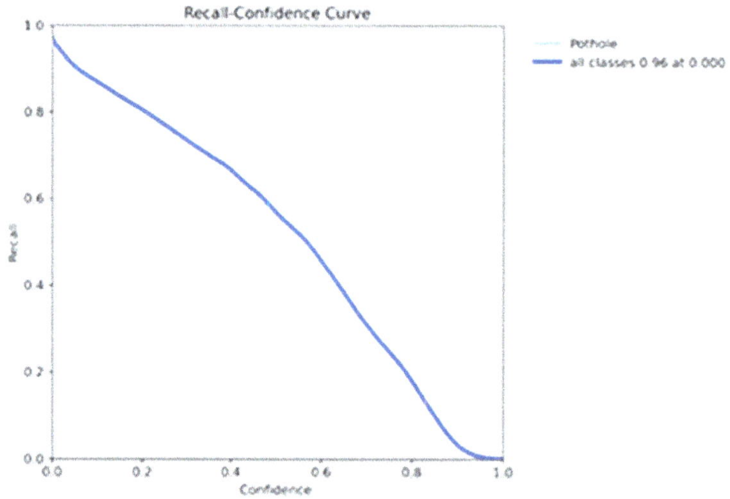

The Recall against Confidence graph in Figure 15 for YOLOv8 plots recall as a function of the confidence threshold. As the confidence threshold increases, recall decreases, meaning fewer actual positives are identified as the model becomes more stringent. This graph is vital for understanding how changing the confidence threshold affects the model's ability to detect true positives, allowing for fine-tuning to achieve the desired recall level.

Figure 16. F_1-Score - Confidence graph for YOLOv9ss

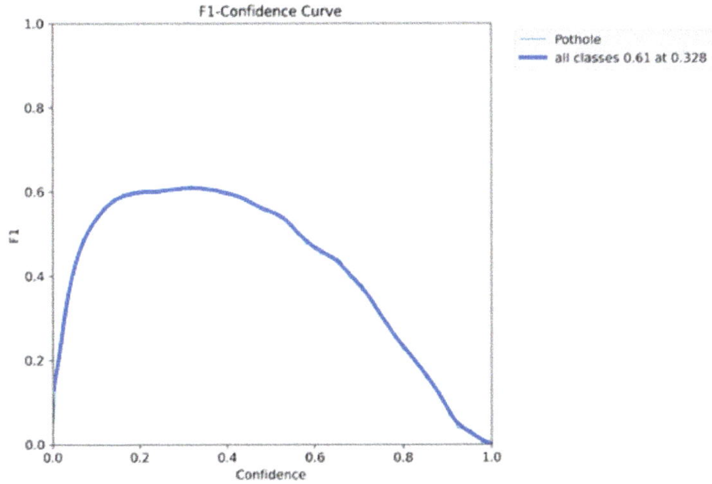

The F_1-Score against Confidence graph in Figure 16 for YOLOv9 shows the harmonic mean of precision and recall as a function of confidence. The F_1-Score peaks at a specific confidence level, indicating the optimal balance between precision and recall. The graph highlights that the F_1-Score reaches its highest value around a confidence level of 0.328, with a peak F_1-Score of 0.61. This information is critical for determining the confidence threshold that provides the best overall performance for the model in terms of both precision and recall.

Figure 17. Precision-Confidence graph for YOLOv9

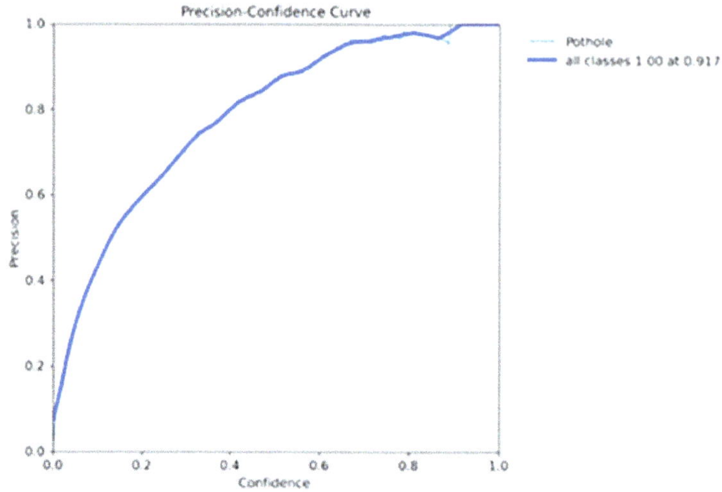

Figure 18. Precision-Recall graph for YOLOv9

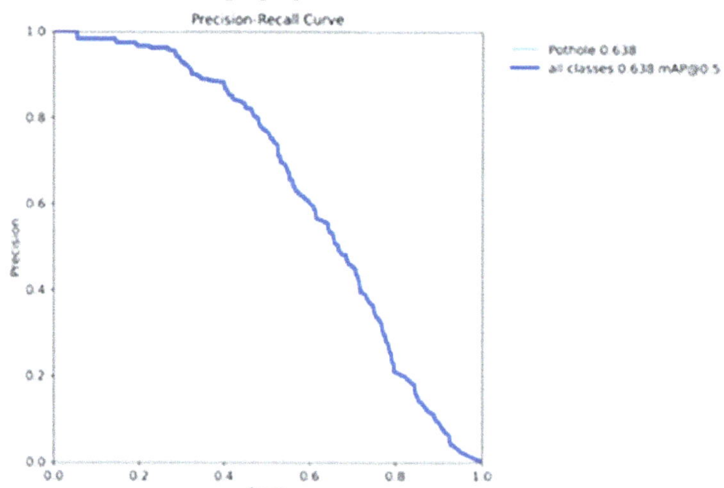

The Precision against Confidence graph in Figure 17 for YOLOv9 illustrates the model's precision as a function of its confidence threshold. The graph indicates that as the confidence level increases, the precision also increases, reaching a high value near the upper end of the confidence scale. This trend demonstrates that YOLOv9 is highly precise when the model is confident about its predictions, with precision nearing 1.0 at higher confidence levels. This behaviour is crucial for applications where high precision is necessary to minimize false positives, particularly in detecting potholes.

Figure 19. Recall-Confidence graph for YOLOv9

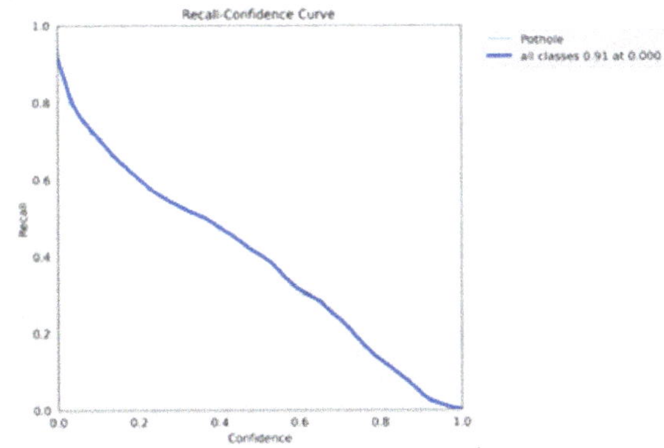

The Precision-Recall curve in Figure 18 for YOLOv9provides a comprehensive view of the trade-off between precision and recall. The graph shows that as recall increases, precision decreases, which is typical behavior in classification models. The average precision (AP) metric for all classes is noted as 0.638 mAP@0.5, indicating the model's overall performance in balancing precision and recall. This curve is particularly useful in evaluating the effectiveness of YOLOv9 in identifying potholes across various recall levels, helping to optimize the model for different use cases.

The Recall against Confidence graph in Figure 19 for YOLOv9 depicts the relationship between recall and confidence. As confidence increases, recall tends to decrease, illustrating the model's ability to retrieve relevant instances. At higher confidence levels, recall drops significantly, suggesting that while the model becomes more precise, it also becomes less sensitive, capturing fewer relevant instances. The recall at maximum confidence is noted to be 0.91, indicating the model's ability to maintain a high level of recall at specific confidence thresholds.

Table 3 shows the inference of YOLOv7 for different test cases. The study infers that YOLOv7 efficiently detects potholes through its advanced object detection capabilities. Leveraging a single neural network, YOLOv7 processes the entire image at once, enabling real-time inference. Its deep architecture efficiently learns features of potholes, allowing it to accurately identify them amidst diverse road conditions. Additionally, YOLOv7's training process incorporates a large and diverse dataset, enabling robust detection across various scenarios commonly encountered on roads, contributing to its efficiency in pothole detection.

Table 3. Inference for YOLOv7

	Input	Expected Output	Actual Output
1			
2	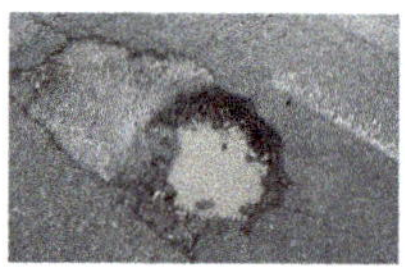		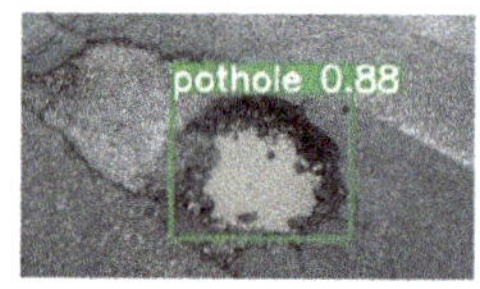
3			

Table 4 shows the inference of YOLOv8 for different test cases. While YOLOv8 demonstrates efficient detection of potholes leveraging its advanced object detection capabilities, it may encounter limitations when identifying roads without potholes. In scenarios where the road surface is uniform and devoid of potholes or other significant features, YOLOv8's detection performance may be less precise. This could result from the model's tendency to prioritize regions with distinguishable features, potentially leading to false positives or inaccuracies in identifying pothole-free areas. Despite this challenge, YOLOv8 excels in accurately detecting potholes amidst diverse road conditions, leveraging its deep architecture and robust training on large and varied datasets. Continued refinement and adaptation to specific road environments can further enhance YOLOv8's effectiveness.

Table 4. Inference for YOLOv8

	Input	Expected Output	Actual Output
1		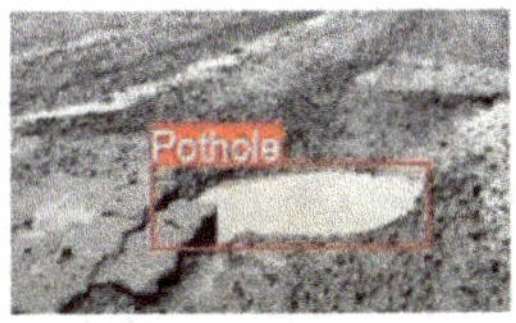	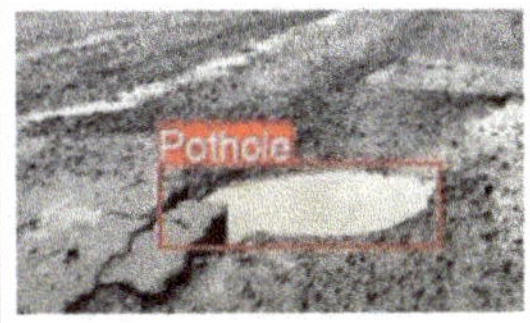
2	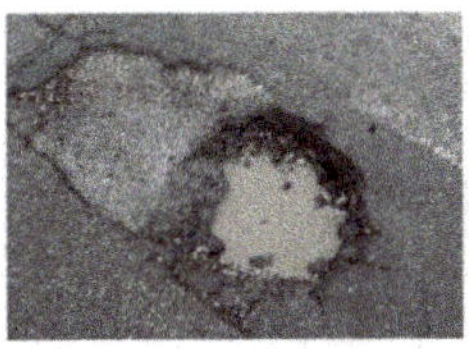	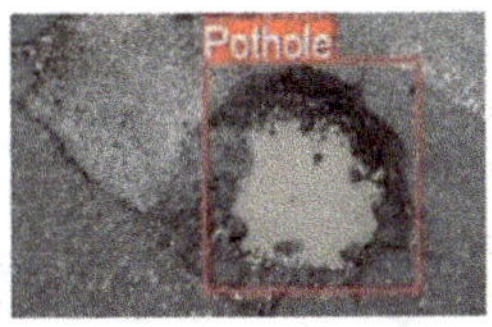	
3			

Table 5. Inference for YOLOv9

	Input	Expected Output	Actual Output
1			
2	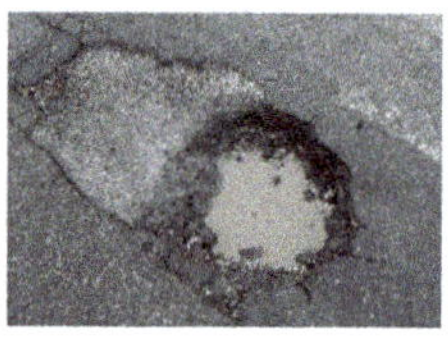		
3			

Table 5 shows the inference of YOLOv9 for different test cases. The study infers that YOLOv9 exhibits the ability to recognize potholes, albeit with lower confidence scores compared to other objects. Despite its capability to detect potholes, YOLOv9 often assigns lower confidence scores to these identifications, indicating a reduced level of certainty. However, YOLOv9 faces challenges in accurately recognizing negative cases, where objects are incorrectly classified as potholes. This limitation may stem from factors such as imbalanced training data or difficulties in distinguishing subtle features in non-pothole areas. While YOLOv9's performance in negative case identification may be less optimal, its proficiency in detecting potholes underscores its potential utility in road inspection and maintenance tasks. Continued refinement and optimization efforts could further enhance YOLOv9's accuracy and confidence in both pothole detection and negative case recognition.

The test cases include images captured under different lighting conditions, such as daytime, nighttime, and varying weather conditions like rain or fog. This diversity ensures that the algorithms can detect potholes and manholes accurately under different environmental conditions commonly experienced on Indian roads.

The test cases encompass images featuring a range of road surfaces and textures, including highways, urban roads, rural roads, and unpaved roads. By incorporating this diversity, the algorithms' performance can be evaluated across different types of road infrastructure prevalent in India, ensuring their applicability in various settings.

Additionally, the test cases include images with different levels of road congestion and traffic density, ranging from sparse traffic to heavy traffic conditions. This variation allows for assessing the algorithms' performance in detecting potholes and manholes amidst different traffic scenarios, which is crucial for real-world deployment where road conditions can change dynamically.

Furthermore, the test cases may include images containing occlusions, such as vehicles, pedestrians, or other objects obstructing the view of potholes and manholes. Evaluating the algorithms' performance in detecting obscured hazards provides insights into their robustness and reliability in challenging traffic conditions.

Moreover, the test cases incorporate images depicting different types and sizes of potholes and manholes, including shallow and deep potholes, as well as various sizes of manhole covers. This variation ensures that the algorithms are capable of accurately identifying different types of road hazards commonly encountered on Indian roads.

Overall, the selection of diverse and representative test cases is essential for conducting comprehensive evaluations of the algorithms' performance in pothole and manhole detection. By encompassing a wide range of real-world scenarios, these test cases provide valuable insights into the algorithms' effectiveness, robustness, and suitability for deployment on Indian roads to enhance road safety and infrastructure maintenance efforts.

Analyzing the performance metrics of YOLOv7, YOLOv8, and YOLOv9 provides deeper insights into their effectiveness for pothole detection on Indian roads. While each model offers unique strengths, understanding the implications of these metrics is crucial for informed decision-making.

The precision and recall trade-off isa critical aspect to consider. Precision and recall often exhibit a trade-off relationship, where improving one metric may come at the expense of the other. YOLOv8 strikes a balance between precision and recall, as reflected in its high F_1-Score. However, in applications where false positives or false negatives have varying degrees of impact, one metric may need to be prioritized over the other. For instance, in pothole detection, a higher recall may be preferred to ensure no potholes are missed, even if it results in more false positives.

The impact of the true positive rate (TPR) is another essential consideration. The TPR reflects the model's ability to correctly identify potholes. YOLOv7's high TPR suggests its effectiveness in accurately detecting potholes, which is crucial for road safety. However, it's essential to also consider the consequences of false negatives (missed potholes), as they pose significant risks to both drivers and the integrity of road infrastructure.

In conclusion, YOLOv7 is the best-performing algorithm, excelling in recall, TPR, and mAP50, despite its moderate overall accuracy. YOLOv8 follows closely, with the highest precision and mAP50-95, demonstrating a balanced yet slightly less effective performance. YOLOv9 lags in most metrics, showing lower effectiveness in both precision and recall, as well as overall accuracy.

CONCLUSION

This paper represents a significant advancement in deep learning for pothole detection on Indian roads. While prior studies used YOLOv2, v3, and v3-tiny, we focus on enhancing performance with YOLOv7, YOLOv8, and YOLOv9. The objective of the paper is to achieve a higher mean Average Precision (mAP), indicating improved accuracy. This metric is crucial for assessing reliability in pothole detection, addressing road safety concerns. By leveraging advanced deep learning, this paper aims to reduce accidents and fatalities by providing timely pothole alerts, benefiting Indian commuters.

Methodologically, the research uses annotated datasets from Kaggle for training and evaluation. This research employs key metrics like mAP, precision, and recall for robust model evaluation. This paper not only enhances pothole detection accuracy but also provides insights for road maintenance stakeholders. In summary, this paperuses deep learningto address pothole detection challenges. By leveraging technology and data-driven approaches, this research aims to reduce accidents and fatalities, contributing to Indian road users' safety.

Furthermore, this paper underscores collaboration between academia, industry, and government in addressing road safety challenges. By fostering partnerships, this research develops innovative solutions with a meaningful impact on public safety and infrastructure. Looking ahead, this paper lays the groundwork for future intelligent transportation systems research. Continued advancements in deep learninghold promise for safer, more sustainable transportation networks. Interdisciplinary collaboration is key to creating smarter, safer transportation for future generations.

While YOLO-based models demonstrate strong performance in pothole detection, false detections present both practical and ethical challenges. False negatives (missed potholes) can lead to severe consequences such as vehicle damage, accidents, or increased maintenance costs, especially on busy Indian roads where infrastructure safety is critical. False positives (incorrectly identifying potholes) may result in unnecessary repair allocations, inefficient use of public funds, or disruption of traffic flow during unwarranted roadworks. From an ethical perspective, over-reliance on automated systems without acknowledging these risks may compromise public safety and erode trust in intelligent transportation solutions. Therefore, ensuring model transparency, conducting rigorous field validation, and incorporating human oversight in decision-making are essential to minimize harm. Addressing these issues not only strengthens the reliability of YOLO-based systems but also aligns their deployment with broader societal responsibilities in road safety and infrastructure management.

FUTURE WORKS

Future works in the field of pothole detection on Indian roads through advanced deep learning techniques should focus on addressing existing challenges and expanding the scope of research to further enhance road safety. Continued efforts should be directed towards collecting high-quality, diverse datasets that accurately represent the various road and lighting conditions encountered on Indian roads. This includes gathering data from different regions, seasons, and times of day to improve model generalization and robustness. Model optimization, further optimization of deep learning model architectures, hyperparameters, and training strategies is essential for improving detection accuracy and efficiency. Techniques such as data augmentation, transfer learning, and ensemble methods should be explored to boost performance and adapt models to real-time deployment scenarios.

Efficient deployment of pothole detection models in real-time applications, such as onboard vehicle systems or municipal monitoring platforms, requires addressing computational resource constraints and minimizing latency. Future research should focus on developing lightweight models and optimizing inference processes to enable seamless integration into existing infrastructure. Integration of pothole detection systems with existing infrastructure, such as road maintenance vehicles and municipal monitoring networks, can facilitate proactive maintenance and timely repairs. Future research should explore ways to seamlessly integrate detection systems into existing workflows to optimize resource allocation and improve road safety.

Combining data from multiple sensors, such as cameras, LiDAR, and GPS, can enhance the accuracy and reliability of pothole detection systems. Future research should explore the integration of multi-sensor fusion techniques to improve detection performance, especially in challenging environmental conditions. Leveraging pothole detection data to automate maintenance planning and scheduling can optimize resource allocation and minimize disruption to traffic flow. Future research should focus on developing intelligent algorithms that prioritize repair activities based on detected pothole severity, traffic patterns, and infrastructure condition. By addressing these future research directions, researchers can further advance the field of pothole detection and contribute to creating safer and more resilient road infrastructure in India and beyond.

DATA AVAILABILITY STATEMENT

The data that support the findings of this study are openly available in Kaggle at https://www.kaggle.com/datasets/andrewmvd/pothole-detection, reference number (Gupta et al., 2020).

REFERENCES

. Chavan, K., Chawathe, C., Dhabalia, V., & Sankhe, A. (2022). Pothole detection system using YOLO v4 algorithm. *International Research Journal of Engineering and Technology (IRJET), 80*, 25171-25195.

Du, J. (2018, April). Understanding of object detection based on CNN family and YOLO. [). IOP Publishing.]. *Journal of Physics: Conference Series, 1004*, 012029. DOI: 10.1088/1742-6596/1004/1/012029

Faramarzi, M. (2020). Road damage detection and classification using deep neural networks (YOLOv4) with smartphone images.

Gupta, S., Sharma, P., Sharma, D., Gupta, V., & Sambyal, N. (2020). Detection and localization of potholes in thermal images using deep neural networks. *Multimedia Tools and Applications, 79*, 26265-26284.

Khan, M., Raza, M. A., Abbas, G., Othmen, S., Yousef, A., & Jumani, T. A. (2024). Pothole detection for autonomous vehicles using deep learning: A robust and efficient solution. *Frontiers in Built Environment*, *9*, 1323792. Advance online publication. DOI: 10.3389/fbuil.2023.1323792

Liu, Q., Huang, W., Duan, X., Wei, J., Hu, T., Yu, J., & Huang, J. (2023). DSW-YOLOv8n: A new underwater target detection algorithm based on improved YOLOv8n. *Electronics (Basel)*, *12*(18), 3892.

Ma, M., & Pang, H. (2023). SP-YOLOv8s: An improved YOLOv8s model for remote sensing image tiny object detection. applied sciences, 13(14), 8161.

Safyari, Y., Mahdianpari, M., & Shiri, H. (2024). A review of vision-based pothole detection methods using computer vision and machine learning. *Sensors (Basel)*, *24*(17), 5652. DOI: 10.3390/s24175652 PMID: 39275561

Sathvik, M., Saranya, G., & Karpagaselvi, S. (2022, December). An intelligent convolutional neural network based potholes detection using yolo-v7. In *2022 International Conference on Automation, Computing and Renewable Systems (ICACRS)* (pp. 813-819). IEEE.

Seraj, F., Van Der Zwaag, B. J., Dilo, A., Luarasi, T., & Havinga, P. (2014, April). RoADS: A road pavement monitoring system for anomaly detection using smart phones. In *International workshop on modeling social media* (pp. 128-146). Cham: Springer International Publishing.

Shaghouri, A. A., Alkhatib, R., & Berjaoui, S. (2021). Real-time pothole detection using deep learning. arXiv preprint arXiv:2107.06356.

Wang, X., Gao, H., Jia, Z., & Li, Z. (2023). BL-YOLOv8: An improved road defect detection model based on YOLOv8. *Sensors, 23*(20), 8361.

Yurdakul, M., & Taşdemir, Ş. (2025). An enhanced YOLOv8 model for real-time and accurate pothole detection and measurement. arXiv preprint. arXiv:2505.04207. https://arxiv.org/abs/2505.04207

Zanevych, Y., Yovbak, V., Basystiuk, O., Shakhovska, N., Fedushko, S., & Argyroudis, S. (2024). Evaluation of pothole detection performance using deep learning models under low-light conditions. *Sustainability (Basel)*, *16*(24), 10964. DOI: 10.3390/su162410964

Chapter 5
Design and Implementation of IoT-Based Sensing Systems for Infrastructure Monitoring

Zaid A. Mohammed
Technical College of Engineering, Al-Bayan University, Baghdad, Iraq

G. Vijayakumar
V.S.B. Engineering College, Karur, India

Mohanad Hatem Shadhar
Department of Civil Engineering, College of Engineering, Al-Iraqia University, Baghdad, Iraq

Balaji Shesharao Ingole
https://orcid.org/0009-0004-6763-2528
Independent Researcher, USA

S. B. Sridhara
https://orcid.org/0009-0005-1929-3303
Amruta Institute of Engineering and Management Sciences, Bidadi, India

ABSTRACT

The complex nature of the infrastructure systems and also the infrastructure aging have demanded innovations, which are more efficient and reliable methods of monitoring. Manual inspection methods, however, are expensive, they are time-consuming,

DOI: 10.4018/979-8-3373-5248-0.ch005

and they have little capability of providing real-time data. The emergence of the technology of Internet of Things (IoT) has transformed the sector of infrastructure monitoring systems due to the capacity to implement cost-effective, in real-time, and automatic sensing networks. This chapter also examines designing and deploying of Sensing systems based InT based on the specific domain of tracking the health of infrastructure like bridges, roads, dams, and buildings. Issues surrounding the implementation of such systems such as choice of sensors, protocols of communication, power supply and data processing are addressed as important factors taken into consideration. IoT in combination with cloud- and edge analytics is analyzed in order to improve data gathering and decision-making processes.

INTRODUCTION

Context and Importance of Infrastructure Monitoring

Effective monitoring of infrastructure is imperative to supporting the management and maintenance of the critical infrastructure ranging between bridges, roads, buildings, dams and tunnels among others. Such infrastructures forms part of smooth running of societies, in transportation, energy distribution, water management, and urban developments. Their security and long life are of utmost consideration to the fact that breakdown of any of the structures would also have devastating economic, social and environmental effects. Infrastructure monitoring entails monitoring and evaluating the state of these constructions in a systematic manner as a way of identifying any problem that may arise before resulting to disastrous consequences, (Abdelraheem et al., 2021). Structural damage, wear and stress monitoring can be utilized at an early stage where precautionary measures can be put in place to carry out repairs and maintenance, which play a critical role in averting accidents, lowering the cost of repairs and increasing lifespan of infrastructures. Such traditional methods of infrastructure monitoring as manual inspections have limited capacity and performance. These methods usually depend on visual inspection that requires experts to visit the facilities at intervals and the errors that it entails are resourceful, time-consuming, and are subject to imperfection in man. Not all defects are easy to detect by inspection, and may be impossible to notice at all; thus, cases of non-obvious or hidden defects like internal cracks or corrosion may not be detected, and the inevitable serious structural failure occur later, (Naraharisetty, 2021). Moreover, the coverage provided by traditional techniques is poor in case of large and/or remote structures and the feedback on the state of the structure is usually delayed thus limiting the proactive action capability. Considering that prompt and accurate data is very crucial so as to guarantee safety and reliability of the infrastructure, there is a rising necessity of higher and efficient monitoring systems, (Bisio, 2022).

IoT-Based Sensing Systems: An Overview

The advent of the Internet of Things (IoT) technology has revolutionized infrastructure monitoring in the sense that the solutions provided by the technology are more automated, scalable and can even deliver real-time information. IoT is the system of the interconnected and interrelated gadgets, sensors, and systems that can interact and exchange information via the internet (Costa Monobi et al., 2024). These sensors are sometimes called smart sensors and are capable of measuring different parameters in real time i.e. temperature, humidity, strain, vibration, pressure etc. Such systems have the capability to collect and analyze data continuously, thereby ensuring the remote monitoring as well as preemptively managing the width and health of infrastructure (F. Zonzini, 2020). The sensing system implemented on the basis of IoT consists of various parts to gather, transmit, and process data. Its major elements are sensors installed on the infrastructure to record different environmental and structural aspects; connectivity infrastructure so that data collected by sensors can be transferred to distant destinations (Kawale, 2023); cloud computing where large data can be stored and handled; and data analytics tools to support the interpretation of data and drawing of actionable knowledge. The systems can be implemented on all forms of infrastructures such as bridges, roads, dams or skyscrapers and produce constant and accurate data that can be used to make more informed decisions. IoT integration into infrastructure monitoring has various main benefits compared to the conventional way. IoT systems automate the process of collecting data, which results in the absence of manual inspections and infection of human error, giving access to very comprehensive data about the condition of the structure of an infrastructure (Poongothai, 2018). The sensors aboard send real-time feedback over a continuous basis, allowing the operators to watch the structure statuses even at distant offices. With cloud computing the massive amounts of sensor data can be compiled and stored and utilizing advanced data analytics and machine learning methods, trends, patterns, and anomalies on the data can be found and used to offer highly insightful information into a predictive maintenance strategy and early warning on faults (Kumar, 2023).

The Role of IoT in Transforming Infrastructure Monitoring

The sensing systems of IoT are having an active role in revolutionizing the process of infrastructure monitoring and management. Current monitoring of the infrastructure through the incorporation of IoT has made it possible to enjoy several advantages that older means are unable to compete with. Most of the major benefits offered by the Internet of Things are related to continuous monitoring, which is available in real-time (Santoso, 2024). The infrastructural sensors can get data constantly and transmit it to a centralized system which interprets the information. This real time monitoring enables detecting any changes in the condition of the infrastructure immediately they are experienced thus leading to immediate indication of the health of the structure. Early fault identification is the other major advantage of the monitoring systems offered by IoT. Through ongoing sensor output analysis, IoT systems are able to detect abnormal patterns or signs of deviation of expected behavior, which could be early warning signals of worn out parts and wear, fatigue or damage (Sepanosian, 2024). This is possible so that possible problems can be detected ahead of time and therefore maintenance and repairs can be done promptly before they develop into serious damages that can cause huge losses. As an example, with bridges, IoT sensors can uncover too much vibration, strain, or movement that can be an indication of a possible failure that will have to be intervened as soon as possible, just to be sure that the population will not be endangered. The another great benefit of the IoT-based sensing systems is predictive maintenance (M. V. Ramesh, 2017). Rather than using scheduled inspection or reparative efforts, the real-life applications of IoT systems allow predicting the future maintenance based on analytical data tendencies. Machine learning algorithms allow such systems to study past data, warn against wear in the future, and suggests the best maintenance regimes. This does not only increase the life time of infrastructure but also save on cost of maintenance since it makes it such that repair is only conducted where necessary and avoids costly repair of other failures through emergencies. Along with these operation advantages, IoT will create smart cities and sustainable infrastructure. The areas where smart cities are implemented through the use of IoT include roads, bridges, public transportation systems, as well as energy grids. The use of the data gathered by the means of IoT systems enables optimal resources distribution, energy savings, easier management of the traffic, and safer living conditions. Second, by using infrastructure monitoring based on IoT, cities and regions might become more adaptable against the effects of disasters linked to the environment (R. Zgheib, 2017), i.e. floods or earthquakes, thanks to early warning systems and possible real-time responses to threats. The monitoring systems through IoT are also important in dealing with disasters. As an illustration, IoT sensors installed in dams or reservoirs may sense the variations in the level of water, pressure, or the state of integrity to forewarn about the potential flood or dam bursts. In areas of earthquakes, the IoT systems are capable of detecting movements of the ground, vibration and building, therefore such data will be important information in the process of earthquake prediction and mitigation. In summary, IoT is transforming the process of monitoring infrastructure since it is more efficient, cost-effective, and reliable to monitor whether our critical infrastructure is healthy (de Oliveira et al., 2023). The potential to accumulate live data, identify emergent defects, and carry out forward-looking maintenance is overhauling the administration of infrastructure, enhancing security, and lessening value, and allowing the creation of assets all the more enduring and dynamic urban circles.

DESIGN CONSIDERATIONS FOR IOT-BASED SENSING SYSTEMS

Design and execution of an IoT-based sensing platform to monitor infrastructure are sensitive issues that need to consider a number of factors in order to make the system effective and reliable and also scalable. These features are the choice of sensors, connectivity standard, power control, data storage plus cloud linkage as well as superior data processing. All the parts of the system are crucial in monitoring infrastructure without errors but in real-time, besides tackling issues relating to energy usage, data safety, and data mining.

Types of Sensors for Infrastructure Monitoring

Sensor choice is perhaps the most important element in creation of an IoT-based infrastructure monitoring system. Depending on the needs of the infrastructure to monitor (e.g. kinds of data that are to be measured and environmental conditions) the sensors may differ in type.

Structural Health Monitoring Sensors

Bridges, buildings and dams are some of the infrastructure that needed constant monitoring to ensure that the structure is standing as time goes by. The physical deformations and strain are normally monitored using sensors such as vibration sensors, strain gauges, accelerometers and rail tilts. Sensors including vibration sensors and accelerometers are used to detect anomalies like unusual vibrations or seismic events that will indicate internal stress in the process and a possibility of failure (N. Kumar, 2023). Strain gauges and predisposing detect the deformation and the displacement of structural elements and give good information about the condition of the infrastructure. Constant observation of such physical changes will enable them to be aware of the most likely problems that could come up such as cracks, sagging, or wearing out of the materials early enough to correct.

Environmental Monitoring Sensors

Operating conditions, like air quality, humidity or temperature, may have a great influence on infrastructure stay and health. Changes in temperatures that may result in expanding, contracting or cracking of substances are measured by temperature sensors. Humidity sensors aid in monitoring the moisture content which is very instrumental in the detection of corrosion or development of mold especially in underground facilities such as tunnels. The air quality sensors control the amount of pollutants above a state; carbon dioxide, nitrogen dioxide and particulate matter; among which can be harmful to the infrastructure and environment (Rajaei, 2023). Such sensors are useful to get data on the conditions in the environment that can cause the deterioration of materials or raise the possibility of damage.

Load and Pressure Sensors

Buildings, tunnels and bridges need to be constantly checked on the aspects of load-bearing capacities. Sensors such as load sensors are placed to estimate the amount of modeling force or weight a structure can handle. Take the bridges as an example, as the load sensors shall be able to warn of an overloading condition when too many vehicles are driving on it thereby causing structural weakness due to overloading. Use of pressure sensors also comes in handy when checking underground structures like tunnels where there is internal pressure of surrounding earth or water and this pressure should be closely monitored (Srilakshmi, 2019). The usage in dams is also made through these sensors as the slightest variation of the pressure in the dam would mean that there is a danger like tapering or even the collapse of the dam.

Visual Sensing Systems

Visual monitoring is a very strong mechanism to determine physical condition of infrastructure. High resolutions imaging systems fitted in the cameras are able to generate detailed images or videos of the structural components, irrespective of whether it is the cracks, corrosion, or wear and tear. These images are analyzed using sophisticated image processing methods, designed to detect the presence of defects, like cracks, abnormal shape or other defects that can not be characterized by eye. Visual inspection of these systems can be automated thus eliminating the number of times one has to make manual interventions and ensuring that visual checks are carried out frequently.

Connectivity and Communication Protocols

To make any IoT-based infrastructure monitoring system work, it also needs a strong communication infrastructure. The system should be able to transfer big quantities of sensor information at long distances and in many cases in real time. A number of communication protocols and technologies are used towards this end:

LPWAN (Low Power Wide Area Networks)

LoRaWAN and NB-IoT are well-known protocols of IoT systems which need to be long-range connected. LoRaWAN finds plenty of application where the devices have to be able to communicate over a long distance (up to 10 km) and using very little energy. It is quite suitable in remote infrastructures such as a bridge or dam where sensors should not transmit data using regular battery replacements especially in long distances (Parimala, 2021). The comparable long-range, more reliable and efficient utilization of resources can be found with NB-IoT operating in cellular networks. Such networks make it possible so that there is constant flow of information that the sensors convey even in regions where there is no provision of communication.

Wireless Sensor Networks (WSN)

The common used protocols are Zigbee and Bluetooth Low Energy (BLE) in wireless sensor networks (WSN) at the local communication. The application of these protocols involves short-range and low-power operations hence it is suited in smaller infrastructures, including monitoring of an individual building or a certain portion of a building. Zigbee is good specifically in the area of development of meshed networks enabling reliable interconnections among many sensors through other sensors. BLE has a reputation of low power consumption and it is often utilized in simple medium scale infrastructure monitoring systems at short distances over which the devices are communicating (Emdadul Hoque, 2022).

Edge Computing

In order to make the necessary real-time decisions and lower latency, edge computing is becoming an essential part of the IoT-based infrastructure monitoring systems. Unlike simply using cloud servers, the edge computing facilitates data processing closer to the sensor or the monitored infrastructure than before. In this way, critical information (good advance warning on structural damage or environmental hazard), can be processed faster and results received in real-time. Edge devices may filter and preprocess the data and only send the relevant parts to cloud, so there is no wasted bandwidth thus making the whole network more efficient.

Power Management

The quality of power management is an important factor during the design of IoT-based sensing systems, especially since most of the sensors will most likely be installed in inaccessible or inaccessible locations, and would not effectively be maintained and new batteries provided.

Battery-Operated vs Solar-Powered Sensors

Numerous IoT based sensing system around have battery-powered sensors, which are easy to deploy but can eventually have short life cycles due to the sensor power consumption. In the case of outdoor or distant infrastructure, the use of solar-powered sensors means an economically viable solution that makes use of sunlight as energy resources to maintain battery life and prevent the necessity to conduct regular maintenance. Solar panels that will incorporate the sensors will mean that one will be self-sustaining in power production and hence suited to long term infrastructure tracking where it is sunny enough (T. Sepanosian, 2024).

Energy Harvesting Solutions

The use of piezoelectric and thermoelectric generators, also known as energy harvesting technologies, is being adopted as a powerful way of powering the IoT sensors in the infrastructure monitoring systems. Piezoelectric sensors can convert mechanical energy with vibrations or pressure into electrical energy and thermoelectric generators convert heat difference into power. Such solutions may come in handy in case of situations where conventional power cannot be utilized and can be used to ascertain the sustainability of the IoT based monitoring system.

Data Storage and Cloud Integration

Cloud implementation and data access are crucial to processing and processing the large capacity of data that IoT sensors collect. The use of cloud platforms like AWS IoT, Microsoft Azure, and Google Cloud allows storing sensor data because of the scalable systems that can help monitoring in a real-time environment. Such platforms provide data analytics tools into it where the state of the infrastructure is visualized and analyzed, and predictive maintenance patterns are created. The security of data and privacy is however the most important especially when handling sensitive infrastructure data. The security and privacy of the information should be ensured by providing high encryption algorithm, and access control and data privacy policies.

Data Processing and Analytics

IoT sensors make massive amounts of data and to derive insightful information, it is necessary to use advanced data processing and analytics devices. The machine learning algorithms used to manipulate sensor data allow detecting anomalies, early fault identification, and predictive maintenance. Such algorithms would allow demonstrating trends in past information to know when one would need maintenance, the allocation of resources, and decision-making. Real-time analytics will also be able to give immediate input to the condition of the infrastructure, reducing the risks and enhancing safety.

REAL-TIME APPLICATIONS OF IOT IN INFRASTRUCTURE MONITORING

Monitoring of infrastructures using IoT-based sensing systems has resulted in huge improvements in the capacity to identify faults on-time and react to unexpected threats in a proactive manner. The IoT sensors have been installed on a broad realm of infrastructure, whether it is bridges and roads, dams, buildings, and tunnels. The systems will allow constant observation and gives a more work-effective solution to protecting the safety and integrity of critical infrastructure which is more efficient and cost-effective. This section will explain some real-time examples of the treatment of IoT in infrastructure monitoring, the application of IoT sensors in various types of infrastructure, as well as case studies to demonstrate the usefulness of the same.

Bridges and Overpasses

Bridges and overpasses are the major part of the transportation system, and its security sustains the safety of people. With time, these constructs encounter a substantial burden, weather, and natural activities such as wind, earthquakes, and thermal expansion which may wear them out and most likely lead to failure. The conventional bridge inspection processes also mostly depend on visual inspections that may be time-consuming, and only fewer defects are identified. Systems based on IoT are monitoring in place and are branded by strain gauges, accelerometers and other gadgets that allow the bridge health be monitored continuously in real-time. Strain gauges are placed to measure deformation of bridge parts that are under a load, and they give data concerning the bridge behavior when subjected to different loads. When a change of the dynamic behavior is to measure vibrations and movements, accelerometers are used. The sensors assist in the detection of possible problems like exhaustion, wear or unwanted vibrations, which could signal an upcoming breakdown. An excellent case of IoT-based bridge monitoring is with reference to a seismic area where an IoT-based load and vibration sensor is attached to the highway bridge. These sensors observe the bridge reaction against the daily loads and seismic events and offer useful information about the behaviour of the bridge. Real time data about the accelerometers can be run during the occurrence of an earthquake to allow an assessment of the effect the earthquake had on the bridge and make immediate decisions about the safety of the bridge. The proactive action plan allows noticing the appearance of stress factors and possible risks of failure early, which may result in timely repair and the prevention of devastating accidents.

Smart Roads and Pavements

Pavement condition is another key area of interest to transport authorities because a damaged pavement in the form of cracks, potholes and others may result in irreparable traffic accidents and damages on vehicles and associated very expensive repairs. Manual surveys or certain visual inspections that are performed periodically are the traditional road inspection methods that might not detect smaller and emerging problems. Using IoT-based solutions, it is now possible to have a better and more efficient solution that is automated when it comes to road condition monitoring. Cameras and IoT sensors will be installed along the roads allowing them to constantly check the status of the pavement. The mechanisms applied in these systems include crack detection and identification of potholes and the cameras on the road capture the surface of the road and additional processing of the images using powerful algorithm detects the anomalies. Another area where the vibration sensors may be used is in detecting the impact of heavy traffic on a road surface; an area which might need repair. These sensors report in real-time data that enables the authorities to know current conditions on the road so they can assign.", A good example is that damaged or cracked section of the road may have to be flagged as an area of immediate need whereas less urgently needed areas can be handled in the future. This is based on data and this would assist in ensuring that available resources are utilized efficiently, time out of service is minimized and road safety in general would be enhanced.

Dams and Water Reservoirs

Water reservoirs and dams are important in the control of water resources, water management, flood prevention and also as hydroelectric stations. Nonetheless, such buildings tend to be established in the areas that are prone to natural calamities like floods or earthquakes and thus it is necessary that they be watched at all times so that structural integrity remains intact. Monitoring systems developed using IoT can supply real-time information about water levels, destruction of dams, and the environment that allows assessing the risk in real-time and making decisions. The sensors used in the dams have the capacity to check the level of water, the integrity of the structure and the environment in terms of temperature and pressure. Flood prediction models may also be incorporated in IoT systems and they are based on the data obtained through environmental sensors in order to determine the risk of its overflow or dam breaking. The example of this would be that IoT based systems can sense the level of water increasing in real-time and sound alerts when water rises to a certain point. Closely placed pressure sensors in the construction of the dam can indicate when its integrity is strained or the dam is developing cracks, and temperature sensors can show when ice is forming in its structure that can also worsen flood dangers. An IoT-based monitoring may be an example of a case study which can be considered in flood-prone locations where dams are fitted with IoT sensors to forecast the potential overflow of water and the possibility of structural failure. In another example, water level, rainfall data and dam pressure sensors were continuously monitored by analysing the sensors on a dam on a flood prone area. The system automatically sounded flood alerts when sensors observed the presence of an increase in the levels of water to possibly critically dangerous levels that enabled the local authorities to take preemptive measures to evacuate the vulnerable locations. This IoT system offered priceless early prediction that could save lives and diminished the economic repercussion of flooding.

Buildings and Skyscrapers

The maintenance of buildings especially those of the higher rise nature such as skyscrapers needs to be done continuously in order to ensure the safety and comfort of the occupants. A diverse set of stresses are encountered to these structures either in the form of vibrations such as, vibrations due to wind, traffic and earthquakes, fluctuations in temperature and humidity. Systems based on IoT enable monitoring of essential parameters including vibration, temperature, pressure, which may cause compromisation of structural integrity and safety of the building in real-time. Vibration sensors are suitable to sense undesired oscillations or vibrations triggered by external forces i.e. wind or earthquake shocks. Temperature sensors can observe variations in internal and external environment of the building and this can affect the behavior of material and systems. As an example, high temperatures may result in thermal expansion of some materials, which may cause some structural conflicts. It can monitor the fluctuations in the air pressure of the building especially tall structures, wherein the pressure variations among levels may cause discomfort to the occupants or destruction of the building systems. The fire and smoke detection systems can also be incorporated in to IoT systems that offer even more safety. The sensors are able to detect the heat, smoke, or gas emissions and activate the alarms alerting real-time to the emergency response teams. The systems enhance the safety of the facility in the sense that it will be able to prevent any risk associated with fire. *Tunnels and Underground Infrastructure:* Formation of tunnels and other infrastructure systems that are found underground has extreme conditions of locating and monitoring them. Some of the risks that are associated with tunnels include ingress of water, ground movements, and emanation of harmful gases, which may interfere with the safety of the facility and its occupants. The IoT-based monitoring systems offer great help as they continuously produce data regarding the most important parameters like stress in the walls, their level of humidity, and gas emissions. Tunnels have devices that measure humidity and this is important in ensuring that there is no probability of water ingress which may cause flooding or corrosion of the structural components. There are gas detectors such as carbon monoxide detectors which can detect the occurrence of foul gases which would be a pointer to poor ventilation or the presence of harmful substances. The structural integrity of the walls of a tunnel can be checked with wall stress sensors that sense cracks or other deformation that might point to the onset of a collapse. The use of IoT systems is applicable also to identify the possible risk of flooding, cave-ins, or poisonous gas leaking. An example would be sensors in a tunnel which would give a preliminary warning of an increasing water level or deteriorating air quality and action could be taken quickly to avert an accident or evacuate the personnel in the case of an accident. In an underground mining operation, the system based on the concept of the IoT could be used to detect structural conditions in real-time and make sure that the integrity of a tunnel could be kept and the employees would not be exposed to hazardous conditions.

CASE STUDIES

Case Study 1: IoT in Bridge Monitoring

The purpose of the case study is to discuss how IoT technologies can be utilized in terms of bridges health and safety control especially in the areas with earthquakes.

The crucially important role that bridges take in the transportation network and the increase of risks to them in certain areas prone to the effects of seismic shocks presupposes the necessity to employ highly enhanced monitoring system in order to guarantee the integrity of the structures. Conventional observation procedures do not always reveal the presence of defects or problems that might result in life-threatening failures in case of a seismic activity. The case study examines ways that the IoT-based solutions will bring constant, real-time monitoring and enhance the use and utilization of such crucial structures. The bridge monitoring system designed on the basis of the IoT applied in the case study encompasses the installation of various sensors on the major portions of the bridge (strain gauges, accelerometers, and displacement sensors). The following parameters that are important to determine the structural health of a bridge are measured using these sensors strain, displacement, and vibration. Strain gauges are used to detect any alteration notions in the material characteristic of the bridge i.e. in the form of stretching or decrease that can depict the initial symptoms of fatigue or injury. Accelerometers record the vibrations due to traffic loads or to seismic activities that are essential in determining the stability of the structures. Also, the displacement sensors are used to detect a change in the alignment of the bridge, thus they could detect subsidence or deformation that may arouse a threat to the safety of the bridge. The sensor data received is relayed in real-time to cloud-based platform where they are processed and analysed immediately in order to take action. Machine learning models are integrated into the cloud platform, where they can learn through past and present data to foresee possible failure points and what should be repaired. The system will be able to predict the times of structural failure using the predictive analytics, which allows interventions before the actual structural failure and minimizing the chances of severe disasters. The constant glimpse by such IoT system allowed structural problems that had not been discovered by physical examination to be identified early enough. In one scenario, cracks that could not be seen during routine check ups on a bridge deck were detected by the sensors. This detection at early stage enabled timely repair which would not lead to more damages and reduce the cost of repair. Predictive maintenance functionality of the system aided in streamlining the maintenance schedules so that repairs can be done when it is deemed necessary and at most cost-effective times. The information provided by the IoT sensors also helped to comprehend the reaction of the bridge to seismic shaking at a better level. In the case of an earthquake, the system advocated real-time responses of the bridges on how it survived and engineers evaluated its structural strength and determined whether they were to carry out urgent repairs or not. The use of predictive models as part of the monitoring system allowed decreasing the amount of downtime by providing more specific interventions to ensure the prolonged life of the bridge and save on the expenses of maintenance over a long-term perspective. Another issue that is an essential element in determining the effectiveness of this bridge monitoring system based on IoT is the location and setting of sensors. This is because the sensors must be well adjusted and placed at the right positions to get the right values. Particular attention should be paid to placing sensors in the problematic areas, which are stressed in the joints and support beams in order to identify the problem at the early stages before the critical situation occurs. Moreover, sensors alignment should be repeated as time goes by to consider any environmental or operational factors that could influence their accurateness. The incorporation of existing infrastructure with IoT was also one of the major issues that were instrumental in this case study. Most of the bridges in the earthquake prone areas were constructed in a way that they did not have modern sensor systems in mind and thus installing the sensor systems would be difficult. The process of equipping bridges with monitoring systems based on the IoT was rather carefully thought, and the design of the bridge had to be changed in some cases. Another challenge was to make sure that the sensor systems would be available to properly communicate with the existing infrastructure. Nonetheless, the advantages of utilizing real-time monitoring information were by far more significant than the initial challenges since it allowed gaining essential information about what was happening upstream and enabled active maintenance.

Case Study 2: Smart Road Monitoring for Pavement Condition

This case study aimed to implement a solution based on the IoT technology that would facilitate setting the state of pavements and roads, allowing planning the maintenance of them and ensuring better road safety. Roads experience a steady state of wear and tear because of the traffic load, climatic conditions and age. Eventually, such decay is characterised by cracks, potholes, and surface degradation that may cause accidents, damage vehicle and result in expensive repair. The conventional road maintenance processes are likely to be made in irregular time frames where there is either a regular check up or corrective maintenance which proves more inefficient hence causing unwarranted expenses. It also shows how IoT systems can speed up on problem identification, minimize costs of repairs, and enhance road safety, which is also reflected in the case study. The IoT system in this case was meant to observe the main highways that were characterized by intensive traffic. Sensors that measure vibrations were installed on the road surface to identify deviations that might suggest the existence of a damaged road, in a form of a crack or unevenness resulting in a breakdown of the material. High resolution image capturing cameras were installed to take snapshots in real time of what is on the road surface. These pictures were studied through computer vision and picture processing to figure out breaks, holes, or any other surface damage. The system would have the capability of detecting the presence of the defects that may not be imminently evident to the human eye and the system would have given a clearer and complete picture of the state of the road. Any data obtained by the sensors and cameras were sent to cloud-based platform to be analysed and decisions made in real-time. The sensor information was manipulated in the platform to provide usable information such as alerts to areas that need urgent action. This system may also be teamed up with the predictive maintenance models which will allow repairing the system according to the level of the road conditions, traffic trend, and resources. With the introduction of the IoT-based system the cost of repair was considerably reduced. This constant monitoring capability permitted the system to identify cracks and potholes in early stages so that the maintenance teams could work on the problems before they escalated. Such proactive approach contributed to the reduced number of expensive emergency repairs since in this case the maintenance work could be performed at the time when there was no traffic or weather conditions were more suitable. Moreover, the system enhanced the safety conditions on the road as it was possible to detect the dangerous locations sooner. An instance case is a pothole that was noticed by the sensors; the sensors marked the pothole and the concerned wondered to get the pothole fixed before becoming dangerous to the drivers. Capacity to book repairs using real time data also minimized the chances of accidents due to road conditions hence the safety of the roads users was increased. Another of the difficulties encountered in this case study was the management of data needed to deal with data generated by the sensors and cameras that is in large quantities of information. Real-time processing of the sensor data and its quality proved to be one of the challenges. As the system was constantly gathering information, the platform needed to be able to accommodate high frequency sensor readings thus making the data viable and usable. The cost-benefit analysis of using IoT sensors versus old maintenance was another significant matter. Although the likely cost of installing the sensors and setting up the system was high, the long term advantage of installing sensor exceeded the expenditures. Through early detection of problems, the system significantly saved the company in terms of repair expenses; saving the company in terms of reactive and emergency repairs. In addition, the increased capacity to make decisions on prioritization of repairing the road using information in real time permitted better allocation of the available public funds.

RESULTS AND DISCUSSIONS

Effectiveness of IoT for Real-Time Monitoring

The building of IoT-based systems in infrastructure monitoring has resulted in the significant improvements in regard to the performance, operational efficiency, and decision-making. A critical infrastructure, like bridges, roads, dams, and buildings, can be monitored in real-time through the use of IoT technology which is a noteworthy innovation. Among the greatest enhancements, one can mention the detecting rate. Manual inspection used to take time as sometimes, the problem was detected late especially in expansive infrastructure. With IoT, data gathered using sensors has a constant flow, thus irregularities and structural anomalies could be reported in real-time. This shortening of detection time is essential, particularly, in those systems of infrastructure that can potentially challenge the safety of the population in the very near future like bridges or dams in the area during floods. In addition, IoT-driven systems minimize the reliance on the manual checks. The use of manual inspection techniques is usually based on periodical review and the experience of a limited number of specialists, which can be considered the presence of loopholes in the detection of emerging problems. By contrast, other IoT systems deliver an automated and real-time flow of data, offering a larger picture of the infrastructure status. As an illustration, Figure 1 graph shows vibration measurements of some bridge monitoring system and one could see that sensor outputs are able to monitor even slightest vibrations and stress points that are otherwise invisible during a manual check. The second major benefit is a predictive capability of IoT systems. Case studies like smart road monitoring systems and preventing bridge failures through detection of possible failures with the use of machines learning have given us an idea of how the machine learning algorithms can be used to analyze the collected data and be able to predict a failure before it happens. As the crack size could be detected over time in Figure 2, it is possible to make a timely forecast of where and when it is necessary to make a scheduled repair, thereby eliminating even more costly repairs in the future. This type of predictive model is based on historical data together with a real-time monitoring that allows the operators of the infrastructure to carry proactive measures and hence increase the life cycle of the structure in effect lowering the cost of maintenance.

Figure 1. Bridge vibration monitoring

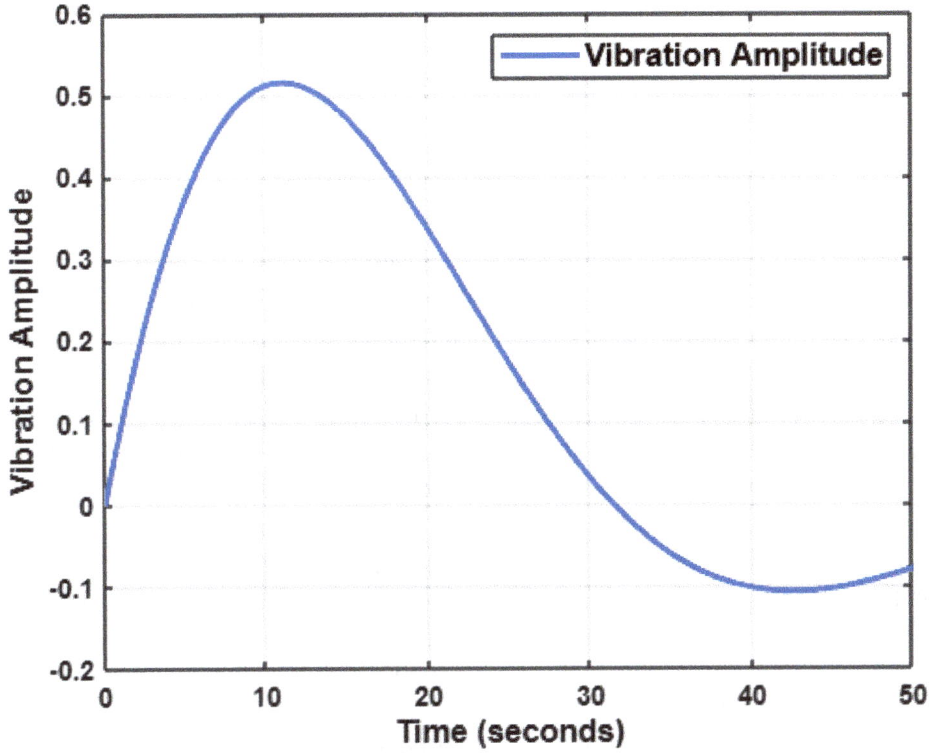

Challenges in Implementation

Although the IoT has many advantages in monitoring of infrastructure, there are some challenges that have to be addressed to make the system successful. Such challenges include technical ones, environmental and economic. One of the most influential obstacles to the ubiquity of IoT-based monitoring systems is presented by the technical issues. Sensor calibration is one of the major problems. To obtain quality data collection, the sensors are required to be properly calibrated in order to take into consideration the change of the environmental conditions, or physical structure of the infrastructure over a period of time. Calibration errors are likely to disrupt the reliability of data thereby compromising the effectiveness of the monitoring system. In addition, IoT systems have a strong linkage with power management. A lot of sensors are at remote or otherwise difficult-to-access locations, which will not be connected to a stable power source. Although other answers, like solar power-sensing or energy gathering (e.g. Piezoelectric) have been tried, they have limitations when it comes to efficiency and longevity as evident in Figure 3, which measures temperature and humidity over a period. Network connectivity is yet another technical challenge. In a mass infrastructure monitoring infrastructure the data output of a broad range of sensors is required to be sent in real-time to a central platform to be analyzed. Nonetheless, weak communication especially in the outskirts may interfere with the smooth flow of data. Such low power wide area networks (LPWAN) solutions as LoRaWAN and NB-IoT have been suggested to solve this problem, yet the network saturation and the signal corruption of extreme environmental conditions may still be problematic. One more adverse factor that does not simplify the implementation of IoT systems is environmental issues, particularly, outdoor or infrastructure support that is subject to harsh weather conditions. The goal behind the design of sensors would be to ensure that they are capable of withstanding harsh weather conditions including high elements of humidity, extreme temperatures, and heavy rainfall especially in infrastructural works such as the dams, roads as well as bridges. There are also the parts of the infrastructure that can become exposed to such elements of the environment as corrosive conditions and, hence, may be in need of the devices that are special sensors and also protective enclosures that will avoid the failure of such solution. Figure 4 modeling repair costs over the time also indicates that harsh environmental conditions are what causes increased wear of the infrastructure, which requires overall durable and long lasting sensors to work with. Lastly, there is the cost of reading and absolute preparation of an IoT framework that is inhibitive. Implementing sensors, connecting them, dealing with cloud platforms, and how a system would be reliable can be quite expensive, especially in large-scale infrastructure. The cost savings in the long-term maintenance and efficiency of operation are however normally worth the cost of the initial investment. This is very pronounced where real time monitoring of the road inspection has been done which has cost a lot of money on emergency repairs in case studies of the smart road monitoring systems.

Figure 2. Road crack detection over time

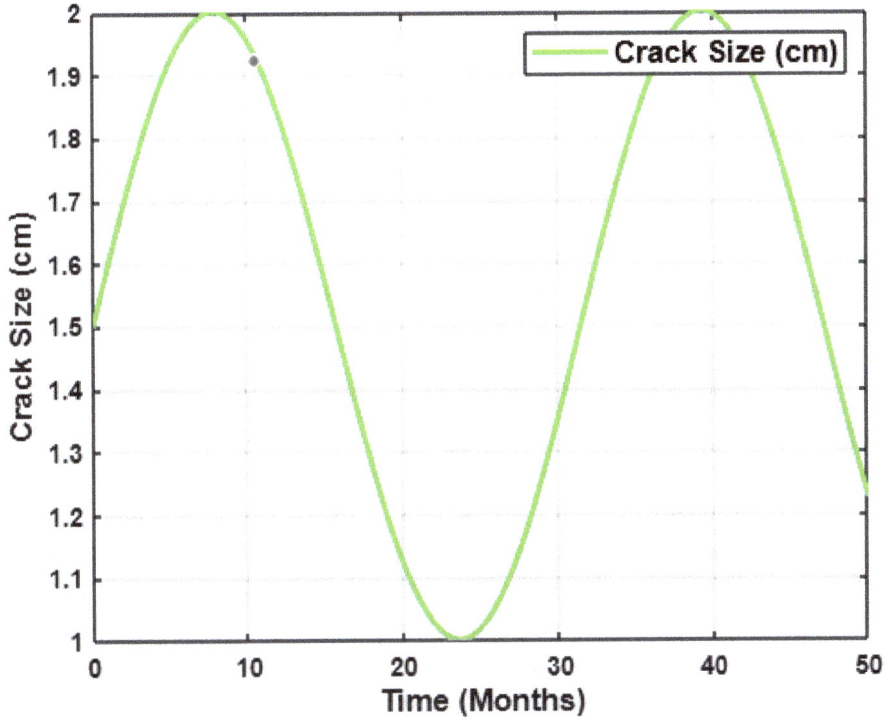

Scalability and Cost-Effectiveness

Scalability is one of the main benefits of the IoT systems; hence, it is perfect to implement it in large-scale environments. One of the main features is the ability to scale these systems to oversee more infrastructures, including the entire cities, or regions. As an illustration, the Internet of Things sensors may be installed in citywide transportation networks, combining information in the roads, bridges, and tunnels with the aim of capturing the whole picture of the urban infrastructure. As the technology evolves, the industry may expect that IoT systems are capable of scaling more efficiently, and cloud platforms and edge computing help to manage large proportions of data that are encoded by thousands or even millions of sensors. Another significant aspect is the economics of using IoT systems. The first costs of setting up may prove to be very expensive, but the benefits of the long run can prove much more than the costs. The Common infrastructure monitoring models are based on The manual inspection and an occasional maintenance that are costly and ineffective. Conversely, systems supported by IoT offer constant observation where predictive maintenance is possible and this can help eliminate expensive emergency fixation. Predictive maintenance also contributes to a cheaper maintenance cost in the long run as the problem is damaging structures are handled before they worsen hence causing major damages. In addition, the fact that data collection and analysis could be automated saves a lot of labor costs and trained personnel. When assessed with the use of IoT-based systems compared to the old methods, periodic manual inspection can incur huge expenses in terms of labor cost, traveling expenses and possibilities of unnoticed problems. Comparatively, the initial study of IoT systems will be more costly but the overall ownership cost is cheaper due to real-time access of data, decreased downtimes, and improved maintenance schedule.

Future Directions

Further development of IoT technology may be linked to the integration of artificial intelligence (AI) and machine learning with the technology, which provides great potential in terms of improving predictive maintenance. Machine learning algorithms have the capacity to read big amounts of sensor data, finding the patterns and making predictions of when and where the maintenance is necessary. This will result in smarter and more efficient maintenance approaches and one will be able to predict failures with still a greater level of precision. The future of infrastructure monitoring is so far using AI to automate the data collection and analysis processes and even the decision-making process involving repairs and replacements. Lastly, due to the integration of drones and robotic systems in visual monitoring and maintaining infrastructure, it is estimated that this will revolutionize how infrastructure is maintained. The unmanned airliners fitted with high resolution cameras and sensors may help to inspect, bridges, dams and such other difficult structures, enabling the operators to get real-time visual data. This will minimize the risks that may occur during human inspection of dangerous or inaccessible places. They can also use a robotic system to perform repair and maintenance services (cleaning, painting, and sealing cracks), which will also help cover the unnecessary usage of humans in unsafe areas. To sum it up, the monitoring of infrastructure with the help of the IoT can be expanded at a fast rate and there is potential of sustained improvement in AI, robotics and the sensor technology. With further maturity of the technologies, they are bound to be even more precise, affordable as well as efficient in addressing the issue of health and safety of the critical infrastructure. Future smarter and greener infrastructure systems will be cleared by the further competence of the IoT.

Figure 3. Temperature and humidity monitoring

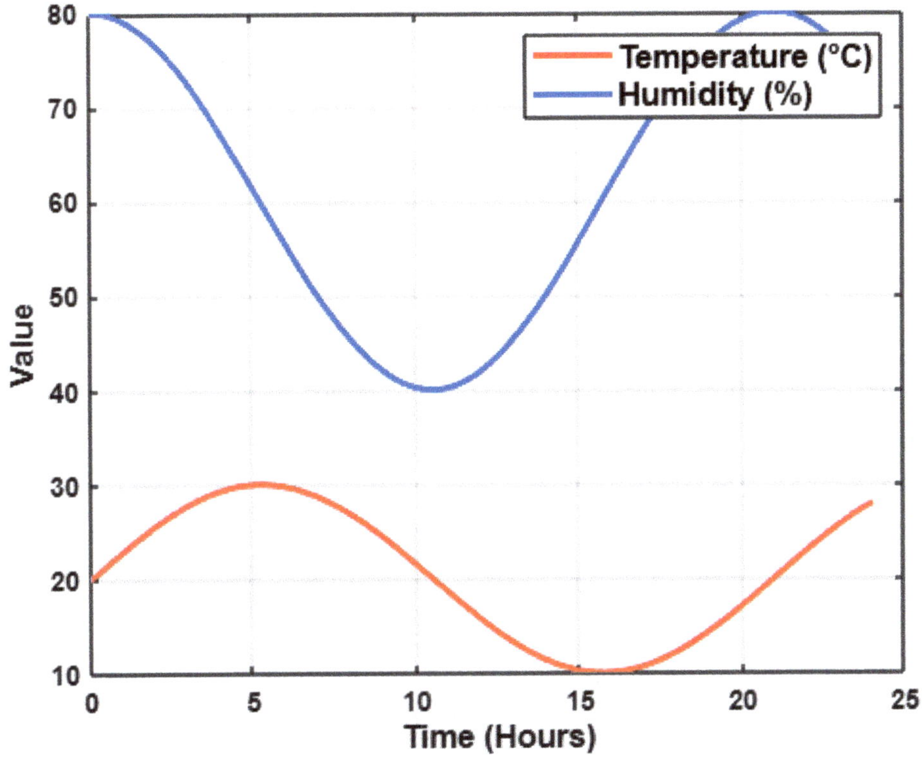

Figure 4. Predictive maintenance for road condition

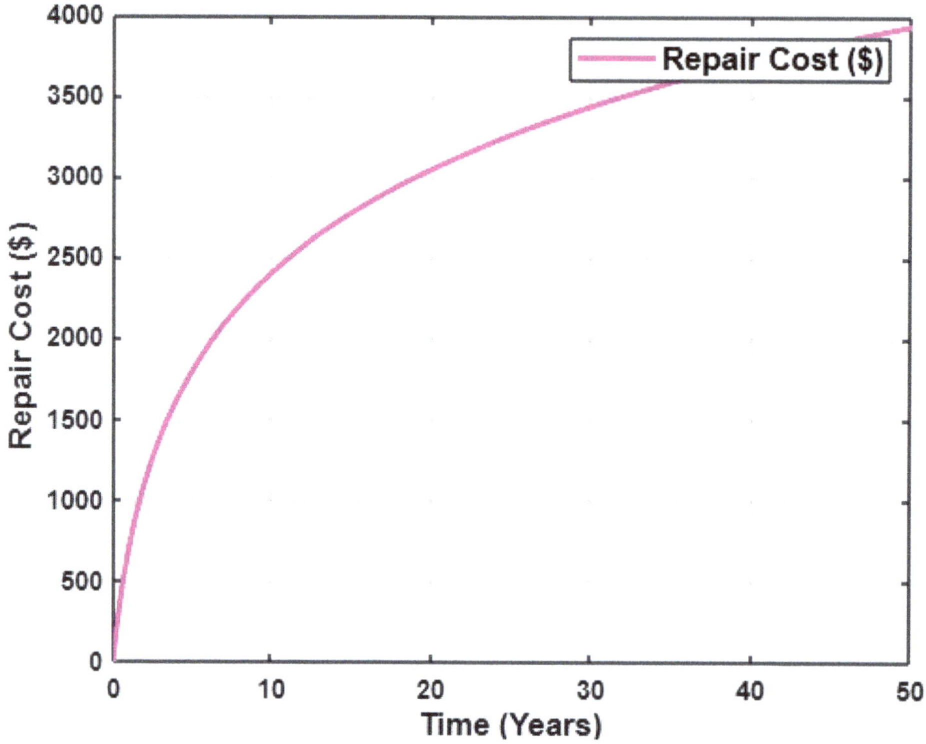

CONCLUSION

Sensing systems using IoT have been seen to revolutionize infrastructure monitoring, meaning that it has a lot of benefits when it comes to the collection of real-time data, real-time predictive maintenance, and budgetary impact. Such systems can help to keep a constant eye on the critical infrastructure, which would lead to early warning about possible problems and eventually a minimal number of downtimes and costs of their maintenance. Improving the infrastructure management efficiency, and via predictive maintenance, IoT systems can lower the number of expensive emergency repair services, thereby prolonging the service life of infrastructure resources. But further research and developmental activity are necessary to deal with issues like the upgrading of the sensor technology, data analytics, network protocols in order to improve the system reliability and its scalability. IoT will have its performance optimized by improvements made in sensors that can work in harsh conditions and development done in edge computing and machine learning. Moreover, the development of communication procedures will assist in external networking challenges in distributed or distributed infrastructures. IoT technologies will play a bigger role in the world of infrastructure in the future, and the prevalence of smart infrastructure will guarantee the design of highly resilient and sustainable infrastructures. The built environment will also be more efficient and intelligent because integration of AI, robotics and advanced analytics will facilitate autonomous monitoring and maintenance.

REFERENCES

Abdelraheem, M., Abdelhafeez, M., & Nassr, A. (2021). IoT-Based Interdigital Capacitance Sensing System for Damage Detection in CFRP-Concrete Structures. *IEEE Access : Practical Innovations, Open Solutions*, *9*, 138658–138667. DOI: 10.1109/ACCESS.2021.3118299

Bisio, C. Garibotto, F. Lavagetto, and A. Sciarrone, (2022) "A Novel IoT-based Edge Sensing Platform for Structure Health Monitoring," IEEE INFOCOM 2022 - IEEE Conference on Computer Communications Workshops (INFOCOM WKSHPS), New York, NY, USA, 2022, pp. 1-6.

Costa Monobi, F. M., Martins, A. C., Cerento de Lyra, P. H., Roberto, J. M., del Bel Sonoda, L., & Coelho Leite, B. C. (2024) "Development of an IoT Based Sensing System for Structural Health Monitoring Using Low-Cost Accelerometers," *2024 Symposium on Internet of Things (SIoT)*, Rio de Janeiro, Brazil, 2024, pp. 1-5. DOI: 10.1109/SIoT63830.2024.10780766

de Oliveira, L. D., De Araújo, J. V. S., Da Silva, J. H. B., Villanueva, J. M. M., de Souza Filho, C. A., & Ochoa, M. N. (2023) "Smart Water Management: a Self-Sufficient IoT-Based Application for Pressure and Flow Monitoring in Water Distribution Systems," 2023 7th International Symposium on Instrumentation Systems, Circuits and Transducers (INSCIT), Rio de Janeiro, Brazil, 2023, pp. 1-6. DOI: 10.1109/INSCIT59673.2023.10258484

Hoque, E., Saha, D. K., & Rakshit, D. (2022). "IoT-Based Intelligent Solar Energy Harvesting Technique with Improved Efficiency," in Artificial Intelligence and Internet of Things for Renewable Energy Systems. *De Gruyter*, *2022*, 279–302.

Kawale, S. R., Prasad, K., Palanikkumar, D., & Mary, P. A. (2023) "A Novel IoT Framework and Device Architecture for Efficient Smart City Implementation," 2023 7th International Conference on Trends in Electronics and Informatics (ICOEI), Tirunelveli, India, 2023, pp. 420-426. DOI: 10.1109/ICOEI56765.2023.10125677

Kumar, N., Haque, M. J., Mittal, A., Singh, A., Rayal, A., & Oli, S. (2023) "IoT Based Smart Weather Monitoring System," *2023 Annual International Conference on Emerging Research Areas: International Conference on Intelligent Systems (AICERA/ICIS)*, Kanjirapally, India, pp. 1-6.

Kumar, N., Haque, M. J., Mittal, A., Singh, A., Rayal, A., & Oli, S. (2023) "IoT Based Smart Weather Monitoring System," *2023 Annual International Conference on Emerging Research Areas: International Conference on Intelligent Systems (AICERA/ICIS)*, Kanjirapally, India, pp. 1-6.

Naraharisetty, V., Talari, V. S., Neridu, S., Kalapatapu, P., & Pasupuleti, V. D. K. (2021) "Cloud Architecture For IoT Based Bridge Monitoring Applications," *2021 International Conference on Emerging Techniques in Computational Intelligence (ICETCI)*, Hyderabad, India, pp. 39-42. DOI: 10.1109/ICETCI51973.2021.9574044

Parimala, M., Rani, M. J., Sai, K. B., & Sunitha, M. (2021) "IoT Based Intelligent Fertigation through Drip Irrigation," *2021 Third International Conference on Intelligent Communication Technologies and Virtual Mobile Networks (ICICV)*, Tirunelveli, India, pp. 410-414. DOI: 10.1109/ICICV50876.2021.9388433

Poongothai, M., Subramanian, P. M., & Rajeswari, A. (2018) "Design and Implementation of IoT Based Smart Laboratory," 2018 5th International Conference on Industrial Engineering and Applications (ICIEA), Singapore, 2018, pp. 169-173. DOI: 10.1109/IEA.2018.8387090

Rajaei, H., & Upendra, S. (2023). *"Intelligent Fall Detection Monitoring System," 2023 Congress in Computer Science, Computer Engineering, & Applied Computing*. CSCE.

Ramesh, M. V.. (2017) "Water Quality Monitoring and Waste Management Using IoT," *2017 IEEE Global Humanitarian Technology Conference (GHTC)*, San Jose, CA, USA, 2017, pp. 1-7. DOI: 10.1109/GHTC.2017.8239311

Santoso, K., Nurba, H. P., Sarief, I., & Munastha, K. A. Kusmadi, and R. Gunawan, (2024) "Performance Evaluation of Early Warning System for Transformer Faults at 20KV Distribution Substation," 2024 18th International Conference on Telecommunication Systems, Services, and Applications (TSSA), Bali, Indonesia, 2024, pp. 1-5.

Sepanosian, T., Küpers, X., Joza, P., Massah, F., & Bemthuis, R. (2024) "An IoT-Based Architecture for Real-Time Emission Monitoring at Construction Sites," 2024 26th International Conference on Business Informatics (CBI), Vienna, Austria, 2024, pp. 139-148. DOI: 10.1109/CBI62504.2024.00025

Sepanosian, T., Küpers, X., Joza, P., Massah, F., & Bemthuis, R. (2024) "An IoT-Based Architecture for Real-Time Emission Monitoring at Construction Sites," 2024 26th International Conference on Business Informatics (CBI), Vienna, Austria, pp. 139-148. DOI: 10.1109/CBI62504.2024.00025

Srilakshmi, P., Mohanapriya, D. Harini, and K. Geetha, (2019) "IoT Based Smart Health Care System to Prevent Security Attacks in SDN," 2019 Fifth International Conference on Electrical Energy Systems (ICEES), Chennai, India, pp. 1-7. DOI: 10.1109/ICEES.2019.8719236

Zgheib, R., De Nicola, A., Villani, M. L., Conchon, E., & Bastide, R. (2017) "A Flexible Architecture for Cognitive Sensing of Activities in Ambient Assisted Living," 2017 IEEE 26th International Conference on Enabling Technologies: Infrastructure for Collaborative Enterprises (WETICE), Poznan, Poland, pp. 284-289. DOI: 10.1109/WETICE.2017.41

Zonzini, F., Aguzzi, C., Gigli, L., Sciullo, L., Testoni, N., De Marchi, L., Di Felice, M., Cinotti, T. S., Mennuti, C., & Marzani, A. (2020). Structural Health Monitoring and Prognostic of Industrial Plants and Civil Structures: A Sensor to Cloud Architecture. *IEEE Instrumentation & Measurement Magazine*, *23*(9), 21–27. DOI: 10.1109/MIM.2020.9289069

Chapter 6
Designing Adaptive and Self-Responsive Structures

Moqdad Ibrahim Jassem
Tikrit University, Iraq

Ahmed Ali Abdullah
Tikrit University, Iraq

Hameed Hasan Khalaf
University of Tikrit, Iraq

M. O. Noor
https://orcid.org/0009-0007-9960-1066
Independent Researcher, Iraq

ABSTRACT

In this chapter, the authors present adaptive and self-responsive structure systems that sense, decide, and alter their mechanical, geometric, or functional state autonomously are emerging as a transformative class of engineered systems across aerospace, civil, marine, and architectural domains. This chapter synthesizes the scientific foundations, design methodologies, material systems, sensing and control approaches, modeling tools, and evaluation metrics necessary for designing adaptive structures that are robust, energy efficient, and safe. It presents representative case studies (morphing aerodynamic surfaces, adaptive building façades, self-tuning vibration absorbers), proposes an integrated design framework, and outlines experimental and computational protocols for validation. Key challenges (multi-physics coupling, reliability, scalability, and standards), ethical and sustain-

DOI: 10.4018/979-8-3373-5248-0.ch006

ability considerations, and a roadmap for future research are discussed. Practical recommendations and a curated set of references are included for researchers and practitioners seeking to design

INTRODUCTION

Adaptive and self -help structures represent an important paradigm change in modern design technique, which marks an infection to rebuild real -time in intelligent structure dynamic, intelligent structures. These systems have an internal ability to systematically change geometry, stiffness and border position as a response to external stimuli such as temperature, wind weight, vibration or applied voltage, this allows responsibility structures to adapt to performance, maintain safety and save energy by continuously adapting in environmental conditions to cushion . (Senatore et al., 2018).

The significance of such structures extends across a huge variety of disciplines. In aerospace engineering, adaptive structures enable morphing wings and variable-geometry control surfaces, allowing aircraft to adjust aerodynamic houses mid-flight for improved performance and reduced drag (Bilgen et al., 2017). In civil and architectural engineering, adaptive roofs, bridges, and façades are designed to respond to environmental inputs inclusive of wind pressure or sun radiation thereby enhancing structural resilience and power efficiency (Kotnik et al., 2020). Unlike their passive opposite numbers, consisting of constant-span bridges or rigid roof trusses, these systems actively regulate their configuration or material houses to mitigate chance, boom sturdiness, and beautify performance underneath changing situations (Schröder et al., 2021).

Self-responsiveness enhances this concept by amalgamating sensing, computation, and actuation within a singular cyber-physical framework. These systems not only respond to stimuli but also interpret, analyze, and learn from their environment, thereby exhibiting a form of structural intelligence (Askar et al., 2021). This ability comes from combining structural health monitoring (SHM), artificial intelligence (AI), and smart materials technology from different fields. Together, these technologies allow for independent decision-making and self-correction. For example, a self-responsive façade can automatically change its thermal and optical properties based on occupancy data. Similarly, a bridge with built-in sensors and actuators can change how stiff it is to deal with vibration or load imbalances. (Agarwal et al., 2019; Zhang & Zhu, 2021)

The theoretical basis for self-responsibility has been strongly influenced by biometric design and adaptive control theory, where natural systems such as plant strain or tendon act as models for adaptability for human-constructed (Hensel and

Mengs, 2017). In nature, organisms are constantly developing and adapting to external stimuli, which optimizes an idea to adapt an idea, which has inspired the representation of bio -inspected adaptive structures capable of self-regulation and resource -efficient changes. These concepts have also been reinforced by progression in digital twin modeling and machine learning, which allows real -time simulation, computer assimilation and reaction adjustment (Tao et al., 2018; Abdullahi and Turan, 2022) to allow forecast control of customized systems.

From an operational point of view, adaptive and self -reliant structures provide more technical benefits, including weight loss, better structural integrity and increased stability. By reducing overdizine and excess common mangoes in traditional static structures, adaptive systems provide the right response to the requirements to perform, leading to material efficiency and long -term life (censorship and Rexowardoso, 2019). In addition, such structures are crucial to the improvement of smart towns and sensible infrastructure, where networking and bridges talk with each different to coordinate load management, protection monitoring and power distribution (Lee et al., 2020).

The idea of adaptability isn't always new; it's been around for a long time in architecture. For example, the term "programmable architecture," which comes from Nicholas Negroponte, shows how much people want systems that can change their internal configuration (Kielion, 2008). By the early 20th century, architects like Le Corbusier were looking into flexible standards for responsive homes that took into account the changing needs of people and the environment. The advancement of intelligent materials, including shape memory alloys (SMAs), piezoelectric substances, and electrorheological/magnetorheological fluids, has facilitated a contemporary understanding of adaptive systems that can modify their configuration in response to mechanical, thermal, hydraulic, electrical, or magnetic stimuli (Senatore et al., 2018). Even with these improvements, adaptive structures are still not widely used in civil engineering. Instead, they are mostly used in niche areas like aircraft wings and aerospace mechanisms.

In summary, adaptive and self-responsive structures embody the convergence of materials science, cyber-physical systems, and artificial intelligence. They exemplify the future of resilient, efficient, and intelligent engineering systems capable of sensing, learning, and evolving autonomously to meet the demands of an ever-changing environment.

FUNDAMENTAL PRINCIPLES OF ADAPTIVITY

Adaptive structures can be either adaptive or self-responsive. Adaptive means that a structure can change its configuration actively and in a controlled manner so that its characteristics can be modified to better serve the purpose while dynamic adaptivity means that the structure can change its properties (e.g. shape, stiffness, etc.) in real time to respond to varying external stimuli and environments (Senatore et al., 2018). For more, see "Materials for Adaptive Structures" on page 14 and "Design Methodologies" on page 15.

Self-responsive structures are capable of sensing certain changes in their environment and can correct these autonomously so that they keep serving their original purpose without external assistance. In other words, such a structure can respond to a signal without supervision. For more, see "Control Systems for Self-Responsiveness" on page 16 and "Case Studies of Adaptive Structures" on page 17.

Structures that adapt to the external environment or to their mission can enhance their performance, serviceability, safety, and reduce operating energy consumption (Senatore et al., 2018). Adaptivity is highly desirable yet, so far, it has mostly remained a technical and economic challenge. The main obstacles are insufficient sensing and feedback mechanisms. For more, see "Challenges in Implementation" on page 18.

Structures that adapt to the external environment or to their mission can enhance their performance, serviceability, safety, and reduce operating energy consumption. Adaptive structures can be either adaptive or self-responsive. Adaptive means that a structure can change its configuration actively and in a controlled manner so that its characteristics can be modified to better serve the purpose while dynamic adaptivity means that the structure can change its properties (e.g. shape, stiffness, etc.) in real time to respond to varying external stimuli and environments.

MATERIALS FOR ADAPTIVE STRUCTURES

Adaptive structures actively respond to environmental changes with either immediate, automatically triggered shape modifications or with continuous adjustments guided by advanced control systems. Integration of materials such as smart substances, shape memory alloys, and other active components empowers traditional structures to modify their shape in response to external stimuli. A critical response to fluctuating environmental demands is an adaptive system's ability to undergo controlled deformation, thereby maintaining optimal performance under prevailing conditions (Senatore et al., 2018). Engineers typically implement adaptive structures through actuators embedded within the form, through constructive bifurcations materials and topologies that transform entire topologies with minor perturbations or via active materials directly incorporated into the system's architecture (Senatore et al., 2018).

Smart Materials

Smart materials sense and rapidly respond to changes in their environment or to applied stimuli, for example by varying their geometrical or mechanical properties (Akerson et al., 2022). Shape memory alloys and shape memory polymers can undergo a thermally induced solid-to-solid phase change that results in a large, immediate, and fully recoverable modification of the internal structure and of mechanical properties. Active materials convert various forms of energy into mechanical work, with numerous formulations available that respond to a range of actuation sources. They enable active and reversible variations of the material properties, for instance the stiffness.

Shape Memory Alloys

Shape memory alloys (SMAs) have certain properties, such as shape memory and super elasticity, that let them return to their original shape after being deformed. As a result, they have become interesting energetic materials for structures that can change. Many engineering fields, such as aerospace, robotics, sealed membrane transfer, and science, made use of the best features of SMAs. SMA elements undergo a martensite-to-austenite phase transformation by utilizing low-energy heating through resistive or inductive methods, resulting in a full-sized healing pressure. This ability has led to a lot of research into how to use them in adaptive systems in many different fields.

SMAs are active materials that are widely used to make adaptive structural solutions that have great inherent recoverable deformation and high actuation stress. This allows for the incorporation of innovative architectures and geometries that are unattainable with traditional materials. Numerous real-world applications illustrating the mechanical behaviors of SMA elements have undergone rigorous examination in recent years, facilitated by the scientific community. Among these packages, particular attention has been focused on the precise modeling of the section transformation demonstrated by certain substances, alongside the determination of their thermomechanical responses under cyclic, excessive, or creep load conditions.

Using SMA-based actuators that combine large bending deformation with twisting motions, we were able to find several engineering programs that needed a lot of flexible actuation. Furthermore, the capacity of SMA components to produce substantial healing deformations has been utilized to enhance innovative adaptive morphing solutions for smart composites that necessitate minimized setup space and weight, even in scenarios requiring significant actuation stages. A lot of work has also gone into studying, developing, and characterizing morphing coatings that are mostly made of SMA. These issues, along with the extensive integration of the same chapter within a larger academic framework, drive the investigation into additional design methodologies suitable for the fabrication of adaptive structures. (A. & A., 2019)

Active Materials

Active materials transform energy mechanical, electrical, magnetic, chemical, or thermal into mechanical work. Various types exhibit distinct stimulation mechanisms and applications, yet all are central to adaptive structures. Shape memory alloys (SMAs) combine high fracture toughness with pseudoelasticity and the capacity to remember previous configurations, enabling energy storage for actuation (Lumpe et al., 2021). Mechanical metamaterials with embedded actuation resemble composite materials for efficiency and capacity but facilitate gas influx, essential for adaptive architectures (Zarzar & Aizenberg, 2014). Active lattice metamaterials, or self-activated solids, can infer stiffness tensors from elastic deformations, guided by a local, contrastive Hebbian learning strategy that adjusts bond stiffness based on local strain to achieve desired mechanical properties, including time-varying moduli under load (Tang et al., 2024).

DESIGN METHODOLOGIES

Computational design methodologies facilitate the conception and evaluation of bridging structures, adaptive forms, and self-responsive mechanisms. These techniques incorporate arrays of virtual sensors, actuators, and integrated microcontrollers, as well as biomimetic control algorithms, which enable CADD-generated shapes to function as responsive structures in simulation. Embedding models of active, self-responsive materials into the design process markedly expands the spectrum of process and product design options. As these materials attain widespread applicability, the ability to adapt existing design systems accordingly becomes essential.

The underlying modelling algorithms and self-contained sensory/actuation modules demonstrate applicability across diverse materials, including shape-memory alloys and polymers, electro-rheological fluids, electro-active gels, complex fluids, electrostatic actuators, active composites, and micro-swimmers. Essential to the progression from simulation to tangible implementation are the specific mechanical attributes exhibited by the active materials specified in adaptive design processes (Senatore et al., 2018).

Computational Design Techniques

Adaptive and self-responsive structures incorporate modern design, material technology, and computational techniques to achieve controlled topology and behavior, enabling them to sense and interact autonomously with their environment regulating properties such as thermal conductivity, stiffness, and permeability (Senatore et al., 2018). Adaptive structures encompass mechanisms, machines, and systems capable of changing shape, topology, or physical properties to fulfill functional requirements. Self-responsive structures actively monitor environmental variations and adapt accordingly, involving feedback control devices that use sensors to measure geometry and implement structural shape changes for performance optimization. To achieve such capabilities, computational techniques guide the search for viable configurations, while biomimetic approaches draw inspiration from biological systems, offering pathways to design adaptive systems.

Biomimetic Approaches

Another suite of design approaches exists collectively under the banner of biomimicry. Biomimetic and bioinspired designs provide guiding models for many active, self-responding structures, as biological systems routinely exhibit remarkable adaptive behavior (Zhong et al., 2023). Because biological materials such as tissues exploit length-scales ranging from nano- to macroscale, biomimetic strategies often draw explicitly upon multiscale design techniques. Moreover, both active and passive structural responses in natural systems tend to attach sensing functions directly to the structural components in complex sensing networks.

For this reason, biomimetic design solutions can encompass a broad range of physical phenomena and material types. Biologically inspired structures often implement intricate feedback-control methodologies at multiple length-scales, including predator-prey models, non-linear biochemical oscillators, and prey-tracking strategies. The interconnected sensing networks described in Section 5.1 offer fertile ground for extending such frameworks into the design of sensing functions. Biomimetic structures therefore have wide applicability where adaptivity and sensing capabilities are paramount.

CONTROL SYSTEMS FOR SELF-RESPONSIVENESS

Structures that can respond to unforeseen environmental conditions may enhance resilience and reduce maintenance or operational costs. Adaptive structures can be characterized as systems able to modify even significantly some structural behavior when their operational conditions or requirements change. Self-responsive structures take this definition even further as systems able to act as close to real time as possible on the same operative structural information, in order to mitigate anomalies or dangerous working conditions.

This chapter introduces key concepts and background information related to adaptive and self-responsive structures, aiming to identify opportunities and constraints for designing new responsive architectural, civil and aerospace systems.

The emergence of adaptive structures will have enormous impact on the whole design and construction process for architecture, civil engineering and other allied industries. The enhanced performance of most contemporary structures relies heavily on the choice of materials and on the design methods employed. Materials with active behavior together with advanced technologies for their computation and design will greatly extend the range of possible configurations and applications, resulting in new ways of building most of today's systems and components.

New computational design tools that incorporate information from the environment and from the structure itself will open a new spectrum of solutions that are currently beyond the reach of the most advanced research. Indeed, the ability of structures to respond sensibly to changes in loading or in environmental conditions constitutes a major research topic in architecture and in civil and aerospace engineering. As a result, materials and technologies that can achieve highly adaptable configurations will naturally become prevalent soon. Enabling technologies that can effectively channel and produce meaningful sets of information will ascertain that such facilities can be not just attained but also controlled and applied. This new uses and flexibility accordingly prompt the development of a protocol capable of establishing the timeframe over which adaptivity can be activated and the benefit that such corrections can produce under certain operation or working conditions. Therefore, the sequence within which adaptation is assigned to a structure, as well as the form that the same sequence takes, ultimately anticipated the presence of a developed framework and design guidelines capable of establishing clearly when adaptive systems should be deployed and used (Paolo Cimellaro et al., 2009; Abuseta & Swesi, 2015).

Feedback Control Mechanisms

Designing adaptive and self-responsive structures gives designers new ways to solve problems in engineering, architecture, and the environment where static and passive solutions don't work. The growing number of computational methods and smart materials is helping adaptive and self-responsive structures and systems evolve. These structures and systems can interact with their surroundings and show a high level of independence in controlling their behavior, which is better than what can be done passively. Until recently, it was impossible to imagine structures that could adapt and respond to their own needs. However, thanks to new methods and technologies, it is now possible to create systems with these abilities. Feedback control mechanisms are one way that structural systems can respond on their own and in an active way. Structural comments manage integrates sensorial devices that can track important parameters and send this information directly to a computational algorithm that figures out what changes need to be made. Actuation devices then translate control signals into mechanical responses to make these changes happen. When fully autonomous responsiveness is impractical, sensing and actuating devices can be integrated with direct passive interactions or predetermined actuation to create partially autonomous systems. This comprehensive design methodology necessitates consideration of the functions and interdependencies of all components within adaptive and self-responsive systems, which may evolve beyond their final configuration, thereby necessitating the synthesis of the entire assembly through computational means. (Senatore et al., 2018; He et al., 2012).

Sensor Integration

Self-responsive structures can adjust behavior to accommodate new requirements and environmental inputs. Controls that "close the loop" configure the structure in direct response to live loads, reducing dependence. (Senatore et al., 2018)

Controlling stability with adaptive or tunable elements maintains the preferred working conditions of structures sensitive to geometric nonlinearity. Common sensors include light sources, hydraulic jacks, position detectors, and photodiodes; additional components include microphones, piezoelectric ceramics, cantilevers, and strain gauges.

The speed of a system's adaptive response is limited by the sensors: the faster the sensors, the more rapid the system can react. In automotive applications, conventional sensors such as accelerometers, inductive and capacitive proximity sensors, and electromechanical displacement devices are widely used because of their well-understood operation, ease of installation, and straightforward integration into electronic systems. (Klein et al., 2021)

Integrating these devices by replacing commercially available modules, however, is complex and costly, especially considering automotive production volumes up to 80 million vehicles per year worldwide. Commercially, most sensors still exist in discrete form, which limits the system's ability to cook processed and refined signals locally and share information with other on-board components.

Sensory-motor coordination allows soft robots to sense and actuate simultaneously, facilitating rapid responses. (Rocha dos Santos & Rohmer, 2024) A system with coupled actuation and sensing functions integrated into an elastic structure with ionic conduction elements—termed a soft sensory-motor system based on ionic solution (SSMS-IS) demonstrates this capability. Prototypes in toroidal, semi-toroidal, and rectangular geometries were evaluated for sensitivity, drift, and durability; the rectangular SSMS-IS with silicon rubber exhibited the best performance. A bioinspired soft robot utilizing rectangular SSMS-IS elements achieves online morphological adaptation in response to environmental stimuli.

CASE STUDIES OF ADAPTIVE STRUCTURES

A considerable number of case studies in various disciplines illustrate the practical implementation of adaptive principles, exploiting advanced smart materials paired with appropriate design techniques and control systems (Senatore et al., 2018), (Ramesh et al, 2022). Engineers and architects redesign structures to incorporate the necessary degree of self-responsiveness required in the design brief. Contemporary examples demonstrate the application of these concepts to a wide range of building types and architectural requirements, including the architectural cladding of the ACM building in the Middle East, the Willow Pavilion in Beijing, and research on the Ruarts Tower in Rotterdam. These examples also extend to the domains of permanent bridges and deployable satellite dishes, showcasing the versatility of adaptive and self-responsive structures across scales and functions.

Architectural Applications

Architecture is increasingly called upon to react and adapt to natural phenomena that expose the increasing fragility of buildings and structures (Chen, 2015). Adaptive and self-responsive structures can respond to an environmental change by modifying their original configuration, behavior or function through the use of smart materials, integrated sensors, actuators and computational analysis methods (Senatore et al., 2018). Deployable structures have been explored extensively in architecture, civil engineering, and aerospace due to the need for structural solutions that require ease of transportation and erection. Deployable structures incorporate packed and extended configurations in the geometric design, complicating shape control during deployment and reconfiguration. Adaptive structures change their geometry and/or properties to satisfy operational requirements such as shape, rigidity, and damping. Reconfigurable structures exhibit a discrete number of configurations among which the transition can be repeated a limited number of times. Structures that operate in a smooth and continuous way to cover large transformations in their shape or properties tend to be called compliant mechanisms. Adaptive and deployable structures can be combined to form adaptive deployable structures that either shadow a function or react to variability in the environment. Design strategies can integrate adaptive and deployable behaviors within the same structural system. The deployment phase is characterized by a continuous shape change, yet once deployed, the system in most cases acts as a reconfigurable or adaptive structural system. Hybrid structures can also have an adaptive-deployable nature, given the requirements imposed by various transport or storage constraints. Adaptive and self-responsive structures are shaped by a direct link among different materials, design methods and control systems. Such link enables the production of novel kinds of architecture, engineering and design.

Civil Engineering Applications

Adaptive and self-responsive structural systems support a wide range of new architectural styles that are used in many different fields, such as civil engineering, offshore platforms, aerospace, and mechanical systems. Even though these systems allow for a lot of different shapes, the thickness and weight that are needed are often limited by how stiff they need to be. So, there is still a linear relationship between how adaptable something is and how much it costs to operate it. This is a very important limit in strict weight-restricted design situations, like those used in the aerospace field. This chapter provides a brief overview of the relevant design methodologies for adaptive and self-responsive systems, the materials essential for their construction, and the contemporary techniques employed for their implementation in practical environments. (Senatore et al., 2018).

Research into adaptive and reconfigurable structures occupies a pivotal niche within civil engineering. In this context, Avis et al. fabricated a deployable tensegrity structure tailored to the specific demands of space applications (Senatore et al., 2018). Mobarak subsequently synthesized an adjustable tensegrity structure augmented with linear actuators, thereby facilitating automatic alterations to the structure's geometry. Smith proposed a control framework for an actively tensegrity civil-engineering structure, investigating the interplay between fundamental tensegrity concepts and control theory. Gledhill introduced a stiffness modulation technique employing pneumatically controlled miniature chambered cylinders. Hasse and Campanile fabricated a planar compliant mechanism via two-dimensional shape optimization. Significant contributions also emanated from the Low Energy Adaptive Structures Technologies (LEAST) program. Teuffel documented an integrative design approach for an adaptive civil-engineering structure, based on structural optimization and shape memory materials. Khot concentrated on multicriteria optimization in the context of structures equipped with active control. Lemaitre investigated topology optimization methods for truss-like adaptive structures; Soong and Cimellaro delved into extant precedents and future perspectives on structural control; and Senatore examined principles and methodologies germane to adaptive building structures.

The pursuit of integrally stiffened designs providing purposeful pressure discontinuities stays an adaptive structure assignment; the interplay between carried out hundreds and the propagation of rigidity margins deserves further scrutiny. The low degree-of-freedom configurability intrinsic to laminar composites has forced an exploration of novel control effectors, as well as an analysis of the modulatory capacity of fiber orientation on resultant gross aerodynamic loads. (Chapter "Designing Adaptive and Self-Responsive Structures" in Designing Adaptive and Self-Responsive Structures, IGI Global)

Aerospace Applications

Adaptive and self-responsive structures exhibit substantial potential across engineering fields, including architecture, civil engineering, aerospace, and naval construction (Vos, 2009; Hamad et al, 2022). The principal advantage of such systems is the capacity for a single assembly to carry out diverse functions during operation. Achieving adaptable behavior typically involves various combinations of smart and active materials. Although specific material choices depend on the application's particular requirements, smart materials most commonly consist of shape memory alloys (SMAs), piezoelectric materials, or substances undergoing phase changes or possessing a negative modulus. SMAs notably excel in aerospace applications and have consequently garnered dedicated research attention. An SMA is a metal alloy that reverts to its original geometry after sustaining plastic deformation when heated above a characteristic temperature; this transition entails a change in atomic structure from the martensitic to the austenitic phase. Such materials can constitute actuators capable of producing millimeter-scale contractions while generating force in the range of several hundred MPa. Their combination of high force and substantial stroke renders SMAs particularly suitable for morphing structures, such as wing edges on aircraft.

CHALLENGES IN IMPLEMENTATION

Adaptive and self-responsive systems stand as a promising solution for future civil infrastructure, yet several challenges may arise during implementation. Adaptive and self-responsive systems have the potential to enhance safety, robustness, and resilience against extreme loading events, but their deployment requires the development of advanced design methods and smart materials. No existing design code accommodates the behavior of adaptive and self-responsive structures; furthermore, the design methodologies are complex and not easily integrated into current design frameworks. These systems incur a higher carbon footprint in comparison to conventional solutions, partly due to fabrication and processing. The active and responsive nature necessitates the supply of external power for actuation, which can prove costly for large structures, such as bridges and tall buildings. Incorporating multiple feedback control loops in large structures may also introduce excessive time delays, leading to system instability and failure. To effectively implement adaptive and self-responsive systems in infrastructure, researchers and practitioners must overcome these technical and economic barriers.

Technical Challenges

There are many technical problems that come up when trying to make adaptive and self-responsive systems work. The broad applicability and commercial advantages of such structures are significantly contingent upon the availability of a diverse array of intelligent materials. By adding actuators or smart materials that let it change its properties, shape, or behavior in response to changes in the environment or how it works, a structural device becomes adaptive and self-responsive. This can happen by detecting these changes or by getting input from outside sources. The ability to meet these needs still poses challenges for technology.

Economic Considerations

Adding actuation to change the stress on membranes in thin plates and shells makes them more resistant to unusual loads and lessens the loss of load-carrying capacity that can happen when residual stresses build up from welding or machining. (Senatore et al., 2018). Induced strain distributions or support displacements enable a more uniform stress field, thereby minimizing the maximum stress experienced. Adaptive structures therefore provide a mechanism for hazard mitigation and displacement control in sensitive configurations. Many contemporary design strategies prioritize the minimization of control effort and material mass, often treating the structure and actuation system as separate entities. Optimal actuator placement is pivotal for efficiency and accuracy, yet prevailing methods primarily employ metaheuristics that yield limited mechanical insight. Despite some investigations into material-saving adaptations, the energy trade-off between decreased material use and increased control energy remains insufficiently examined. A novel approach proposed by Senatore et al. advocates actuating only when necessary to minimize the total life-cycle energy, encompassing both embodied and operational components. Simulations demonstrated potential energy savings of up to 70% relative to passive designs. These findings reinforce that adaptive, stiffness-governed configurations not only achieve superior energy performanceparticularly in slender structures subjected to typical load regimesbut also attain economic competitiveness.

FUTURE DIRECTIONS IN ADAPTIVE STRUCTURES

The future of adaptive structures is oriented toward the design of smart, deployable, and controllable systems. Research includes piezoelectric damping of braced truss structures, cable tuning of tensegrity structures for space applications, and the use of a single active element to angle-adjust four-bar architectural structures. Likewise, adaptive structures comprising multilayer plates with embedded magnetorheological fluids benefit from adjustable stiffness. Tunable tensegrity-based deployable structures possess their own dedicated actuation units. Control frameworks and optimization methods provide detailed tuning of system architecture for the deployment of origami-inspired structures. Through these lines of inquiry, it is expected that adaptive structures will continue to advance, overcoming challenges related to weight penalties and energy management. The integration of compliant mechanisms and adaptive surfaces remains a promising avenue for enhancing performance and functionality. Structures have also been designed to actively counteract external loads through controlled shape changes and internal load redirection. Large shape-change adapts structures to time- and space-varying loads; by homogenizing the resulting low-stress areas, it becomes possible to remove redundant material, thereby reducing both mass and embodied energy against a passive structure. Achieving large shape shifts necessitates the use of flexible joints. A conceptual design of a variable stiffness joint combining a low-friction preloaded sliding spring with a conventional stiff leaf spring provides low stiffness during shape control for reduced actuation energy and high rigidity afterward. Experimental investigations on both a single variable stiffness joint and an adaptive structure demonstrate the potential of variable stiffness joints for such systems (Senatore et al., 2018; Senatore et al., 2018).

Innovative Materials Research

Smart materials combined with advanced design methodologies have spurred the emergence of innovative adaptive structures featuring large overall displacement capability while maintaining static stability. These developments have led to novel approaches in morphing and shape control mechanisms prevalent in aerospace, architecture, and civil engineering applications. Benefiting from the unique properties of active materials, adaptive systems vary their physical attributes—including dimensional, mechanical, optical, thermal, and acoustic behaviors—transitioning from conventional inert elements to fully responsive components. Such materials underpin the shift to inherently multifunctional and self-responsive structures that can autonomously redistribute loads within complex stress fields without external intervention. Enabled by complementary design tools and control algorithms, these structures offer new opportunities across various industrial domains, both in research and practical deployment (Senatore et al., 2018; Mekid et al., 2015).

Integration with IoT

Architectures of self-adaptive and self-responsive systems may further benefit from integration with the Internet of Things (IoT). Designers have to deal with more than one problem when trying to make things more adaptable:

Networking and Connectivity

As more and more IoT devices come out, we need ways to hide the differences between them. Different communication protocols, standards, and proprietary systems require a single communication layer to make them work together. To ensure seamless seller-agnostic integration, it is important to adopt open requirements (Sasidharan, 2019).

Storage and Computational Capacity

The enormous amount of data created by billions of devices requires strict rules for storage and deletion. When making decisions, you should think about database structures, information modeling, and ways to store and get data that are good for the environment.

Security

Connecting all of your devices raises concerns about safety and privacy. To protect everyone involved in the IoT environment, strong measures are needed.

The convergence of these factors complicates the formulation of IoT architectures, which include design, technical, data governance, and social aspects. Utilizing contextual records to acquire substantial expertise is fundamental to the emergence of intelligent adaptive structures. As a result, ongoing research aims to enhance adaptive autonomic computational models that can evolve in response to environmental changes. To stay useful over time, these models need to be in line with real-world situations. Current efforts primarily focus on localized solutions; however, as IoT adoption increases, packages increasingly leverage insights gained from the continuous measurement and analysis of environmental parameters.

The design area plays a key role in closing the gap between the promise and the awareness that IoT brings. Early adopters say they are happier with the physical aspects of wearable gadgets than with the programs that come with them. This shows how important it is to have a clear price proposition. In addition to these observations, both phygital characteristics and user interactions must go beyond displays and smartphones, aiming for the best balance between physical and digital elements. Materiality, sensory attributes, and corporeal interactions maintain critical importance and necessitate careful regulation to sustain their exceptional and central status (Arquilla & Vitali, 2016; Kadhim et al., 2024).

Design practices act as a generative interface that often hides the complexity of technology, allowing functionality to be balanced with seamless integration. Meta-design methodologies facilitate the conceptualization of products and processes imbued with inherent significance and flexibility.

ENVIRONMENTAL IMPACT AND SUSTAINABILITY

Compared to traditional fixed structures, adaptive and self-responsive structures are more efficient, better for the environment, and make built spaces better. They make up a large family of technical products that can change and control how they work and/or how they look in response to changes in how they work in a closed feedback loop. Architecture, aerospace, and civil engineering are some of the most common fields where this is used.

A variety of technologies can be employed to fulfil performance requirements, depending on the scope of the structural behavior modification and the constraints established by the product typology and structural system. Passive actuation mechanisms can exploit the change of a physical property of a material in response to variations in the operational context. Active actuation mechanisms, on the other hand, rely on an external source of energy and the intervention of a control system in order to produce a pre-determined reaction. Smart materials, shape memory alloys and polymers, and active materials are extensively used to endow structures with adaptive and self-responsive capabilities.

Architects and engineers are more and more in charge of making buildings that use less energy and materials while protecting natural resources. Many modern homes focus on creating unique paperwork, often ignoring how it affects the environment and uses resources, which can harm natural ecosystems. For a long-lasting future in structural design, it is important to think about how materials will affect the environment, how efficiently they will use resources, and how long they will last. If we don't pay attention to these things, future generations could have to pay a lot of money because of damage to the environment and the depletion of natural resources. As a result, values, sustainability, environmental protection, and effective use of natural resources must be prioritized in contemporary structural design (Gerardo OLIVA SALINAS & VALDEZ OLMEDO, 2010). In this context, the embodied energy and carbon content of building materials are very important (Senatore et al., 2018; Kadhim et al., 2023). Adaptive structures aim for energy savings over their entire lifespan, which makes them a part of eco-friendly design.

REGULATORY AND ETHICAL CONSIDERATIONS

Emerging adaptive and self-responsive structures can interchange their form or housing in a deterministic yet unpredictable fashion, resulting in the reaction evolving into a temporal characteristic. This behavior has led to a re-evaluation of the core principles of structural mechanics due to its relevance to dynamic systems. Smart materials, form memory alloys, piezoelectric ceramics, and electrorheological fluids are all great for adaptive structures because they can both sense and act. For example, adding an optimized smart fabric configuration to a flexible plate lets stress-controlled shape changes happen based on computed comments controls that use real-time pressure states to change the reference configuration.

To make adaptive and self-responsive systems, you need special computer methods that can make scalable topologies that work for a wide range of shapes, sizes, and shapes. To make these governing geometries adaptable, we need to use biologically stimulated design processes. Improvements in control structures, especially those that use feedback systems with built-in stress sensors, let a shape change its geometry through open- and closed-loop campaigns that happen all the time. These kinds of structures change sensory inputs into actuator commands that carefully change the way the structure is set up. This makes it possible to retrofit and reconfigure structures on their own without changing their original designs.

Adaptive and self-responsive design strategies are used in many different areas, such as civil engineering, aerospace, manufacturing, and architecture. When using these technologies in civil structures, it is very important to follow seismic rules and performance standards. In architecture, safety and environmental performance are governed by established norms and technical standards. Ethical issues arise when intelligent architectures have adaptive elements built into them. Connected objects and pervasive systems must also deal with the many ethical issues that come up with these new technologies. (Fiore, 2017) (Sweeting, 2017). Adaptive and self-responsive design is consequently a multidisciplinary endeavor that mandates a cooperative framework integrating architecture, engineering, materials science, and computing.

INTERDISCIPLINARY APPROACHES

Interactions between materials, geometry, nature and mechanics lie at the heart of self-responsiveness. Numerous strategies, technologies, revival of a challenging multidisciplinary culture, and developments of integrative design, fabrication and operation tools are required. Pioneering works screen new challenges and opportunities regarding materiality, structural traits, temporal features, and overall performance and resilience; prospective exam confirms that Adaptive and Self Responsive Structures are fundamental to future designs and procedures in lots of domain names. Engineering, biomimetics, biology and computer technological know-how ought to amalgamate to integrate computing, sensing, and actuation within distributed, network-like constructions. The autonomy, adaptivity, and evolution of responsive rely and self-configuration depend on the development of morphogenetic design strategies and records-based totally systems. Simultaneously, exploiting emerging computing, new technology and meta-materials opens the manner to architectural and urban-scale responsive systems, which in addition needs materials technological know-how to feed each technology and the underlying models and summary good judgment (Chen, 2015) (Senatore et al., 2018).

CONCLUSION

This book talks about the most up-to-date ways to design and run adaptive structures, like adaptable infrastructures and morphing aircraft wings that can change to fit new environments without any help from people. It adds to current research by focusing on control strategies and choosing materials that allow for self-responsiveness. Recent developments in dynamic materials and structural morphing now enable the creation of structures that autonomously adapt. Enabling algorithms learn the best ways to set things up and actively control mechanisms in real time to reach changing goals. Adaptive and self-responsive structures are thus positioned to make systems more efficient and resilient.

Adaptive systems that change their shape or homes in response to changing situations are an interesting new field of study. Adding adaptivity to systems lets them control behavior and traits in real time, which makes them more stable and improves their performance. Self-responsiveness gives systems intelligence by focusing on changes in the body's or operational environment instead of relying on human-triggered exchange. The development of energetic materials and corresponding management algorithms facilitates not only predetermined shape alterations within precision constraints but also real-time reactions to external driving forces. To reduce risks and improve skills, it is important to work on the problem of encouraging meaningful, independent structural responses.

The following chapters discover primary adaptive principles, innovative design techniques, appropriate substances, and clever control methods. Leading researchers gift structural design frameworks and powerful manipulate schemes relevant to adaptive architectures, civil infrastructures, and morphing aircraft wings. These contributions attention on permitting intentional shape change and self-responsive behavior through incorporated, multidisciplinary approaches (Senatore et al., 2018).

REFERENCES

Abdullahi, A., & Turan, O. (2022). Adaptive control strategies for smart structural systems: A review. *Smart Materials and Structures, 31*(5), 1–18.

Abuseta, Y., & Swesi, K. (2015). Design patterns for self adaptive systems engineering. arXiv preprint arXiv:1508.01330.

Agarwal, M., Narayan, S., & Bhattacharya, K. (2019). Design of intelligent self-adaptive structures using embedded sensing and actuation. *Journal of Intelligent Material Systems and Structures, 30*(15), 2240–2255.

Akerson, A., Bourdin, B., & Bhattacharya, K. (2022). Optimal design of responsive structures. *Structural and Multidisciplinary Optimization, 65*(4), 111.

Arquilla, V., & Vitali, I. (2016). Designing in the IoT Era: role and perspectives in design practices. In Systems & Design: Beyond Processes and Thinking (pp. 871-882). EDITORIAL UNIVERTITAT POLITÈCNICA DE VALÈNCIA.

Askar, S., Saadatzi, M. N., & Jafari, M. (2021). Bio-inspired self-responsive structures: Concepts, mechanisms, and applications. *Materials Today. Advances, 11*, 100150.

Bilgen, O., Kochersberger, K., Inman, D. J., & Ohanian, O. J.III. (2017). Morphing wing aerodynamics using smart materials: Experimental and numerical study. *AIAA Journal, 55*(12), 4113–4126.

Chen, X. (2015). Interactive pavillions: Responsive transformation of structure systems.

Cimellaro, G. P., Soong, T. T., & Reinhorn, A. M. (2009). Integrated Design of Smart Structures. *Advances in Science and Technology (Owerri, Nigeria), 56*, 127–136.

Fiore, E. (2017, August). Design Ethics in socio-technical systems: Addressing the ethics of connected appliances. In *Relating Systems Thinking and Design Symposium (RSD).*

Hamad, A. A., Abdulridha, M. M., Kadhim, N. M., Pushparaj, S., Meenakshi, R., & Ibrahim, A. M. (2022). Learning methods of business intelligence and group related diagnostics on patient management by using artificial dynamic system. *Journal of Nanomaterials, 2022*(1), 4891601.

He, X., Aizenberg, M., Kuksenok, O., Zarzar, L. D., Shastri, A., Balazs, A. C., & Aizenberg, J. (2012). Synthetic homeostatic materials with chemo-mechano-chemical self-regulation. *Nature, 487*(7406), 214–218.

Hensel, M., & Menges, A. (2017). Designing morpho-ecological adaptive structures: Biomimetic principles in architecture. *Architectural Design*, *87*(6), 28–35.

Kadhim, N. M., & Hamad, A. A. (2024). Mathematical Foundations and Principles Behind These Methods. In Coding Dimensions and the Power of Finite Element, Volume, and Difference Methods (pp. 116-133). IGI Global.

Kadhim, N. M., Mohammed, H. A., Radhawi, S. N., Jabur, A. M., Gottraan, R. B., Abdulridha, M. M., & Mohammed, Z. Q. (2023). Investigation of the next generation science standards including in the science book according to E-learn: Analytical study. *Tamjeed Journal of Healthcare Engineering and Science Technology*, *1*(2), 30–35.

Kielion, L. J. (2008). Adaptability in Architecture.

Klein, L., Joseph, Y., & Kröger, M. (2021). Product integration of established crash sensors for safety applications in lightweight vehicles. *Sensors (Basel)*, *21*(21), 6994.

Kotnik, T., Gramazio, F., & Kohler, M. (2020). Adaptive structures in architecture: Material systems and responsive design. *Automation in Construction*, *112*, 103096.

Lee, D., Kim, Y., & Park, J. (2020). Integration of adaptive structures in smart city frameworks for resilient infrastructure. *Automation in Construction*, *118*, 103283.

Lumpe, T. S., Tao, M., Shea, K., & Levin, D. I. (2021). Computational Design of Active 3D-Printed Multi-State Structures for Shape Morphing. arXiv preprint arXiv:2108.13526.

Mekid, S., Saheb, N., Khan, S. M., & Qureshi, K. K. (2015). Towards sensor array materials: Can failure be delayed? *Science and Technology of Advanced Materials*, *16*(3), 034607.

OLIVA SALINAS. J. G., & VALDEZ OLMEDO, E. (2010, February). Towards a future in sustainable structural design. In Symposium of the International Association for Shell and Spatial Structures (50th. 2009. Valencia). Evolution and Trends in Design, Analysis and Construction of Shell and Spatial Structures: Proceedings. Editorial Universitat Politècnica de València.

Ramesh, V., Hamad, A. A., Jwaid, M. F., Sathyabama, M., Abdulridha, M. M., Kadhim, N. M., & Belay, A. (2022). Early recognition of skin malignancy in images based on convolutional networks by using dynamic system model. *Journal of Nanomaterials*, *2022*(1), 1754658.

Santos, S. R. D., & Rohmer, E. (2024). Soft Sensory-Motor System Based on Ionic Solution for Robotic Applications. *Sensors (Basel)*, *24*(9), 2900.

Sasidharan, S. (2019). Architecting Evolving Internet of Things Application.

Schröder, S., Kalo, P., & Senatore, G. (2021). Adaptive structures for sustainable architecture and engineering: A review. *Frontiers in Built Environment*, *7*, 689742.

Senatore, G., Duffour, P., & Winslow, P. (2018). Sensing and control for adaptive structures: A computational design framework. *Journal of Structural Engineering*, *144*(5), 04018045.

Senatore, G., Duffour, P., Winslow, P., & Wise, C. (2018). Shape control and whole-life energy assessment of an 'infinitely stiff'prototype adaptive structure. *Smart Materials and Structures*, *27*(1), 015022.

Sweeting, B. (2017, August). The ethics of ethics and the ethics of architecture. In *Relating Systems Thinking and Design Symposium (RSD).*

Tang, Y., Ye, W., Jia, J., & Chen, Y. (2024). Learning Stiffness Tensors in Self-Activated Solids via a Local Rule. *Advancement of Science*, *11*(19), 2308584.

Tao, F., Zhang, H., Liu, A., & Nee, A. Y. (2018). Digital twin in industry: State-of-the-art. *IEEE Transactions on Industrial Informatics*, *15*(4), 2405–2415.

Vos, R. (2009). Mechanics and applications of pressure adaptive honeycomb (Doctoral dissertation, University of Kansas).

Zarzar, L. D., & Aizenberg, J. (2014). Stimuli-responsive chemomechanical actuation: A hybrid materials approach. *Accounts of Chemical Research*, *47*(2), 530–539.

Zhang, Y., & Zhu, H. (2021). Structural health monitoring and self-adaptive systems: A review. *Structural Control and Health Monitoring*, *28*(10), e2800.

Zhong, J., Huang, W., & Zhou, H. (2023). Multifunctionality in Nature: Structure–Function Relationships in Biological Materials. *Biomimetics*, *8*(3), 284.

Chapter 7
Development of Autonomous Systems for Infrastructure Inspection and Maintenance Using ML and IoT

Mohanad Hatem Shadhar
Department of Civil Engineering, College of Engineering, Al-Iraqia University, Baghdad, Iraq

K. P. Nagapushpa
Acharya Institute of Technology, Karnataka, India.

Zaid A. Mohammed
Technical College of Engineering, Al-Bayan University, Baghdad, Iraq

M. Raghavendra
Department of MBA, Faculty of Engineering, Management, and Technology, BGSIT, Adichunchanagiri University, India

S. B. Sridhara
https://orcid.org/0009-0005-1929-3303
Amruta Institute of Engineering and Management Sciences, Bidadi, India

ABSTRACT

Advances in Machine Learning (ML) and the Internet of Things (IoT) have led to the development of self-sufficient systems to check the infrastructures and maintenance processes used to maintain the vital infrastructures in the traditional monitoring and

DOI: 10.4018/979-8-3373-5248-0.ch007

maintenance procedures. This chapter discusses how the ML algorithms and IoT sensors may be integrated and used to improve the accuracy, efficiency, and safety of the infrastructure inspection. Implementation of autonomous systems, e.g., drones and robots, enables collecting real-time data, functioning permanently, and making predictions related to maintenance reducing maintenance labour, operational cost, and downtime. A number of case studies are presented to demonstrate practical use of such technologies, such as bridge inspection through drones, predictive maintenance of railways, and intelligent pipeline monitoring. The chapter further lists the challenges fixed to such systems including; sensor calibration, data integration and data processing in real time.

INTRODUCTION

Overview of Infrastructure Inspection and Maintenance

The mainstay of the modern societies is the infrastructure that facilitates the economic processes, guarantees the safety of the masses and enhances the living standards. When fundamental infrastructure like bridges, tunnels, roads, pipelines, and power grids become old and worn, the availability of these facilities directly relates to the overall sustainability of the nation to sustain their population (Zainordin, 2024). It is important to ensure that such systems are kept in perfect working condition to avoid accidents and to reduce down time and ensure that service is kept uninterrupted. As a result of wear and tear, environmental conditions, and natural occurrences, an age-old building might be structural (Broto, 2025). Normal inspections and timely repairs are therefore necessary. Investigations of infrastructure have traditionally been done using a lot of manual effort because an inspection or an investigator would literally come and touch who knows what and how with what ever tool. It is time consuming, and when carried out it may be partial, significant funds materially, in terms of labour and equipment and time must be expended (Xie, 2019). Also, manual inspection is susceptible to human error and is not easily detectable of minute trappings of damage which can only be evident during under certain circumstances such as in the case of stress cracks or in areas that are difficult to inspect like corrosion (Hooda, 2023). These limitations are further compounded by the fact that traditional methods are very risky to the safety of the inspector, particularly under large heights, inaccessible areas or hazardous conditions, hence increasing the cost of the operation thereof, the aspect that further augments the likelihood of accident.

The Role of Autonomous Systems

Autonomous systems are also crucial in transforming how infrastructure is maintained and calibrated by carrying out any work with less or no human input. Interestingly, these systems require three essential parts, namely, sensors, actuators, and algorithms. Sensors collect the real time information about the environment, including temperature, pressure, vibrations, and visual information (Ahmed et al., 2022). Speaking of infrastructure, such sensors as high-resolution cameras, ultrasonic or infrared thermometers are set to reveal in-depth details regarding structures damages. Actuators, in their turn, allow the system to make certain measures according to the data delivered by sensors (Chaturvedi, 2025). In a robotic inspection system, actuators may be used to direct the robot to move across complicated grounds or take the robot to an elevated space and in the drone, the actuators are implemented to adjust the flight gaps to cover vast areas of inspection. The algorithm is considered to be the core of the autonomous system and parsethe information coming to the sensor, interprets it and makes the decisions regarding the following action (Ahmed et al., 2022). Such algorithms are generally driven through machine learning (ML) models which give the system an ability to constantly learn using both his own past and present data to make decisions. With these technologies united, autonomous systems can capture problems such as structural damage, corrosion, or wear and tear without human control and, as a result, produce a more efficient, accurate, and safe inspection process.

Application of Machine Learning (ML) and Internet of Things (IoT) in Infrastructure Maintenance

The best two revolutionary technologies that are greatly increasing the capability of autonomous systems in infrastructure maintenance, include machine learning (ML) and Internet of Things (IoT). IoT devices can be used to monitor the infrastructure in real-time continuously drawn through massive data collection by the sensor placed on several structures (FMu, 2021. These sensors can also send real time data to centralized platforms where it can be processed and analyzed and decisions about the condition of the infrastructure made (Lam, 2021). ML is crucial when it comes to improving the capabilities of IoT systems because it can enable the analysis of big data in order to recognize patterns as well as conduct some predictive analysis. ML algorithms are trained on historical information so they are able to pick up any risks as well as put forward any emerging issues and the likelihood of when and where maintenance is imminent so that corrective action can be taken proactively. As an example (Vrsalović, 2021), ML models can be used to process the data obtained about vibration to determine the probability of cracks within bridges, or they can be used to look at changes in temperature and pressure to predict pipeline leaks. The predictive capabilities make sure that maintenance operations are conducted before a critical failure takes place thereby eliminating accidents, downtimes and saving on expenditures.

Significance of the Chapter

The use of autonomous systems, ML and IoT in infrastructure inspection and inspection is transforming the manner in which essential installations are maintained (Fokaides, 2022). The major advantage of the technologies is to make the process of inspection much more efficient. Autonomous systems are able to run inspections much quicker in an area compared to human beings and as well as in areas that require expertise to reach, the autonomous systems can reach the areas in a significantly short time than humans do. In addition, such systems can work 24 hours providing constant observation without the necessity of breaks and shifts. Autonomous systems also cost less to operate besides being faster (Mahmud, 2018). The usual method of inspecting the traditional infrastructure involves large numbers of people, equipment and spaces of time, which are all a source of high labor costs. This can be significantly decreased with the usage of autonomous systems because fewer human resources are involved, thus, reducing the time of inspections. Moreover, predictive maintenance possibilities provided by ML and IoT decrease the chance of costly emergency maintenance and lead to the infrastructure lifespan extension, therefore, saving money in the long run. The other important value of autonomous systems is safety (Chang, 2019). Dangerous or high-risk conditions that may cause accidents, like the need to go above a bridge and inspect it, or needing to check the pipeline or pipes in the ground, human presence is hazardous and should not be the one that puts their lives at a risk-unless when such inspection is done by an artificial intelligence system, potential accident risk is reduced to a large extent. These inspections can be performed safely by the use of autonomous systems, thus eliminating any possible accidents resulting in injuries or even death when conventional methods are applied (J. A. Jasim, 2024). Altogether, this chapter highlights the question of autonomous systems and possible ways of their use in transforming the practice of infrastructure inspection and maintenance. By bringing together ML and IoT, these technologies can present a way of making the management of critical infrastructure more efficient, safe and cost effective, which will result in more reliable, sustainable more intelligent infrastructure systems in subsequent years.

TECHNOLOGY OVERVIEW

Machine Learning (ML) in Infrastructure Inspection

Machine learning (ML) is significant in the change of infrastructure inspection due to precisely and accurately analyzing data automation and predictive functionalities. Infrastructure monitoring incorporates several ML algorithms such as supervised learning, reinforcement learning and unsupervised learning. Both of them are used to solve certain problems of infrastructure inspection. The most widely used ML approach in monitoring the infrastructure is supervised learning. In this method the algorithm is trained using labelled data where the input is combined with the preferred answer (Wang, 2019). As an example, pictures of a surface of the bridges may be marked with the absence or presence of cracks. The model will acquire the ability to find patterns and make educated guesses with new data based on the learnt features in the training data (Alzubi et al., 2022). Supervised learning is commonly applied to completing a task like image recognition and can be deployed to classify a structural state or detect an abnormal stress in one of the infrastructure parts, i.e., a crack in concrete or holes in the metal component. The reinforcement learning, in its turn, is an effective type of decision-making optimization in a dynamic world. Regarding infrastructure inspection, reinforcement learning may be directed to explain autonomous systems how to move and accomplish duties in situations which might include hindrances or changes across the lifetime of the situation. As another example, an inspection of a bridge by drones or other robots might come across unexpected obstructions (Chen et al., 2010), which reinforcement learning will assist the system to re-route itself to enable the best possible coverage of the bridge. The system is also a learner, as it acts, based on the feedback it is receiving, then rewards or punishments are used to encourage a desired behavior or admonish a negative behavior, respectively. The other technique is called unsupervised learning which allows autonomous systems to identify patterns or anomalies within data without labeled training sets. The approach is especially applicable where a wide amount of data is gathered and it is non-practical to identify each data point. As another example, previously unknown abnormalities in the behavior of the parts of infrastructure, including unusual vibrations or temperature changes could be detected by unsupervised learning (Rosati, 2020). Clustering algorithms, e.g. k-means or hierarchical clustering, can be used to classify similar data points and thus discover patterns that may be evidence of structural weakness or imminent failure. Predictive maintenance is another area of special concern to infrastructure inspection through ML as well. With the help of historical data collected by sensors, ML models will be able to conclude a point at which a structure will have to be maintained or at which a failure is about to occur (Scheffer, 2018). As an example, predictive modeling could be applied to a bridge using data collected on its vibration in order to predict when the bridge would have degraded to a potentially dangerous state such that repairs were necessary. By monitoring and predicting the events of failures before they happen, the infrastructure managers can carry out the maintenance at the right time and at the right price which saves them the pressure of doing emergency repairs, and also eliminates downtime (V. I, 2025).

Internet of Things (IoT) in Infrastructure Maintenance

IoT has transformed the maintenance of infrastructure to such an extent that real-time monitoring is possible owing to the use of a large network of connected devices. Sensors, cameras and drones that make up IoT devices are placed along the infrastructure to continuously capture data giving a clear view of the health status of the infrastructure and the environment in which assets are located (Rinaldi, 2018). The devices have different sensors that can detect parameters such as temperature, humidity, pressure and motion and vibration which play important role in gauging integrity of elements of infrastructure including bridges, roads, pipelines and power lines. The use of the sensors embedded in infrastructure through the IoT can deliver real-time data about the wear and tear state of the materials used or the structural integrity of a component (Dziedzic et al., 2021). As an example, during monitoring exposed to constant vibration such as in bridges or buildings, the ultrasonic sensors can scan or detect the cracks on the concrete whereas accelerometers can be used to measure the vibrations that could possibly detect some faults on the bridges or buildings. High resolution cameras, or thermal cameras, or even infrared cameras are created to take pictures of the surface of the infrastructure so the rust, cracks, and other signs of deterioration can be noted and taken care of. The recent use (to complement IoT-based inspection) of drones to take quality pictures and video footage of otherwise hard-to-access sites like high bridges, high buildings or offshore oil platforms is on the increase. Such drones are working with a wide range of different sensors and cameras so that it can retrieve a comprehensive data set, analyzing which you can determine the state of the infrastructure through many different lenses (Gan, 2021). After gaining data through IoT devices, it is sent to different communication network to be processed further. IoT systems today utilize quick and stable networking like Wi-Fi, 5G, and edge computing to ferry information collected in far places or in areas that are difficult to access to central computing systems. The Wi-Fi is a widespread tool in short-distance communication, whereas 5G is a high speed and low delay connection, which is critical in real time monitoring and massive data transfer. Closer processing of data (edge computing) is valuable as a means of improving latency and the speed of decision making in critical applications Latency Latency (shown in red on the figure) refers to delays that may be caused by the network, by remote processing or action to be taken, or by a combination of those factors. Infrastructure managers can also have an understanding of the performance and fitness of assets through the constant stream of real-time presentation of data by Internet of Things devices (Eneyew, 2022). Using such data, they will be able to make the right decisions regarding where and when to conduct maintenance, so that they use resources optimally and minimize downtimes.

Integration of ML and IoT

The combination of ML and IoT forms a synergy where the potential of the autonomous systems in inspecting and maintaining infrastructures is heightened dramatically. IoT devices gather huge amounts of real-time data collected by many different sensors within infrastructure and data are managed and analyzed with the help of ML models to come up with actionable results. These two technologies have the capacity to be integrated so that infrastructure managers will automate decision-making operations, optimize maintenance operations, and predetermine failure. To illustrate, IoT devices attached to a bridge would have a constant check on such parameters as temperature, vibrations, and load-bearing capacities. Using this kind of data, ML can examine it and identify patterns that can show that structural damage is taking place at an early stage (Wu, 2018). In case the pattern of the vibration indicates formation of a crack or any other flaw, the system is capable of automatically raising alarm, which mobilizes the maintenance departments who take necessary action by studying it further. By doing that, ML models make the decision-making process more efficient, as they can foresee in which area and when certain issues will arise, which allows preventive measures to be taken, lowering downtime and extending the infrastructure lifespan. In addition, the combination of IoT and ML allows automating the routine maintenance activity. There can be autonomous solutions in the form of drones or robots that can be used to inspect and perform basic fixes without any human worker. Such systems are based on IoT sensor data to inform them on what to do and are capable of dynamically changing their behavior thanks to ML models. As an example, a drone operating on the pipeline can review the path according to live data acquired by its sensors and concentrate on the areas, which indicate possible wear or corrosion. When coupled to IoT real-time data collection systems and ML analytics, autonomous systems will be able to better inspect and monitor the state of infrastructure and can even build future maintenance requirements onto a scale that has never been seen before. Such integration saves substantially on manual inspection, as well as lowers the cost of operations, boosting the safety and reliability of any infrastructure system. Altogether, ML and IoT technologies in the inspection and maintenance of infrastructure create groundbreaking opportunities when incorporated into the system. IoT devices allow monitoring the work 24/7 and in real-time, the data is processed in ML models to predict failures or identify anomalies and plan the maintenance of equipment. The combination of these technologies will make the management of infrastructure more effective, efficient and proactive and will therefore increase the life expectancy and reliability of critical infrastructure systems.

DESIGNING AUTONOMOUS SYSTEMS FOR INFRASTRUCTURE INSPECTION

System Architecture

Autonomous systems used in the inspection of the infrastructure incorporate multiple elements that interact with each other in order to facilitate effective and precise monitoring. Such elements are sensors and Internet of Things devices, autonomous platforms, and systems to process data. All the components are critical towards having the system that can gather, analyze, and take actions on data in real-time to give actionable outcomes on maintaining the infrastructure.

Sensors & IoT Devices

Sensors are the basics of any autonomous inspection industry, as they allow capturing the data about the surrounding area. Different sensors are applied to observe and measure other variables of infrastructure health. Temperature sensors are often employed to sense changes in temperature and this can indicate such problems as the effect of heat causing expansion, a lack of insulation in or around the building, or faulty electrical connections. The weaknesses, cracks, and corrosion on concrete and metallic materials can also be checked through the use of ultrasonic sensors. These sensors are used in the form of sending high-frequency sound waves and recording the reflections to detect the imperfections at the solid surface. Imaging sensors like high resolution cameras and infrared cameras are used to capture images, footage to inspect the surfaces, detect cracks that can be seen, rust and other types of corrosion, particularly in inaccessible areas. Motion sensors are used to monitor vibrations, and the deviations in the behavior of infrastructure, i.e. how a bridge responds to traffic loading. Sound waves created by stress through acoustic sensors can be used to detect structural weaknesses or fatigue of materials. Taken together, such sensors deliver a detailed and unbroken flow of information on the condition of the infrastructure and help predict any emerging issues in advance.

Autonomous Platforms

Then, another important element in the system is the autonomous platform which makes the carriage of the sensors. Infrastructure that is inaccessible by any given means or one that may cause harm to human beings is often exposed to drones, robots, and autonomous vehicles. Drones fitted with cameras and sensors are particularly helpful in carrying out checks on buildings that are high rise, bridges, power lines and towers because they can maneuver into tight areas and reach places that are inaccessible. Wheeled or tracked robots are used on-ground, as well, to inspect things above or below ground like in tunnels, pipelines, or bridge interior inspection. Different sensors that are built into these robots help in the tracking of the health of a structure as it passes through tight spaces. Autonomous cars are used in large truck inspections, i.e. to monitor highways, or to check railway systems. These are platforms that are programmed to navigate themselves and can therefore easily cover the routine inspection procedures without the human operator hence they cut out the manual labor and the risks that are involved when using the human operated inspection.

Data Processing & Analysis

After being captured by sensors, data have to be processed and analyzed to gain some meaningful insights. Modern infrastructure inspection technologies are based on a cloud environment where tons of information gathered by IoT sensors is stored and processed. The cloud platform enables fast and scalable storage of information and creates a framework in which powerful computation can be used to perform complex analysis, such as predictive models and anomaly detection algorithms. The other important aspect of data processing is edge landscape in situations where a real-time decision is required. As in edge computing, the data remains as close to the source as possible and this causes latency to be reduced or ensured that one can still make some decisions fast like the sounding of an alarm or the sending of autonomous platforms into an area that still needs its inspection. Data analysis uses artificial intelligence (AI) and machine learning (ML), so that algorithms are trained on patterns, failures, and to make real-time decisions. As an example, one can consider ML models that recognize crack patterns on images or analyze trends in a vibration signal that can predict a structural problem. Through these models, infrastructure managers can be provided with the latest and correct information in the decision-making process, which is automated.

Actionable Insights & Decision Making

Among the most prominent benefits of autonomous systems, data analysis can be pointed out to supply actionable insights. ML models are applied to identify the point at which the maintenance should be triggered to correct the situation: schedule a repair, organize a follow-up, or place an area under emergency intervention. As an example, when drone with visual and thermal sensor identifies a crack in a bridge, an ML model can be used to define the severity of the crack, compare it to previous data, and decide where the repair on the bridge must be done immediately and where the problem can be left to see whether more damage could be found. These insights allow infrastructure managers to make decisions based on data alongside maximizing maintenance schedules, which save costs and enhance safety.

Challenges in System Design

Although participating in autonomous systems in the infrastructure inspection presents a bulk of benefits, the development and creation of such systems are fraught with quite a number of problems, mostly concerning the environment, data collation, and real-time analysis.

Environmental Conditions

Inspection of infrastructure usually occurs in harsh conditions, the external situation that may dramatically affect the work of autonomous systems is the climate, landscape, and remote places. Unfavorable weather such as rain, snow or strong winds may compromise the drone or robot capacity to undertake its tasks. Also, hard terrain like steep slopes, or undulating field may pose a challenge to ground-based robots restricting their mobility and accessibility. Some areas which may be far away or inaccessible may pose the challenge of keeping a stable communication network between the sensors and the central control unit resulting to data lag and time in making a decision. It is such key to making autonomous platforms be built in a way that will enable them to bear such challenges, therefore, be reliable and successful in any environment.

Data Integration and Communication Challenges

The other task in the autonomous system design is to combine data of multiple sensors and devices. It is important that the information provided by various sensor types could be integrated easily and analyzed because there is a common tendency to use various sensor types in monitoring the infrastructure. The network systems with which they communicate the data of the sensors to the control center should be strong and have an ability to transfer an abundance of data in real-time. Some of these problems can be solved with the help of 5G or low-power wide-area networks (LPWAN) which can provide high speed, low-latency network which makes it a good choice to achieve remote monitoring and data shared among people. Nevertheless, how to have reliable data flow and compatibility and how different sensors of different kinds can play along in the same system continue to be an enormous challenge.

Real-Time Processing and Decision-Making

The real-time choice is an essential issue of the autonomous inspection systems since the problems of infrastructure usually demand urgent action. Real-time data processing and analysis requires very strong computers, and low-latency communication links. It can be challenging to guarantee that, when certain data is processed, it is within a reasonable timeframe to command the activities that need to take place, like sending a drone to a specific area or even creating an alert to make some urgent repairs. Also, ML models will have to be able to provide rapid decisions regarding the information that was provided, which cannot be achieved when using cloud-based solutions. The dilemma is to make the system quick to respond and precise where downtimes are minimal.

Conclusively, designing autonomous systems to be subjected to infrastructure inspection is not a simple task and requires a thorough deliberation of a number of factors such as: choice of sensors, design of the autonomous platforms, and robotized systems of data processing and analysis. Although there are certain obstacles related to environmental conditions, data integration, real-time processing, developments in technologies are currently made to overcome these issues and bring more productive, safe, and cost-efficient technology to monitor and maintain infrastructure.

CASE STUDIES

Case Study 1: Autonomous Drone Inspection of Bridges

In a practical example of autonomous systems, drones with IoT sensors have been utilized in the process of inspecting bridges and have served as an alternative method to the conventional ways of bridge inspection since they are both cheaper and efficient. The drones will be meant to travel to inaccessible parts of bridges and take real-time information with the help of sophisticated imaging devices. Drones have diverse sensors, which are RGB, thermal cameras, as well as ultrasonics. RGB cameras give high resolution images that can detect visible surface cracks or wear and thermal cameras can detect differences in temperatures which may relate to underlying corrosion or water ingress problems. In situations where the cracks or the weakness of the structure may not be able to be visible to the surface, ultrasonic sensors come in handy to realize such damages or weak areas which can fall down and lead to collapse of the bridge.

The images brought in drones are subjected to machine learning (ML) models, which recognize and categorize the structural damages that include, but are not limited to, cracks, corrosion, and surface wear. The Image recognition algorithms examine the digitized images to be automatically read and cracks identified, classified as per the severity and give a quantitative measure regarding the structural health of the bridge. It makes inspections much faster and accurate. The models can be also trained to identify other kinds of damage in accordance with their size, location, and depth which assists in prioritizing repair works. Moreover, the system may give predictive information through analysis of past records to predict when certain sections of the bridge would need repair.

The outcomes of this method are that the frequencies of inspections increase, since the drones are able to inspect bridges increasingly and in reduced time lapsed, than in the case of conventional means. It contributes to minimized downtime and increased safety of the inspectors because they are not forced to perform under rather dangerous or troublesome conditions (working on the top of tall buildings or in the traffic, etc.). In addition, drones have shown the capability to reveal unnoticeable damages, which cannot be seen with visual inspections, e.g., hairline cracks or corrosion that have not progressed further. The ability facilitated early detection assists in increasing the life span of bridges and minimizing on expenditures required when carrying out major repairs.

Lot of lessons however were learnt in the implementation of autonomous drone inspections. Among the major obstacles was the Interference with the GPS signal in the urban area, being subject to tall buildings or other surfaces with subsequent blocking of the satellite signal, drones will not have stable holding. A solution to this problem was the introduction of other sensors, e.g. the visual odometry or inertial measurement units (IMU), which allow preserving the position of the drone in a GPS-denied space. The other issue was how to provide battery life needed in long term flights. Since drones have to examine vast parts of the infrastructure, both battery life management and the control of the flight path to reduce the energy costs of the flight path were key to the success of the inspections.

Case Study 2: Predictive Maintenance for Railway Infrastructure

An operator of a railway introduced a predictive maintenance system based on IoT to control the state of tracks and trains of this complex transport system with an essential infrastructure. The system employs a web of vibration sensors, acoustic sensors and temperature sensors installed on tracks and within trains to provide round the clock environment monitoring of operations. These sensors gather information regarding the vibrations generated during movement of trains, temperature of the tracks and the signals in addition to the acoustic signatures given out by the infrastructure. Such real-time data is inputted to machine learning model that reads the trend and predicts if there is a failure before it happens.

The main models of machine learning applied to the case study are predictive models that operate based on historical data and make predictions with the wear and tear or potential failures in mind. As an example, the vibration data recorded in the tracks can be examined to identify deviant shake patterns, which can be an indication of a misalignment problem, a deteriorated track or a worn out wheel. On the same note, acoustic data assists to indicate mechanical issues or loose parts, whereas temperature sensors can indicate overheating signs on essential units e.g. signal boxes or switches. The ML algorithms are then able to indicate when a component is expected to break down so that maintenance can be done in advance before problems develop to be very crucial.

The outcome of such predictive maintenance system would be that the unplanned outages would be reduced, since this system gives early alerts about possible failures and maintenance teams are able to fix the problems before they lead to costly downtimes. More reliability of the rail network has been obtained as well, since the maintenance work is more focused and timely, and no longer reactive. The proactive maintenance strategy has also saved the firm a lot of expenses in terms of maintenance, due to the diminution of emergency maintenance, increase in serviceability of infrastructure materials, and proper speed in exploitation of resources.

Notwithstanding the advantages, a multiple number of challenges were involved in the implementation of the system. Among the obstacles, integration of diverse data sources was noted as one of the most important. Railway infrastructure consists of different types of sensors and systems and obtaining the data across such sources in order to analyze it holistically needed substantial work on data standardization and system integration. The right alignment and processing of all the incoming data in real-time were also essential elements in order to make timely decisions. There was also a need to have real-time analytics, so that predictive maintenance messages were created in real-time, and the actions that need to be performed could be made at that time without delays. This was a bit tough in the aspects of computation capabilities, network capacity and in locations where infrastructures are minimal.

Case Study 3: Smart Pipeline Monitoring System

A major oil and gas corporation introduced an intelligent pipeline monitoring system that allows identifying possible problems like corrosion, leaks, and physical damages throughout the pipeline ways using the IoT sensor networks and machine learning models. The system implies installation of different kinds of sensors (pressure, temperature, and ultrasonic sensors) along a pipeline. These sensors constantly check the state of the pipeline and record values of such parameters as internal pressure, change of temperature and vibrations. These parameters are analyzed by the sensors to find any anomaly which may signify failure in the pipeline.

The data that comes in through the sensors is fed into the machine learning models that correlate these data to identify patterns that mean a failure. The models developed on anomaly detection are used to identify the changes in the operating conditions, i.e. the reduction of pressure or the increase of the temperature unexpectedly. These designs are able to detect the beginning of problems like leaks, corrosion decay or mechanical damage, sometimes before they cause a disastrous failure. Whenever the system identifies faults, it activates alarms, and maintenance teams are able to respond to deal with a problem before it becomes serious.

The outcomes of this intelligent pipeline monitoring system are that it will be able to detect pipeline leaks early before they lead to serious environmental effects of the leaks because it will neutralize the problem before it takes place. The system also enhances the maintenance scheduling in terms of determining at what time or date a certain section of the pipeline would be prone to problems. The concept reduces emergency fixes and therefore expensive downtimes. Moreover, the system further improves the safety of pipeline operations due to prevention of catastrophic failures that will put the lives of the workers, population, and the ecosystem at risk.

There are however various difficulties that were experienced when this system was implemented. Among the key difficulties was calibration of sensor in remote locations where the pipelines pass through inaccessible locations like desert or forest. To maintain reliability in the monitoring system, it was vital to ensure that the sensors used were well-calibrated even in those rough conditions, where temperature fluctuations and the physical conditions might be an issue to the functioning of the sensor. Also, there was the problem of false positives which needed to be overcome whereby, the system needed to be highly adjusted such that no false alarms are raised so as to make the unnecessary maintenance interventions upon the false alarms by some minor or irrelevant anomaly. This needed optimization of the machine learning models that would have high sensitivity and specificity and the system was precise and robust.

Conclusively, these case studies point at the transformational capacity of autonomous systems in the inspection and maintenance of infrastructure. Coupled with IoT sensors and support of machine learning models, such systems are the keys to improving efficiencies, assure safety and cut costs, thereby serving invaluable assets in the management of critical infrastructure. Nevertheless, issues of calibration, data fusion and real-time analytics should be closely tackled to get the most out of the technologies.

RESULTS AND DISCUSSIONS

Performance Evaluation of Autonomous Systems

The autonomous inspection of infrastructure systems has shown major benefits of the autonomous system when compared to the manual inspection systems in speed, accuracy, and cost. Conventional approaches to infrastructure inspection entail the involvement of human beings to perform the inspection as they carry out a manual inspection process which is not only time consuming, costly as well as it has a probability of error on the part of the human being. Such techniques are usually demanding in terms of labor, equipment and travel duration particularly in the remote areas where the construction is being built like the high bridges or deep tunnels. As a comparison, autonomous systems, e.g., drones and robots, can travel through vast distances and do so rapidly and efficiently, delivering real-time data with only a little involvement of people. As an example, the inspection of such objects as a bridge or a giant pipeline system can be done by a drone significantly faster than by an entire human crew. The rate of inspection also becomes much higher as demonstrated in Figure 1 where drones can do constant inspection without rest as compared to manual inspection that has limited working hours and man power. In addition, autonomous systems provide high-quality sensors, and thus they can receive data more correctly and in detail than the traditional approach. As indicated by figure 3, machine learning (ML) models incorporated in autonomous systems result in high levels of accuracy in the identification of cracks and other structural defects, which is much higher than the performance that would be attained through manual inspections, which have their share of likelihood of missing smaller or subtle defects.

Self-governing systems are cost-effective as well. Conventional inspections consume a lot of resources, which involve workers with expertise, travelling, and gear. In comparison, the cost of labor is also minimized upon use of drones and robots, which do not require any human expenditures. Moreover, the early detection of problems, as illustrated in figure-4, facilitates predictive maintenance whereby, infrastructure breaks are predictive to act upon, thus, avoiding the expensive emergency repairs and extend the life spans of infrastructure. This proactive maintenance switch as evidenced by Figure 2 saves a lot of cost in the long-term since unplanned repair costs are minimized.

Real-Time Data Processing

Among the greatest strengths of autonomous systems, one should point out the option to process the data immediately, thus making a more immediate decision related to the maintenance of the infrastructure. The role of real-time data analysis is that it enables constant monitoring of the infrastructure giving a current picture of its state. This will allow instant identification of problems, e.g. cracks, corrosion or wear, in relation to the time less as soon as they materialise permitting swift measures and decreasing downtimes.

Nonetheless, such real-time data analysis also has limitations. The amount of data that will be produced by IoT sensors and cameras may be excessive, and powerful computational systems are needed to handle and process massive data within a brief period. Although cloud-based systems and edge computing are already a popular choice to do so, the transmission and the processing of data may include a delay, particularly in the case of a remote area with low network coverage. These limitations are vital to be solved with the use of edge computing and 5G networks, which were described in the section of future directions.

Reliability and Accuracy of ML Models

Such a success or failure in autonomous systems depends on how the ML models perform under failure prediction or trouble detection. Model-based infrastructural inspection, including visual inspection of image recognition models and algorithms and crack detection models, are very responsible in locating problems in structures. Such models, as illustrated in Figure 3, are extremely accurate in identifying cracks and other defects, whose accuracy is higher compared to that of human inspectors. Nevertheless, similar to any technology, there is no perfection about ML models. False negatives (Failure to identify a defect that exists) and false positives (Identifying a defect that does not exist) will be there. Such mistakes usually depend on training data quality, constraints of sensors and natural conditions that may be lighting or noise in visual data. Nevertheless, ML models see constant improvement, and reducing such errors with time is supposed to depend on the further advancement of algorithms and the data they deal with.

Cost Savings and Return on Investment (ROI)

The adoption of autonomous systems leads to high savings in costs providing a good return on investment (ROI). Figure 2 illustrates that in the long-term outlook, using predictive maintenance with the help of ML models will also help to save substantial costs on the maintenance. More often than not traditional methods entail reactive maintenance where repairs are only carried out after they result to failures which are usually very expensive because of their emergency nature. Autonomous systems, contrary, allow proactive maintenance, as issues can be identified even before they become serious. This strategy, besides cutting down repair expenses, will also cut down the downtime incidence in case of infrastructure failure. Moreover, sensitized hybrid automation of the inspection would result in low labour expenses since we shall use less human resources and also be able to have more frequent inspection time, and the resources become more productive.

Limitations of Current Technologies

Although there are wide-ranging advantages, a number of limitations accompany the existing technologies applied in autonomous infrastructure inspectioning systems. Among the major problems lies sensor limitation. Many factors may influence the capacity of the sensors to achieve range and accuracy and sustainability, including the condition of the environment, weather conditions, and physical barriers. As an example, ultrasonic sensors utilized in detecting the existence of crack or corrosion may not be effective in some materials or may fail to properly work with uneven surface conditions. Also, long-range inspections of large-scale structures such as power lines and bridges cannot be inspected using the autonomous drone because of its battery life.

The other issue is data integration. Autonomous systems can use the data provided by a plurality of sensors, and their smooth combination and proper interpretation is the key aimed at efficient decision-making. The non-uniform data types, malfunction of sensors and the communication problems might reduce the overall functionality of the system. Furthermore, it would be difficult to guarantee real time processing in remote territories or places which are inaccessible, having a reduced network band.

Finally, there are safety consideration issues in harsh conditions like extreme weather condition or harsh terrains. Therefore, autonomous systems should be built to help them fight these obstacles, and seek their operation in conditions harsher than before without compromising reliability and safety of the drones and robots.

Future Directions

In the long term, there is a variety of newer technologies that can contribute to the further elevated level of autonomous infrastructure inspection systems development. 5G networks and edge computing will be helpful in terms of enhancing the processing of real-time data by allowing quicker data transfer and processing. It will serve to eliminate the latency and bandwidth that plagues remote inspection at the moment. Also, if autonomous systems are integrated with Digital Twins, which are virtual mock-ups of physical infrastructure, they will enhance visualization and bring more proper predictive analytics. Dynamic simulation and real-time monitoring can occur with Digital Twins, and this will ensure that decisions are made better and maintenance activities are improved.

Further development of AI is also going to facilitate autonomous decision-making as systems will not only identify the problem, but also decide which action constitutes the best course of action. This may consist of making appointment on repair, alerting or even repairing small damages on its own. Application of autonomous systems to other forms of infrastructure (e.g. power lines, dams and water supply systems) may further optimize infrastructure management by continually monitoring infrastructure and eliminating the possibility of failures.

To sum up, even though the use of autonomous systems to inspect and maintain infrastructures is extremely advantageous due to the fast inspection pace, high accurateness, and affordable cost of implementation, the challenge lies in the inconvenience of the sensors, data collection, and real time processing. Nevertheless, the development of 5G, edge computing, and AI will help to enhance the work of these systems even more every time so they could become a major part of the contemporary infrastructure control.

Figure 1. Inspection frequency vs. traditional methods

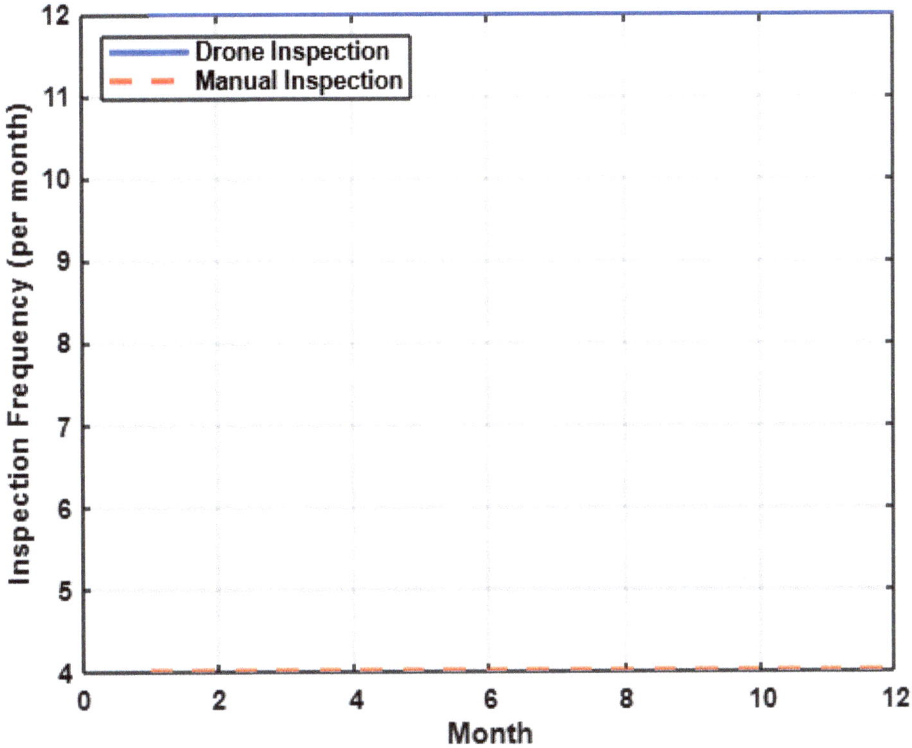

Figure 2. Predictive maintenance - maintenance costs reduction

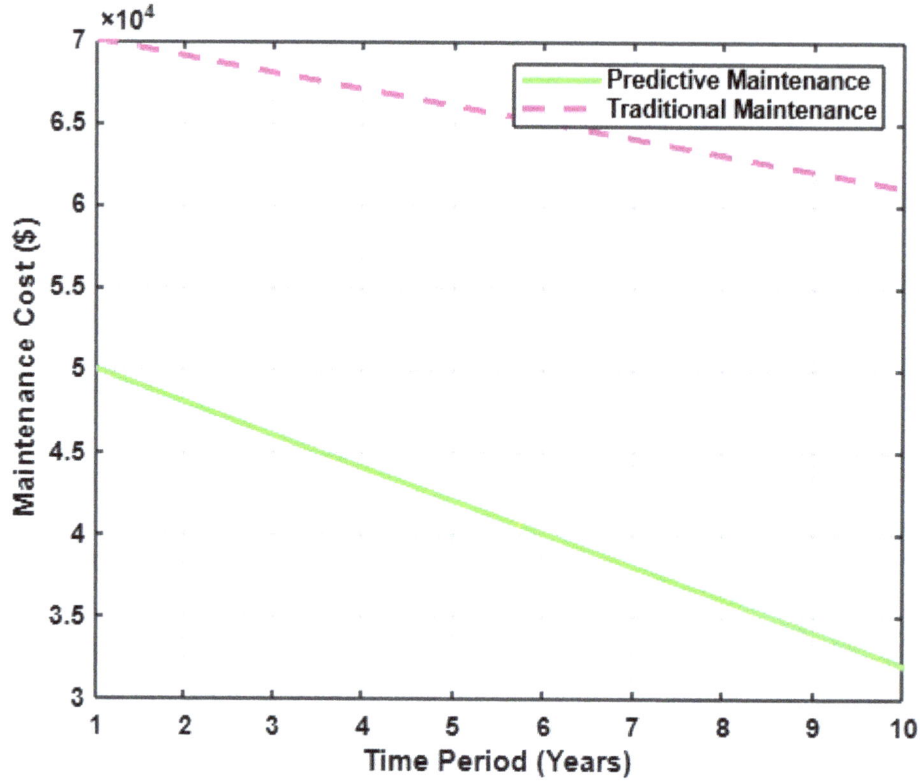

Figure 3. Anomaly detection performance

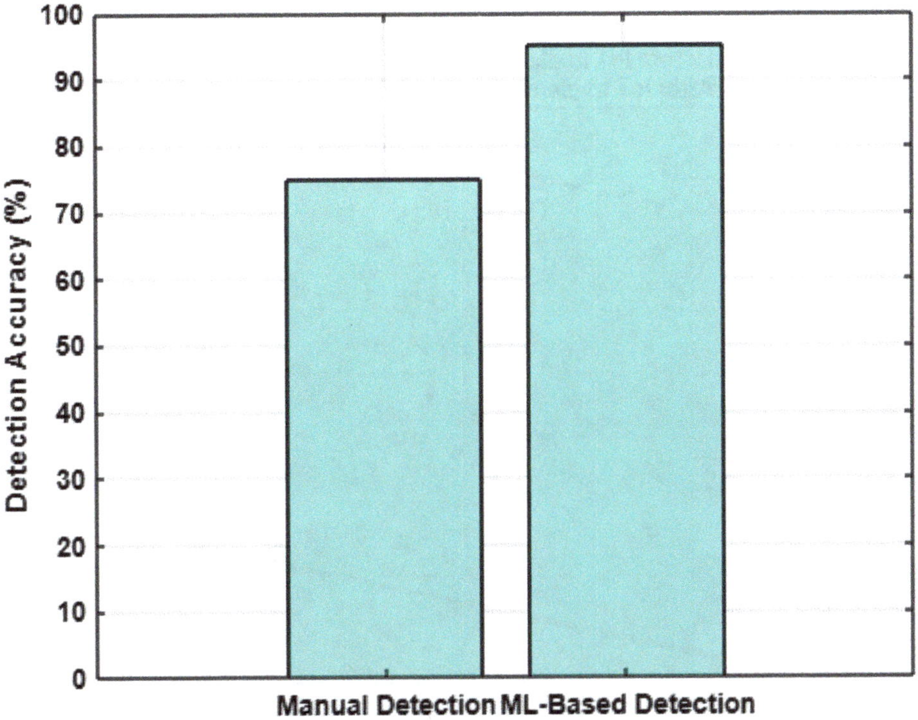

Figure 4. Pipeline leak detection efficiency

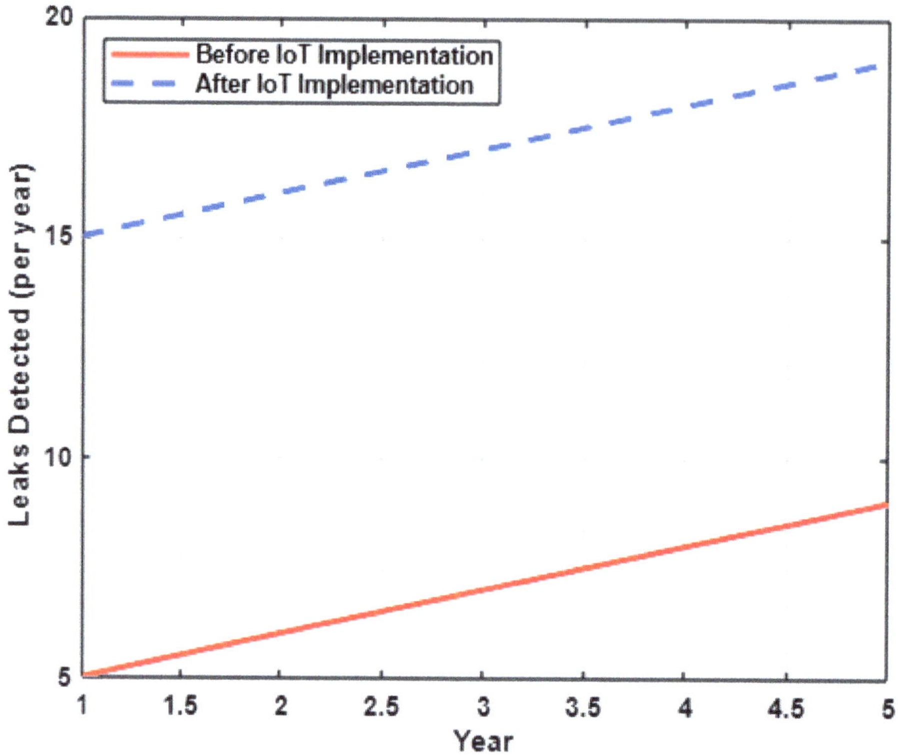

REAL-TIME APPLICATIONS

Smart Cities

In smart cities, autonomous systems are vital to the management, prescription and renovation of the urban infrastructure. It is through such sensor based, artificial intelligence (AI) and machine learning (ML) driven systems that real time monitoring and decision making can be effected toward efficiency, sustainability, and safety of urban environments. The most typical activities conducted through autonomous systems in smart cities are associated with traffic management, streetlights control, and waste management, among the others. Indicatively, the autonomous control of the traffic refers to a set of cameras, sensors, and ML models working together to study the movement of traffic, identify the traffic jams, and adjust the traffic lights timings to minimize delays and enhance mobility. This not only improves the traffic management system as a whole but also helps reduce carbon emission because of less amount of idling.

Streetlights also require autonomous systems in order to be maintained. The IoT sensors will enable enough streetlights to track the health of the streetlights and the energy they spend in real-time, permitting the system to identify the faults and maximize energy efficiency. Such a proactive maintenance minimizes cost of energy and makes urban centers well lit serving the common good of people. Additionally, automatic cars and drones can also be utilized to examine the city infrastructure, disclose problems in the structure of bridges, roads, and tunnels, and deliver immediate feedback to make the required changes. Through automation of these functions, cities will run efficiently and with efficiency the city resources are utilized effectively and the provision of services to the public will not be skipped.

Disaster Management

The future use of autonomous systems in the disaster management will be difficult to avoid, particularly regarding the inspection of the critical infrastructure following the disaster, e.g., earthquakes, floods, or hurricanes. Such systems, such as drones, as well as autonomous robots can be sent out quickly after a disaster to survey damage to inaccessible or dangerous areas where human workers may lack the ability to travel and confidently survey. As an example, the results of an earthquake may be inspected with the help of drones with cameras and thermal sensors to detect cracks, structural vulnerabilities, or possible collapsing. Rapid damage assessment also assists emergency responders in prioritizing their actions and allocating resources, so the most important problems could be addressed at the first stages.

Autonomous robots might come in handy during inspection of waterlogged areas to ensure no damages have occurred to the water treatment plants, dams, and levees during the floods. In the event of a large flood cover, drones can also survey the area to know the areas that are likely to be flooded further, or where other rescue works may be necessary. Thanks to the ability to introduce the information in real time, autonomous systems can greatly increase the rate and precision of the assessment, to limit the possibility of additional harm, and to hasten the recovery processes.

Construction & Mining

Autonomous robots are widely used in construction and mining industry as they help to make the work of employees safer and avoid accidents. Construction safety hazards monitored using autonomous systems include falling objects, unstable scaffolds or intrusion of hazardous areas by unauthorized people. It can be done by using robots that are fitted with cameras and sensors to man the construction sites and are able to monitor them in real life with sensors that give a live output of conditions on the builds this way robots have the capability to patrol areas that need to be gated off where it may be unsafe to walk or areas that are being guarded by police. Such systems are also programmable to identify environmental hazards, e.g. leaks of toxic gases, or an unstable structure, and send automatic alerts to the workers to initiate precaution procedures.

Autonomous robots and vehicles are applied in mining activities whereby the underground or remote sites that may be too dangerous to be manned appear to warrant inspection as well as monitoring. Ground robots and drones can be utilized to oversee mining machinery, identify the signs of failure, and check that the infrastructure of the mine is not damaged. This further eliminates accidents and enhances overall efficiency of mining activities since equipment is adequately maintained and problems that might arise in an event are solved before they occur.

Energy Sector

Robots have been so useful in the energy industry where they are increasingly being deployed to monitor the power plants, wind turbines, and solar panels. Autonomous robots and drones have applied in power generation plants to undertake periodic checks on the equipment like turbines, boilers, and cooling towers to check any wear or failures. Being a difficult location to visit and inspect a wind turbine due to its height, drones loaded with high-resolution cameras with the ability to inspect the turbines by capturing high resolution pictures of the turbines is particularly effective in its use. These autonomous systems are collecting the real-time data that can aid the operators in detecting any problems in the turbine blades like cracks or misalignments and there is less cost spent on manual inspection and hence reduced downtime.

In case of solar panel inspections, the drones installed with thermal cameras are adopted to inspect faulty panels checking hot spots that can characterize internal problems. Autonomous systems are also used to streamline the effectiveness of solar farms, which would mean data collection to understand the condition of the panels and have indicators of maintenance. Automation of such inspections allows the energy sector to work more efficiently, save, and guarantee an increased degree of safety and reliability within the critical infrastructure.

To conclude, the list of industries in which autonomous systems are finding their application is rather long, and they offer real-time data, enhance safety, lower expenses, and serve as efficiency tools. In the field of smart cities, disaster management, construction, mining and energy industry, autonomous systems are revolutionizing managing the infrastructure and also providing answers to age-old problems in these sectors. With the changes in technology, the future use of autonomous systems in the management of complex infrastructures can only increase, which will also improve the measures of safety, productivity, and sustainability.

CONCLUSION

Modernization of infrastructure management through the evolution and fusion of self-regulating systems to inspect and manage infrastructure, aided by progress in Machine Learning (ML) and the Internet of Things (IoT), provides a revolutionary approach to revising infrastructure management. These systems are more efficient, accurate and cost effective than conventional options, and they can monitor and understand in real time and perform predictive maintenance. Use of self-directed drones, robots, and sensors increase the capability of more regular inspections, increased speed of locating the problem and minimize cases of downtime, which ultimately translates to longer infrastructure life span and better operational costs. In addition, ML algorithms combined with the capabilities of IoT devices allow real-time decision-making under the influence of failures forecasted and an optimized effort on maintenance activities. Although the applicability of the current technologies is substantial, there are still issues with sensor constriction, data allure, and real-time analyses that require processing incessantly. New technologies such as 5G, edge computing, and AI will only promote the functionality of these autonomous systems making them even more stable and capable of handling critical infrastructure. With these systems moving to other areas, including smart cities, disaster management, and the energy industry it is just becoming more and more apparent that the benefits in terms of improved safety, reduced risks, and cost savings can be achieved. The future of infrastructure maintenance will be centred around the use of such intelligent, and autonomous technologies to make vital infrastructure systems sustainable and resilient.

REFERENCES

Ahmed, S., Kurnaz, S., Hamdi, M. M., Khaleel, A. M., Khaleel, A. M., & Seno, M. E. (2022) "Study for Buildings with IoT System for Energy Management," *2022 International Symposium on Multidisciplinary Studies and Innovative Technologies (ISMSIT)*, Ankara, Turkey, pp. 53-57. DOI: 10.1109/ISMSIT56059.2022.9932729

Alzubi, K. M., Salah Alaloul, W., Al Salaheen, M., Hannan Qureshi, A., Musarat, M. A., & Mohsen Alawag, A. (2022) "Reviewing the Applications of Internet of Things in Construction Projects," *2022 International Conference on Decision Aid Sciences and Applications (DASA)*, Chiangrai, Thailand, pp. 169-173. DOI: 10.1109/DASA54658.2022.9765143

Broto, J. M., Cipriano, J., Mor, G., Gavaldà, O., Masoumi-Verki, S., & Eicker, U. (2025) "An interoperable ontology-based information model for better integration of building physics and IoT data analytics models," 2025 IEEE/ACM 7th International Workshop on Software Engineering Research & Practices for the IoT (SERP4IoT), Ottawa, ON, Canada, 2025, pp. 1-8. DOI: 10.1109/SERP4IoT66600.2025.00005

Chang, Y. (2019) "Study of Improving Teaching Materials for Barrier-free Facility Inspector Training in Taiwan," 2019 IEEE Eurasia Conference on IOT, Communication and Engineering (ECICE), Yunlin, Taiwan, 2019, pp. 125-128. DOI: 10.1109/ECICE47484.2019.8942669

Chaturvedi, K. Prasad, A. K. N, B. Ashreetha, and C. H. Priya, (2025) "Optimizing Environmental Monitoring in Smart Urban Areas with IoT-Enabled Wireless Sensor Networks," 2025 5th International Conference on Pervasive Computing and Social Networking (ICPCSN), Salem, India, 2025, pp. 455-460.

Chen, Y., Zhang, J., Mu, X., & Ma, J. (2010) "Study on the theoretical framework of the internet of building energy systems," 5th International Conference on Computer Sciences and Convergence Information Technology, Seoul, Korea (South), 2010, pp. 973-976.

Dziedzic, R.. (2021) "A framework for asset management planning in sustainable and resilient cities," *2021 IEEE International Symposium on Technology and Society (ISTAS)*, Waterloo, ON, Canada, 2021, pp. 1-10. DOI: 10.1109/ISTAS52410.2021.9629158

Eneyew, D., Capretz, M. A. M., & Bitsuamlak, G. T. (2022). Toward Smart-Building Digital Twins: BIM and IoT Data Integration. *IEEE Access : Practical Innovations, Open Solutions*, *10*, 130487–130506. DOI: 10.1109/ACCESS.2022.3229370

Fokaides, P., Jurelionis, A., & Spudys, P. (2022) "Boosting Research for a Smart and Carbon Neutral Built Environment with Digital Twins (SmartWins)," 2022 IEEE International Smart Cities Conference (ISC2), Pafos, Cyprus, 2022, pp. 1-4. DOI: 10.1109/ISC255366.2022.9922513

Gan, V. J. L., Wang, T., Hu, D., & Yin, C. (2021) "Developing an Integrated Data Model based on IFC and SensorML for Post-COVID Facility Management," 2021 IEEE 23rd Int Conf on High Performance Computing & Communications; 7th Int Conf on Data Science & Systems; 19th Int Conf on Smart City; 7th Int Conf on Dependability in Sensor, Cloud & Big Data Systems & Application (HPCC/DSS/SmartCity/DependSys), Haikou, Hainan, China, 2021, pp. 1845-1850. DOI: 10.1109/HPCC-DSS-SmartCity-DependSys53884.2021.00273

Hooda, Y., Singh, H. D., Mehta, J., Passi, A., & Mehta, S. (2023) "An Interdisciplinary Approach for Evaluation of Green Building Index," *2023 International Conference on Advances in Computation, Communication and Information Technology (ICAICCIT)*, Faridabad, India, 2023, pp. 509-512. DOI: 10.1109/ICAICCIT60255.2023.10465795

Jasim, J. A., Hilal Mutlag, A., Othman, N. A., Shareef, H., & Hameed, N. H. (2024) "Innovative Applications of AI and IoT in Smart Urban Infrastructure and Housing Projects," *2024 International Conference on Emerging Research in Computational Science (ICERCS)*, Coimbatore, India, 2024, pp. 1-6. DOI: 10.1109/ICERCS63125.2024.10894956

Lam, R. C. Y., Lam, L. C. H., Li, N. X., & Lee, P. K. K. (2021) "IoT-based Monitoring System to Prevent Collapse of Falsework Systems During Construction," *2021 International Conference on Computational Science and Computational Intelligence (CSCI)*, Las Vegas, NV, USA, 2021, pp. 1454-1459. DOI: 10.1109/CSCI54926.2021.00080

Mahmud, M. A., Bates, K., Wood, T., Abdelgawad, A., & Yelamarthi, K. (2018) "A complete Internet of Things (IoT) platform for Structural Health Monitoring (SHM)," 2018 IEEE 4th World Forum on Internet of Things (WF-IoT), Singapore, 2018, pp. 275-279. DOI: 10.1109/WF-IoT.2018.8355094

Mu, F., & Wang, Z. (2021) "Smart Fire Safety Early Warning System of Large Complex Building based on Data Collection and Processing," 2021 Fifth International Conference on I-SMAC (IoT in Social, Mobile, Analytics and Cloud) (I-SMAC), Palladam, India, 2021, pp. 1201-1204. DOI: 10.1109/I-SMAC52330.2021.9640862

Rinaldi, S., Flammini, A., Tagliabue, L. C., & Ciribini, A. L. C. (2018) "On the Use of IoT Sensors for Indoor Conditions Assessment and Tuning of Occupancy Rates Models," 2018 Workshop on Metrology for Industry 4.0 and IoT, Brescia, Italy, 2018, pp. 123-128. DOI: 10.1109/METROI4.2018.8428327

Rosati, A. Cervo, and C. Fantuzzi, (2020) "Air Quality Monitoring in a BIM model by means of a IoT Sensors Network," 2020 Fourth International Conference on I-SMAC (IoT in Social, Mobile, Analytics and Cloud) (I-SMAC), Palladam, India, 2020, pp. 110-115.

Scheffer, M., ... (2018) "Evaluation of Open Data Models for the Exchange of Sensor Data in Cognitive Building," 2018 Workshop on Metrology for Industry 4.0 and IoT, Brescia, Italy, 2018, pp. 151-156. DOI: 10.1109/METROI4.2018.8428349

V. I., R. K, S. Saranya, B. Shankar, M. Tamilselvam, and S. Venkatachalam, (2025) "The Future Role of Artificial Intelligence and Internet of Things Smart Green Homes Buildings," 2025 International Conference on Frontier Technologies and Solutions (ICFTS), Chennai, India, 2025, pp. 1-5.

Vrsalović, T. Perković, I. Andrić, M. A. Čuvić, and P. Šolić, (2021) "IoT Deployment for Smart Building: Water Consumption Analysis," 2021 6th International Conference on Smart and Sustainable Technologies (SpliTech), Bol and Split, Croatia, 2021, pp. 01-05.

Wang, W., ... (2019) "Dijkstra Algorithm Based Building Evacuation Recognition Computing and IoT System Design and Implementation," 2019 IEEE 13th International Conference on Anti-counterfeiting, Security, and Identification (ASID), Xiamen, China, 2019, pp. 229-233. DOI: 10.1109/ICASID.2019.8925068

Wu, M., Liu, H., Huang, L., Lin, J., & Hsu, M. (2018) "Integrating BIM and IoT technology in environmental planning and protection of urban utility tunnel construction," *2018 IEEE International Conference on Advanced Manufacturing (ICAM)*, Yunlin, Taiwan, 2018, pp. 198-201. DOI: 10.1109/AMCON.2018.8615004

Xie, Q., Zhou, X., Wang, J., Gao, X., Chen, X., & Liu, C. (2019). Matching Real-World Facilities to Building Information Modeling Data Using Natural Language Processing. *IEEE Access : Practical Innovations, Open Solutions*, *7*, 119465–119475. DOI: 10.1109/ACCESS.2019.2937219

Zainordin, N., Salleh, S., Farsana, N., Khoo, S. L., Omar, A. F., & Heka Ardana, P. D. (2024) "Stoppage to the Implementation of Internet of Things (IoT) in Construction Projects: from Lens of Civil Engineers in the United Kingdom," 2024 5th International Conference on Artificial Intelligence and Data Sciences (AiDAS), Bangkok, Thailand, 2024, pp. 191-202. DOI: 10.1109/AiDAS63860.2024.10730249

Chapter 8
Future Directions and Emerging Research

Hameed Hasan Khalaf
University of Tikrit, Iraq

Hatem Kareem Ahmed
College of Administration and Economics, University of Samarra, Iraq

Mohammed Ameen Haitham Ahmed Alhababi
College of Administration and Economics, University of Samarra, Iraq

Shihab A. Shawkat
https://orcid.org/0000-0002-9529-2151
University of Samarra, Iraq

ABSTRACT

The study results show that emerging research represents a major driver of technological transformation, particularly in the fields of artificial intelligence (23.5% annual growth), renewable energy (18.2%), and big data (20.1%). Global funding in these fields has reached record levels, with AI alone expected to receive $90 billion by 2024. In contrast, developing countries face a significant gap, with the number of researchers not exceeding 740 per million people, compared to 4,500 in developed countries. The most prominent challenges are a lack of funding (72%), weak infrastructure (65%), and a lack of skilled personnel (60%). The analysis confirms that the integration of technology and civil engineering is essential for building smart and sustainable infrastructure. The study recommends reorienting research policies towards open innovation, scientific foresight, and effective knowledge governance. The conclusion indicates that keeping pace with future trends is no longer an option, but rather a strategic necessity to ensure scientific and developmental positioning

DOI: 10.4018/979-8-3373-5248-0.ch008

INTRODUCTION

This study is situated within the context of profound global transformations in science, technology, and society. In light of accelerating change, understanding future trends in research has become a necessity rather than an option. This work examines how emerging research areas can shape knowledge production and influence societal development. In light of the profound transformations the world is witnessing in the fields of science, technology, the environment, and the economy, it is no longer possible to talk about scientific research as a traditional (Akinola & Emuze, 2020). closed system concerned only with the production of knowledge in isolation from accelerating societal and global changes. Emerging research and future trends have become the cornerstone of shaping the scientific agenda for the coming decades. Researchers, policymakers, and universities must reconsider patterns of scientific thinking, how topics are selected, questions are formulated, funding is directed, and results are evaluated (Bocken *et al.*, 2014). This is in line with new global challenges, such as climate change, the digital revolution, geopolitical developments, epidemics, demographic shifts, and growing development needs.

The nature of science itself has changed. Once confined to laboratories and equations, science today has become a strategic tool for states to extend their influence, enhance their economic security, and lead global competitiveness. Knowledge is no longer an academic luxury (Ibrahim 2024) it has become an integral part of major political, economic, and social equations. Therefore, the importance of understanding future trends in scientific research lies not only in tracking technological developments, but also in recognizing their cognitive structure, societal dimensions, and potential impact on the structure of power, identity, and development. This requires scientific communities today to shift from a reactive to a proactive mode, from a logic of consumption to a logic of production, and from a subordinate role to a leading role (Eastman *et al.*, 2018).

It has become clear that traditional paths of scientific research are no longer able to keep pace with accelerating changes, whether in terms of organizational structure, (Hammad et al., 2024) priorities, or even methodologies. The world is moving toward complex and interconnected knowledge systems that rely on the integration of sciences, cross-sector partnerships, investment in innovation, and the proactive anticipation of risks and opportunities. In this context, there is an urgent need for a deep understanding of the concept of emerging research, which embodies new waves of change and expresses transformations in the human perception of itself, nature, and the tools for understanding and controlling the world (European Commission, 2020), (Fagnant & Kockelman, 2015). Methodologically, the study relies on a comparative analysis of global reports, statistical indicators, and foresight-based analytical tools. This approach enables a multidimensional understanding of emerging research and provides a foundation for assessing the readiness of developing societies to engage with these transformations.

Many global reports indicate that the next two decades will witness a radical change in the map of scientific research (Ali & Ghanem 2025), (Hameed *et al.*, 2020) with some fields disappearing and others emerging. Research funding methods, publishing approaches, and evaluation criteria will change, and there will be fierce competition over who takes the lead in discovering solutions, providing answers, and producing alternatives. Hence, Thus, the preparation of this research is positioned at a pivotal historical juncture, where scientific inquiry intersects with strategic decision-making, when the question of knowledge intersects with the question of destiny, when the scientific and value-based aspects overlap, and when the future of scientific research shifts from possibility to necessity, from option to inevitability (Gao & Pishdad-Bozorgi, 2019), (Ghosh & Chatterjee, 2023). Accordingly, the objective of this chapter is to explore the dynamics of emerging research within the global scientific landscape and to analyze their implications for developing countries, particularly in the Arab world. By situating the discussion within broader epistemological and institutional contexts, the chapter seeks to clarify the conceptual framework, set the analytical boundaries, and highlight the practical importance of linking foresight methodologies to national research agendas.

THE RESEARCH PROBLEM

A review of the existing literature highlights the growing recognition of emerging research as a critical driver of knowledge production and societal transformation. Studies emphasize the shift toward interdisciplinary, foresight-based approaches in scientific research. However, much of the current literature remains concentrated on developed contexts, with limited focus on the readiness of developing countries to engage in this transformation. Furthermore, existing studies often describe emerging research trends without integrating them into a comprehensive framework that accounts for cultural, institutional, and epistemological dimensions. This chapter seeks to address these gaps by providing a multidimensional analysis that links global trends with the specific challenges and opportunities facing developing societies.

Amidst this major transformation the world is witnessing in its cognitive, technological, and social structures, a pivotal research problem emerges: the ability of scientific research to keep pace with the future, rather than merely reacting to its challenges. The issue is not simply about developing research tools or updating methods, but rather fundamentally about reshaping the role of science itself, the limits of its contribution, and the methodologies for positioning it within a complex network of global, digital, and cultural changes. Hence, this study poses a key question: How can we monitor and analyze future trends in scientific research, and what is the nature of emerging research currently taking shape on the fringes of the global scientific system (International Energy Agency, 2021), (Liu *et al.,* 2017).

The problem of this research is to attempt to understand and analyze these emerging trends in their global context, with a focus on the readiness of developing societies, particularly in the Arab world, to engage in this new landscape, in which there is no longer room for stagnation, imitation, or falling behind. The most pressing question is no longer: "What is being produced scientifically?" Rather, it has become: in (Addula et al., 2025) How is this knowledge employed? Who benefits from its results?" We are faced with a global knowledge structure that is changing in terms of funding patterns, disciplinary maps, and publishing systems. This requires every country, university, or research center to seriously and rationally reassess its position and research strategy (Lu *et al.,* 2020), (OECD, 2021).

Numerous sub-questions stem from this problem, such as: What are the most important emerging research areas expected to drive the future of knowledge in the coming decades? What are the potential impacts of this research on economic, social, and educational structures? What is the nature of the research gap between developed and developing countries? Can developing countries transcend the role of reception and accompaniment to become part of global knowledge production? What are the structural challenges preventing these countries from integrating into the future scientific system? How can the role of national policies be activated in supporting future research trends in terms of funding, capacity building, legislative modernization, and linking research to the labor market and development (Qiao *et al.,* 2024), (Salama & Wiedmann, 2013).

The research problem also includes a critical dimension related to how to anticipate these trends. A large part of the gap relates not only to a lack of capacity or weak infrastructure, but also to a lack of vision, the absence of foresight tools, and the reliance on outdated models of knowledge management. Therefore, the core of the problem lies in the inadequacy of institutional understanding of the meaning of "futures research." Innovation is sometimes reduced to technology alone, when it actually requires deeper transformations in research culture, critical thinking, state strategies, the role of the university, and the researcher's relationship with society (United Nations, 2015), (Zhang *et al.,* 2022). Hence, studying this problem is not limited to describing reality or repeating what has been published in global reports. Rather, it seeks to offer an independent scientific reading that goes beyond translation to interpretation, moving from observation to analysis, and from analysis to the presentation of applicable reform models. The problem at hand is not recognizing the existence of emerging research, but rather how to make this research part of the fabric of academic and national decision-making, and part of the actual tools for change in economic, educational, and social life.

Addressing this research problem requires adapting emerging methodologies capable of anticipating societal transformations while maintaining epistemological integrity. Approaches such as participatory foresight, scenario-based modeling, and AI-driven analytics provide valuable tools for projecting future trajectories of science and society. To preserve methodological rigor across cultural and institutional contexts, these tools must be applied with triangulation, cross-validation, and cultural sensitivity, ensuring that foresight does not become detached from the social realities it seeks to influence.

RESEARCH OBJECTIVES

Defining the objectives of scientific research represents one of the fundamental pillars that determine its path, methodology, and outcomes. Objectives also serve as a roadmap that helps the researcher precisely define the focus of work, control the relationship between the theoretical framework and the applied framework, and direct analytical tools toward measurable and evaluable results. In light of the problem posed in this research, which relates to the position of future trends and emerging research in the global scientific landscape, and the extent of developing countries' readiness to absorb and interact with these transformations, the objectives of this research are presented at several integrated levels.

This research seeks to provide an in-depth and comprehensive analysis of emerging research expected to leave its mark on the future of knowledge and technology in the coming decades. This research explores the most prominent advanced scientific fields, such as artificial intelligence, data science, synthetic biology, sustainable energy technologies, and digital social sciences, while examining the factors that led to the emergence of these fields and their impact on the economy, the labour market, and public policy.

The research also aims to assess the structural and cognitive readiness of academic and research institutions in Arab and developing countries to keep pace with these trends. This is achieved by analysing elements such as funding, infrastructure, competencies, legislation, and governance, and by identifying strengths and weaknesses in national scientific research systems. A central objective of this work is to establish a comparative framework between scientific research policies in developed and developing countries, revealing differences in funding priorities, researcher support mechanisms, and publishing and innovation systems. This will allow for the development of realistic recommendations that can be used to develop national systems.

The research also aims to provide a practical vision for anticipating research trends by integrating quantitative and qualitative analysis tools, leveraging global reports and digital platforms, and future-oriented approaches based on predictive models and scenarios. This will help decision-makers build scientific policies based on information rather than estimates. One of the vital objectives of this research is to contribute to guiding Arab universities toward more influential roles in the social and economic environment. This is achieved by proposing mechanisms to link emerging research with societal needs, developing academic curricula to keep pace with these changes, and activating the role of research centres as engines of innovation rather than relying solely on traditional teaching.

The research also seeks to build a new academic awareness among young researchers of the need to engage in emerging fields, acquire tools for future foresight and analysis, encourage a culture of specialized and interdisciplinary critical thinking, and enhance the presence of Arab research in the global publishing landscape. In this context, the central objectives of this research can be summarized in the following points:

1. Analyse future trends in scientific research and monitor the most prominent emerging research fields globally.
2. Examine the readiness of developing countries to engage in this research by assessing the current state of research infrastructure and knowledge policies.
3. Provide a comparative analysis of research systems in the Global North and South, identifying factors of success and failure.
4. Develop a proposed applied model for integrating emerging research into national research and education plans. Enhancing the forward-looking vision in scientific thinking and broadening the horizons of Arab researchers towards the future.

SIGNIFICANCE OF THE RESEARCH

This study derives its importance from a set of cognitive, methodological, and practical considerations, which make its subject not merely a circumstantial response or limited academic interaction, but rather a strategic investment in the knowledge infrastructure upon which societies' scientific, economic, and developmental futures will be built. At a time when human priorities are being reshaped in light of major innovations, environmental and health risks, and geopolitical conflicts, it has become essential for every society to possess a high capacity for foresight and place scientific research at the heart of its civilizational project.

The primary importance of this research stems from its inherently forward-looking nature. It does not merely describe the current reality or diagnose its manifestations, but rather seeks to interpret weak indicators, anticipate future transformations, and understand knowledge trends in their future contexts. This places it within the category of visionary studies needed by academic and national decision-makers alike. In a rapidly changing world, a failure to understand research trends means falling behind the curve, relying on external factors, and retreating to the margins.

The importance of this study also lies in its attempt to transcend superficial classifications of the concept of "emergent research," which is often reduced to the mere use of new technologies or entry into new fields. In reality, emerging research represents a shift in the very philosophy of science itself, in terms of the nature of the questions posed, the relationship between disciplines, modes of production, new actors in the field of knowledge, and the links between science, society, economy, and politics.

The importance of this study stems from its ability to address the challenges facing developing societies in effectively engaging in the global research ecosystem. Lack of vision, limited funding, weak governance, and the absence of clear strategies to direct research toward promising future fields have all contributed to these societies remaining outside the circle of influence on major scientific decisions. This study proposes a set of alternatives and policies based on an understanding of the reality of these countries, rather than replicating models that do not take context into account. The importance of this study lies in its integration of theoretical, analytical, and applied aspects. It does not merely describe phenomena, but rather analyzes leading international examples, draws lessons from them, adapts them to the needs of local environments, and presents a preliminary roadmap that can be built upon by universities, research centers, and scientific policymakers.

The importance of this research is heightened by the existing gap between what is taught in Arab universities and the research produced worldwide, as well as the scientific and technical solutions needed by the market and society. The lack of alignment between research outputs and real development needs represents one of the greatest manifestations of waste in human and knowledge resources. Therefore, redirecting research toward future horizons that open the way to new jobs, emerging markets, and smart solutions to chronic problems is an existential necessity, not just a developmental option.

The importance of this research can be summarized as follows:

1. It provides an in-depth scientific analysis of the concept of future trends and emerging research as a strategic tool, not merely a cognitive one.
2. Addresses the chronic shortcomings in linking research to societal needs and calls for a redefinition of the function of scientific research.
3. Supports decision-making centers with a vision based on foresight and comparative analysis

4. Contributes to building a new research awareness among actors in universities and academic institutions in developing countries.
5. Constitutes a primary reference in the debate on knowledge sovereignty, scientific security, and research positioning in a changing global system.

RESEARCH METHODOLOGY

This section adopts a mixed-method approach that combines quantitative and qualitative analysis. providing statistical indicators on research expenditure, growth rates, and institutional readiness. These datasets were selected for their reliability and global coverage. Qualitative insights were obtained through a critical review of academic studies and foresight-oriented policy reports. Analytical tools included comparative assessment across developed and developing countries, triangulation to validate findings, and scenario-based reasoning to anticipate future trajectories. This methodological framework enhances the credibility and replicability of the results by making explicit both the sources of data and the techniques of interpretation.

RESULTS

This research is built on a multidimensional theoretical framework based on contemporary scientific approaches that address emerging research from a foresight and applied perspective. It is closely linked to technology and civil engineering, two dynamic fields that effectively intersect to solve contemporary challenges and build smart and sustainable cities. Scientific research is viewed here not only as a knowledge tool, but also as an effective element in formulating national policies, driving innovation, and achieving sustainable development through integrated models between theory and practice. The theoretical framework is primarily based on the concept of scientific foresight, which represents the ability to analyse global trends and anticipate future transformations through the use of future prediction tools such as mathematical models, scenario analysis, and artificial intelligence. This concept is particularly important in the field of civil engineering, where foreseeing climate risks, infrastructure transformations, and sustainable energy needs has become essential in the planning of cities and major facilities, directing engineering research toward preventive solutions rather than post-hoc treatments.

Second, the framework is based on the theory of open innovation, which moves beyond the notion that knowledge is produced solely within academic laboratories to include contributions from industry, the public sector, and end users. In the context of civil engineering, this model enables research projects to be linked to real-world problems such as waste management, improving building materials, developing smart construction technologies, or designing smart infrastructure monitoring systems through collaboration between universities and technology companies. Third, the framework embraces the concept of interdisciplinarity, whereby civil engineering cannot be separated from computer science, environmental science, data science, and artificial intelligence. Modern projects relying on digital twin models, 3D simulations, and smart sensors require a multidisciplinary approach that integrates engineering and technical expertise with statistical analysis. Therefore, successful future scientific research will cross disciplinary boundaries and build solutions from the interaction of knowledge, not just its accumulation.

Fourth, the theoretical framework is integrated with the concept of knowledge governance, which addresses how knowledge production can be directed to serve strategic objectives, achieve rational use of resources, and build a transparent system for monitoring scientific impact. In civil engineering, this governance is evident in how research findings are transformed into smart infrastructure policies, how legislation is enacted to stimulate the application of innovations in urban projects, and how technical and environmental risks are systematically assessed. Fifth, the framework highlights emerging technologies as an incubator for civil engineering development. Concepts such as artificial intelligence in structural design, 3D printing in construction, self-healing smart materials, responsive geotechnical systems, and lifecycle management techniques are all examples of how technology can radically transform civil engineering practice and redefine efficiency, quality, and sustainability in infrastructure projects.

This framework also draws on ongoing analysis of reports from international institutions such as UNESCO, the World Bank, and the World Economic Forum, which indicate that applied research related to infrastructure, climate, energy, and natural resources will be at the heart of the research landscape in the coming decades. These studies confirm that technology-based engineering solutions are key to addressing major challenges such as climate change, urban stress, and water and energy shortages, reinforcing the importance of integrating civil engineering with emerging research within flexible and applicable research frameworks. This framework not only provides a description of concepts, but also establishes an analytical lens used to interpret the relationship between scientific research, technological changes, and structural needs. It clearly demonstrates that betting on the future is a bet on a scientific mind capable of integration, on applied research capable of solutions, and on forward-looking methodologies capable of transcendence.

Table 1. Growth rates in emerging research areas

Research Area	Annual growth rate (%)
Artificial Intelligence	23.5
Renewable Energy	18.2
Nanotechnology	16.4
Big Data	20.1
Synthetic Biology	14.9

Table 2. Key challenges in supporting emerging research

The Challenge	Prevalence in studies (%)
Lack of Funding	72
Weak Infrastructure	65
Lack of Skills	60
Policy Gaps	58
Knowledge Gap	55

Table 3. Comparison between developed and developing countries in research support

Indicator	Developed countries	Developing countries
Research expenditure (% of GDP)	2.8	0.5
Innovation centers	3200.0	430.0
Number of patents	125000.0	7800.0
Number of researchers per million inhabitants	4500.0	740.0

Table 4. Most funded fields in the world (2024)

Research Area	Financing (billion dollars)
Artificial Intelligence	90
Renewable Energy	76
Digital Health	58
Robotics	45
Cybersecurity	52

Figure 1. Annual growth in emerging research fields

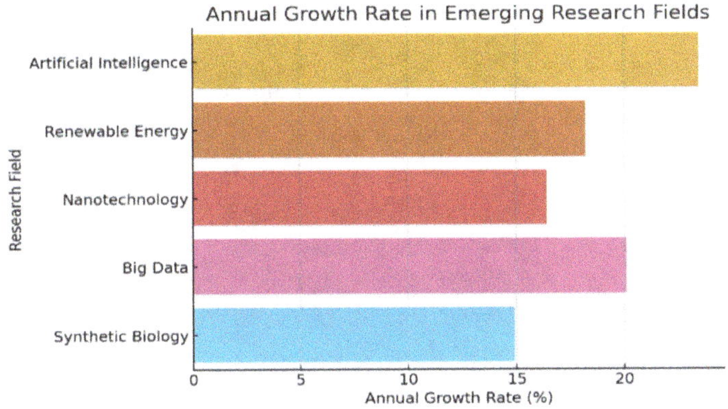

Figure 2. Most common challenges in supporting emerging research

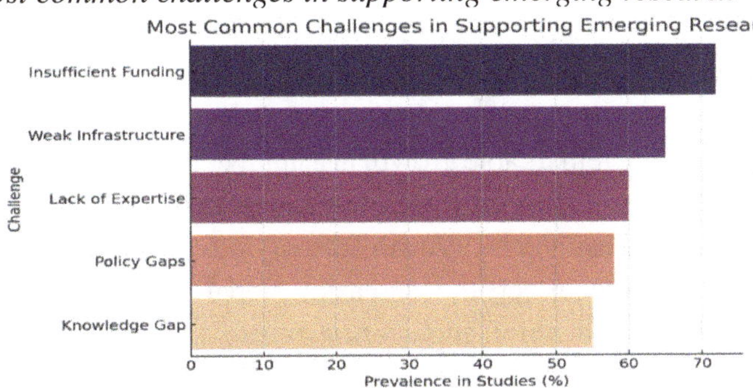

Figure 3. Comparison of research support

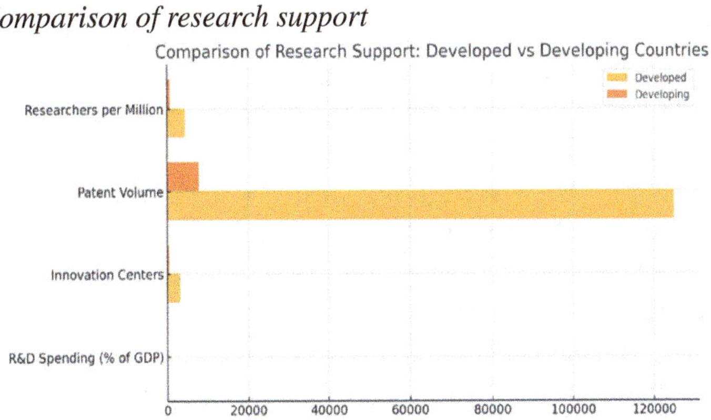

Figure 4. Top funded research areas worldwide (2024)

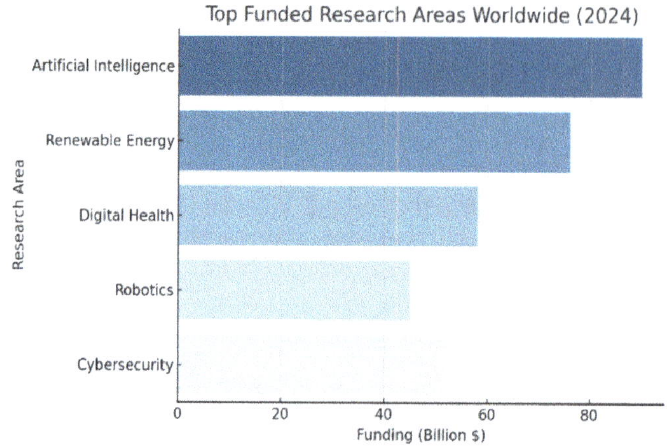

DISCUSSION

Figure 1 reflects the remarkable growth in a number of advanced research fields. Table 1: Growth Rates in Emerging Research Fields shows that artificial intelligence (AI) achieved the highest annual growth rate of 23.5%, outperforming all other disciplines. This increase reflects the increasing global reliance on AI technologies in various sectors, including civil engineering, where deep learning algorithms are used to analyse infrastructure data, predict material behaviour, and manage smart construction projects.

Big data comes in second with a growth rate of 20.1%, reflecting the growing importance of using large-scale computer models to process and analyse information. This is critical in urban planning projects, where data is used to estimate loads, population densities, and plan roads and vital facilities. Renewable energy ranked third, accounting for 18.2% of the total, is a vital field in the context of environmental and civil engineering. Sustainable building designs rely on integrating solar and wind energy sources into the building structure itself, which requires advanced research in engineering materials and storage and distribution systems.

Figure.2 and 3, the Most Prominent Challenges in Supporting Emerging Research, shows that the lack of funding is the most prominent challenge, according to 72% of the studies. This challenge is directly reflected in civil engineering research projects Table 2, as smart building technologies and the development of composite materials require high investments in equipment and laboratories. This is followed by weak infrastructure (65%), reflecting the absence of advanced laboratories or realistic testing environments, which are necessary to simulate earthquake conditions, thermal loads, or environmental corrosion.

The data also shows that the lack of expertise and skills (60%) represents an additional obstacle, especially in developing countries, where few engineers are able to use digital analysis tools or implement AI-based solutions in the design of structures and facilities. The third image, "Comparison of Research Support: Developed vs. Developing Countries," supported by Table 3: Comparison between Developed and Developing Countries, reveals a sharp disparity across all scientific research indicators. For example, the number of patents in developed countries is 125,000, compared to only 7,800 in developing countries. This indicator reflects the depth of innovation in industrialized societies, compared to the limited innovative output in resource-poor countries.

The difference in the number of researchers per million people is significant: 4,500 in developed countries compared to 740 in developing countries. This disparity affects countries' ability to build multidisciplinary teams for major projects, such as intelligent bridge control systems, traffic simulation, or 3D modelling of urban projects. The supported by Table 4 and fig.4 shows that artificial intelligence (AI) leads global funding with $90 billion, which is consistent with its high growth rate in the first table and confirms the growing reliance on it in building the future of digital infrastructure. Renewable energy follows with $76 billion in funding, reflecting the world's transition to design solutions that integrate environmentally friendly construction and technologies.

Digital health ($58 billion), robotics ($45 billion), and cybersecurity ($52 billion) also received significant funding. These fields directly intersect with modern civil engineering, particularly through the development of maintenance robots for high-rise buildings, automatic fault detection systems, and securing industrial control systems from cyberattacks. While the figures and tables illustrate clear disparities between developed and developing countries, a deeper analysis reveals underlying structural causes. The persistence of the skills gap, for instance, can be attributed not only to weak infrastructure but also to limited integration of future-oriented curricula in universities and the absence of incentives for interdisciplinary training. Similarly, the success of certain countries in overcoming funding challenges lies in their adoption of innovation-driven economic policies and the creation of strong links between academia, industry, and government. These examples suggest that disparities in research capacity are not merely resource-driven, but reflect broader policy choices, governance models, and institutional cultures.

CONCLUSION

Analyzing the images and tables shows that investment in emerging research is no longer a cognitive luxury, but a development priority. The integration of technology and civil engineering has become a prerequisite for achieving efficiency, flexibility, and sustainability in infrastructure. Furthermore, the gap between developed and developing countries in research support is clear and cannot be bridged without a comprehensive strategy that begins with research funding, continues with human capacity building, and finally, integrates knowledge into national decision-making.

REFERENCES

Addula, S. R., Mamodiya, U., Jiang, W., & Almaiah, M. A. (2025). Generative AI-enhanced intrusion detection framework for secure healthcare networks in MANETs. *SHIFRA*, *2025*, 62–68. DOI: 10.70470/SHIFRA/2025/003

Akinola, A., & Emuze, F. (2020). Future of construction: An overview of emerging technologies. *Journal of Engineering. Defence Technology*, *18*(4), 813–830. DOI: 10.1108/JEDT-11-2019-0317

Ali, A., & Ghanem, M. C. (2025). Beyond detection: Large language models and next-generation cybersecurity. *SHIFRA*, *2025*, 81–97. DOI: 10.70470/SHIFRA/2025/005

Bocken, N. M. P., Short, S. W., Rana, P., & Evans, S. (2014). A literature and practice review to develop sustainable business model archetypes. *Journal of Cleaner Production*, *65*, 42–56. DOI: 10.1016/j.jclepro.2013.11.039

Eastman, C., Teicholz, P., Sacks, R., & Liston, K. (2018). *BIM Handbook: A guide to building information modeling for owners, designers, engineers, contractors, and facility managers* (3rd ed.). Wiley.

European Commission. (2020). *A new ERA for research and innovation*. https://research-and-innovation.ec.europa.eu

Fagnant, D. J., & Kockelman, K. (2015). Preparing a nation for autonomous vehicles: Opportunities, barriers and policy recommendations. *Transportation Research Part A, Policy and Practice*, *77*, 167–181. DOI: 10.1016/j.tra.2015.04.003

Gao, X., & Pishdad-Bozorgi, P. (2019). BIM-enabled facilities operation and maintenance: A review. *Advanced Engineering Informatics*, *39*, 227–247. DOI: 10.1016/j.aei.2019.01.005

Ghosh, S., & Chatterjee, S. (2023). Artificial intelligence in civil engineering: A systematic literature review. *Automation in Construction*, *148*, 104753. DOI: 10.1016/j.autcon.2023.104753

Hameed, M. H., Shawkat, S. A., & Al-badri, K. S. L. (2020). Multi bands metamaterial absorber optimized by genetic algorithm in microwave regime. In AIP Conference Proceedings (Vol. 2213, No. 1, p. 020055). AIP Publishing LLC. DOI: 10.1063/5.0000075

Hammad, O. M., Smaoui, I., Fakhfakh, A., & Hashim, M. M. (2024). Recent advances in digital image masking techniques: Future challenges and trends—A review. *SHIFRA*, *2024*, 67–73. DOI: 10.70470/SHIFRA/2024/008

Ibrahim, A. A. A. (2024). Features and functional components in the use of artificial intelligence in human capital development (higher education as a model). *ESTID-AMAA, 2024*, 1–6. DOI: 10.70470/ESTIDAMAA/2024/001

International Energy Agency. (2021). *World Energy Outlook 2021*. https://www.iea.org/reports/world-energy-outlook-2021

Liu, R., Ding, L., & Love, P. E. D. (2017). BIM and big data for construction cost management: A systematic review and future research directions. *Automation in Construction*, *72*, 143–156. DOI: 10.1016/j.automatica.2016.10.011

Lu, Y., Xue, F., & Wang, M. (2020). Digital twin-enabled smart construction: Concepts, technologies and future directions. *Automation in Construction*, *112*, 103087. DOI: 10.1016/j.autcon.2020.103087

OECD. (2021). *Science, technology and innovation outlook 2021: Times of crisis and opportunity*. OECD Publishing., DOI: 10.1787/75f79015-

Qiao, Q., Zhang, Y., & Zhang, W. (2024). Integration of sustainable construction practices with AI-based systems: A critical review. *Sustainable Cities and Society*, *95*, 105628. DOI: 10.1016/j.scs.2024.105628

Salama, A. M., & Wiedmann, F. (2013). Demystifying the concepts of sustainable architecture. *Archnet-IJAR*, *7*(2), 94–112. DOI: 10.26687/archnet-ijar.v7i2.121

United Nations. (2015). *Transforming our world: The 2030 agenda for sustainable development*. https://sdgs.un.org/2030agenda

Zhang, X., Shen, L., & Zhang, L. (2022). Predictive analytics and future trends in smart city infrastructure: A review. *Journal of Urban Technology*, *29*(2), 85–102. DOI: 10.1080/10630732.2021.2003032

Chapter 9
Integration of UAVs and Robotics in Infrastructure Inspection and Maintenance

V. Dankan Gowda
https://orcid.org/0000-0003-0724-0333
B.M.S. Institute of Technology and Management, India.

Kottala Sri Yogi
https://orcid.org/0000-0002-3671-7420
Symbiosis institute of Business Management, Symbiosis International University, Pune, India

Ashwini S. Shivannavar
SSIT, SSAHE Tumakuru, Karnataka, India

B. C. Kavitha
https://orcid.org/0000-0001-5775-4901
BGS Institute of Technology, Adichunchanagiri Univeristy, Mandya, India

M. Benila
V.S.B. Engineering College, Karur, India

ABSTRACT

The use of the Unmanned Aerial Vehicles (UAVs) and robotics in infrastructure inspection and maintenance has transformed the old procedures of inspecting essential property. In this chapter, UAVs and robotics are discussed in terms of their potential to enhance efficiency, safety, and accuracy of infrastructure inspection. It

DOI: 10.4018/979-8-3373-5248-0.ch009

is possible to conduct precise aerial inspections of difficult to access locations, i.e. bridges, high-rise structures as well as industrial facilities with the help of UAVs that boast high specification sensors. In the same way, the systems in the form of robotic systems, including ground robotic systems and climbing robotic systems can perform the precise inspection in tight spaces, which has the great benefit in the process of maintenance. Such combination of UAVs and robotics enables shorter inspection period, better data gathering and cheaper operations. Nevertheless, such barriers as regulatory rules, data handling, and technical restrictions are still common.

INTRODUCTION

Modern societies operate hugely on infrastructures such as transportation means such as bridges, highways, energy supply and industrial establishment. These infrastructures are relevant to the development of the communities as well as their economic well-being based on their resilience, safety and efficiency (Ali, 2025). Hence, a practice of routine check up of the infrastructure and repair is required to discover possible problems before they blow up and become serious breakdowns. Nevertheless, conventional ways of inspecting and maintaining infrastructure are usually expensive, time-consuming, and unsafe. Such traditional methods include manual processes, are put into use through scaffoldings and ladders, or through big machines whose access is limited. Although these approaches are commonly in use, there are also limitations attached to these methods which include the nature of cost, accessibility and safety which are the powerful factors that determine the overall efficiency of the infrastructure management (Addula et al., 2025). The use of human intervention in conducting the inspections is another area that is of concern because of the dangers that may come along with working at heights, in small spaces or under any hazardous environmental conditions. In addition, due to deterioration of infrastructure and growing need of quicker and more precise assessments, the issues of these traditional techniques have become more apparent (A. A. A. Ibrahim, 2024). Against this background, the upcoming of Unmanned Aerial vehicles (UAVs) and robotics in infrastructure inspection and maintenance services has become a revolutionary stepping-stone in combating challenges in the sector. Designed to perform airborne inspection and minimizing the exposure of human life to hazardous environments, UAV, more popularly known as drones, have high-tech sensors and advanced imaging systems (Bartlett, 2023). They are also able to take reading photographs, videos and sensor data of important buildings which give an accurate and thorough evaluation within a shorter period. Certain areas that cannot be easily or safely reached by the human hand and feet, e.g. high-rise buildings and bridges, offshore structures can be reached by UAV, giving inspection to be done more efficiently and safer (B. Bartlett, 2024). Furthermore, UAVs also present the benefit of also being able to collect data in real-time and analyze it to determine any wear and tear on the item beforehand (by identifying cracks, corrosion, or deformities) so that it can be maintained in advance rather than automatically going through it and fix it, which is rather reactive (Du et al., 2024).

Robotics has also been taken advantage of in infrastructure inspection and maintenance along with UAVs. Robotic techniques, like ground-based robots and climbing robots can maneuver in the narrow or cluttered environments or along the vertical surface, so it can make every fine check in the places that it is difficult to navigate into with other techniques even with UAVs (Bruggemann, 2022). These are sometimes assigned special tools, sensors, and allowed to undertake risky tasks on behalf of their owners, this includes house cleaning, repair, and inspection in areas that would be otherwise dangerous to a person, e.g., tunnels, pipes, or underground mechanisms. Another advantage of incorporating robotics into the maintenance of infrastructure is the possibility of increasing automation with regard to the performing of routine tasks, which can either be carried out automatically or with not much attention to humans. UAVs and robotics have great potential to not rest with the benefit of making inspections more accurate and efficient than ever before as well as the cost of operations and the safety of people alike in the infrastructure before they have been merely combined (Lynch, 2024). Geotagging technology eliminates the possibility of accidents and injuries since it allows human beings to avoid high-risk situations, such as climbing or entering enclosed spaces. Moreover, those technologies provide this data gathering in a very precise and consistent way, compared with the conventional methods, which makes it possible to arrive at a better decision on matters concerning the management of infrastructure (Yang, 2025).

Besides these important benefits, the UAVs and robotics application in infrastructure inspection and maintenance also have a number of challenges to overcome. Regulations, like airspace limitation of UAVs and robotics safety, as well as the absence of common practices in various industries, are obstacles to embrace large-scale adoption (F. F. Nyboe, 2023). Current limitations of the existing UAVs and robotic systems can be also considered as an issue, such as battery life and payload, and accuracy of sensors. Moreover, the processing and analysis of the big data produced by such technologies may be complicated, and they need specific software and experience to decipher the data in the proposed way (Paul et al., 2023). Finally, the opportunities of UAVs and robots used to inspect and maintain infrastructure are an important step forward compared to the existing practice. The technologies are more efficient, safe, and good in gathering data, and as such are important in infrastructure management field. Nevertheless, the implementation of such technologies will presuppose the solution of technical, regulatory, and operational issues. On-going innovations around the UAV and robotic technology and avenues to develop common protocols and best-practices will go a long way in eliminating these challenges to harness the overall potential of these technologies in the resiliency and long-term sustainability of the critical infrastructure.

TECHNOLOGICAL OVERVIEW

The Unmanned Aerial Vehicles (UAVs) and robotics applications have resulted in improved developments in the infrastructure inspection and maintenance. These two technologies are created to overcome the drawbacks of the traditional technologies, including their expensive nature and safety issues, as well as the problem of accessibility (Car et al., 2020). These sophisticated instruments have now got into high-tech sensors and imaging technologies that help in increasing their applications and in becoming very efficient in inspection and maintaining structures. This part deals with the background about the UAVs, robot or any technology by the description of the features and range of functions, technologies that have been developed and upgraded, sensors and technologies that have been incorporated by these systems to enhance their performance. UAVs started out as a plain flying object, being simple machines. However, with time they have become specialized in precision and flexibility. Some UAVs can be applied in interior inspection where small quadcopters to bigger fixed-type drones would have different tasks (Perez-Jimenez et al., 2020). The most widespread of its kind is the multirotor UAV, highly agile and able to hold stationary in the air, and thus excellent at performing tedious inspection duties in an environment with limited or convoluted topography like bridges, towers, and building facades. The uses of fixed-wing UAVs more clearly lie in large-scale inspections e.g. the monitoring of a long highway or large energy infrastructure because they can travel long distances without requiring any landing due to their flying capabilities.

The most recent technology led to hybrid UAVs, containing the features of both multirotors and fixed-wing UAVs. This gives flexibility of vertical takeoff and landing (VTOL) capability and the capability of long-distance flight and such kind of UAVs suits any infrastructure projects where there is detail inspection required together with large area coverage (Gil-Castilla et al., 2025). The hybrid UAVs usually possess enlarged payloads, which enables them to carry heavier sensors as well as imaging system hence makes them collect more data. Integration of hi-tech sensors and imaging technologies has also facilitated a boost in capabilities of UAVs. In close connection with the RGB (Red, Green and Blue) camera, this type of UAV is also equipped with the LiDAR (Light Detection and Ranging) system, enabling the mapping of infrastructure of a high accuracy. LiDAR is based on laser beams that create an accurate, three-dimensional surface model that assists in identifying structural aberrations including defects, deformations, and surface wear (Chand et al., 2021). The other type of sensor that is implemented on the UAVs is the thermal camera that can detect the differences in the temperature of different parts of the infrastructure and help determine the key areas where the heat is lost, the moisture is accumulated, or the electrical problems can occur. UAVS also have the capability of installing ultrasonic sensors to identify fissures and wall instability mainly in concrete and metal walls where visual examination may not be able to indicate the existence of a problem.

The field of robotics is also very important in regard to the inspection and maintenance of infrastructure especially those tasks which the UAVs are not designed to take up like repair or access to confined areas (Ramon-Soria et al., 2019). There are various kinds of robots that have been invented to perform particular functions in the infrastructure systems. The most prevalent is the use of ground robots in checking of the infrastructure. These robots are traditionally fitted with wheels or tracks, which give them mobility in any garages and on road, railway, floors areas. Ground robots have a potential to carry a range of tools and sensors and thus they have the ability to carry out inspections, cleaning of surfaces, or repairs at areas that human workers cannot access due to their dangerousness or challenges to reach. Robotic arm is another crucial form of robots that are applied to maintain infrastructure (Montambault, 2010). These are very adaptable and accurate robots that are created to do structural repair, maintenance as well as clean-up of areas that are inaccessible. Robotic arms have various tools, grippers, drills, welders to work in order to tighten bolts, clean debris, or fix the parks of a building that are spoiled. Installing robotic arms as a part of infrastructure systems can contribute greatly to the efficiency of processes due to repetitive routine procedures automation and the human laborers no longer having to perform dangerous jobs. Another type of robotics is the climbing robots that are specialized to reach the vertical or inclined surfaces including bridges, towers and wind turbines (Sadik Hassan Hussein, 2024). The robots are able to scale the walls and other vertical surfaces by using mechanism such as magnets, suction cups or even leg types devices. This capability of scaling-up leads to surface usually only subjected to inspection and maintenance by means of a scaffold or a ladder, freeing up time, money, labor and risk.

Combination of hi-tech sensors and imaging with UAVs and robots enables them to take accurate and effective inspections. LiDAR is very important to both ground robot and UAVs in that they would produce such high accuracy 3D models of infrastructure and spot minute differences in the structure that could identify the necessity of a repair (Wu, 2008). A thermal camera is also an essential resource when it comes to the recognition of heat loss, moisture, or electrical faults; especially on the exterior of buildings, their roofing, and in industrial facilities. The thermal sensors are sensitive to changes of temperature not visible to the human eye and provide priceless information in the early sign of problems, which otherwise might have been unknown. Both UAVs and robotics makes extensive use of ultrasonic sensors to inspect the material, mostly in locating cracks, corrosion or other structural degradation of metals, concrete and other building materials (Hament, 2018). These sensors project high frequency sound waves and can detect potential defect in a product even when they are buried inside the product by detecting the reflection of waves on a surface. There is also a rise in the use of radar sensors especially where there is a difficult environment where visual sensors may not perform so well. Radar systems have been shown to be able to overcome rain, fog and dust and hence UAVs and even robots can carry out inspections in areas that humans might find it hard or even dangerous to conduct such checks.

Lastly, UAVs and robots are being equipped with AI and machine-learning algorithms to enhance the quality of data analysis with their help. These technologies are compatible with auto-processing of the data collected in large amounts during inspection, outlining the patterns, anomalies, and possible areas of concern that need to be looked over further (Patsiouras, 2022). The AI has the potential to deliver further by empowering UAVs and robots to learn the inspections and relatively become accurate and efficient. With this, the ability of structure inspection and repair has been greatly boosted due to the technological innovation that has been witnessed in the UAVs, robotics and the various senses that go hand in hand with them. The use of UAVs and robots is currently powered with a combination of various sensors such as LiDAR, thermal cameras, ultrasonic sensor, and radar that can ensure that an inspection can be done with a very high level of accuracy in any conceivable environment. The technologies are not only making operation of inspections much more efficient and precise, they are also making them safer as they allow limiting the activities that have to be done by human workers, which are often dangerous (Petit, 2024). As the technologies keep developing, they will be introduced into infrastructure systems even more critically towards ensuring sustainability and stability of infrastructure structures.

APPLICATIONS IN INFRASTRUCTURE

UAVs and robotics, used in the infrastructure field, have increased the possibilities of engineers and inspectors to find a more effective and safer method of monitoring and stabilizing the declared infrastructure. The benefits of these technologies are vast, among which there can be mentioned speeding up the data collection process, increasing its leanness, lower the cost of hands, and the possibility of using technologies as a measure of increased safety since now humans do not have to perform the high-risk roles. The next sections describe the application of UAVs and robotics in the inspection and monitoring of the different types of infrastructures, such as bridges, tunnels, highways, buildings, towers, and industrial plants, and in gathering data about the hazardous or the hard to reach environments.

Inspection of Bridges, Tunnels, and Highways

Bridge, tunnel and highway inspection are some of the most important uses of UAVs and robotics. These are the parts of infrastructure that must be maintained as they are the parts of transporting system and need to be maintained on the regular basis to keep them safe and durable. Most of the conventional inspection techniques associated with bridges involve visual inspection by people who have to physically go to difficult parts of the bridge through climbing or the use of scaffolders (Liu, 2023). This is not only time consuming but also risky. This can be more effectively replaced by the use of UAVs with high-resolution cameras and LiDAR equipment, making it safe and efficient. The UAVs will be able to fly near to the bridge surface, and provide close in mileage images and 3D models so that it can be analyzed to see any evidence of structural damage by cracking, corrosion or wear down. It limits the possibility of scaffolding or heavy equipments, which makes it much safer and cost-efficient. In the case of tunnels, robot systems are becoming common to inspect confined areas that can only be accessed through narrow passages, which are hard to reach or risky through human inspection. Robotic systems installed with ultrasonic sensors and LiDAR are sent to crawl beneath floors and walls on tunnel surfaces and gather valuable information on the surfaces and record any signs of crack or water entry. The systems have the ability of working fully independently, which limits the dangers of entering confined spaces or hazardous areas by people. When it is related to highway infrastructures, UAVs may be deployed to scan the road surfaces looking at possible potholes, cracks, etc., and examine the situation on the highway or other safety-related items, which will help to keep the transport system intact.

Monitoring of Buildings, Towers, and Industrial Plants

The UAVs as well as robots have been found to be of special use in tracking down their buildings, towers and industrial plants where there are high cases of insecurity and inaccessibility. Other tall buildings like the skyscrapers, communication towers, and wind turbine are considerably difficult by the conventional methods of inspection because they involve the use of workers who are required to scale heights or employing the use of a crane to conduct the inspection. Multirotor drones and other UAVs can easily reach these tall objects and be able to deliver high quality imagery, thermal imaging, and 3D mapping and require no scaffolding or aerial lifts. These

UAVs are able to obtain close details of the exterior and the structure elements, and highlight the problems; such as cracks, corrosions, or other types of damages and such, and this is done without putting the workers into danger. Safety and efficiency in terms of operation in industrial plants are of the essence, including oil refineries, power plants, and factories. There are robotics systems that are implemented to enter in to routine inspections, maintenance services and even simple repairs. Example: robotic arms perform cleaning, inspection and repair work on heat exchangers of a plant, or pressure vessels, which would otherwise need a human operator to go into confined and dangerous spaces. There is also a possibility of robots with gas detectors and thermal cameras detecting gas leak, overheated equipment, or other dangerous situations, which is safer and causes less downtime. The applications assist in ensuring that industrial plants are operated in a safe manner, therefore, reducing the chances of occurrence of accidents and prolonging the service cycle of these equipments.

Use of UAVs and Robotics for Data Collection in Hazardous or Hard-to-Reach Areas

Data collection of places that are dangerous, inaccessible or places that are not suitable to human inspectors is also very effective with the help of UAVs and robotics. These are offshore offshore, distant, platforms structures, power lines and nuclear plants with conditions that are usually quite harsh and chances of being injured are high. UAVs with thermal cameras, LiDAR, and multispectral sensors can deliver information about the state of infrastructure in such surroundings in real-time to avoid maintenance issues in the future without human labor working in hazardous zones. As an example in offshore oil rigs, the UAVs are applied to check the integrity of platforms, locate leakages and also checking the status of the most important equipment. Conditions under which such platforms are subjected to harsh weather conditions usually occur, and using the conventional inspection approach is both expensive and dangerous. UAVs will provide a safer solution where they can have an inspection done swiftly and in detail without exposure of the human life at stake. In the same manner, the power lines that are miles long in some inaccessible lands can be tracked on via UAVs to check any worn parts or any corrosion or breakages thus averting more failures that may be triggered off before connections are cut widely. And robots are also used in risky places, in the situation of nuclear plants or mines where the humans cannot go and work in risky conditions of radiation or poisonous exposure. The inspections and maintenance of cooling towers in power plants is carried out by climbing robots with thermal sensors and ultrasonic testing equipment, the same is used to monitor the condition of the underground mining equipment. The ability of such robots to conduct inspection despite being independent and transferring the information to the main control allows the problems to be identified earlier than they would not otherwise have been noticed.

Case Studies and Examples of Infrastructure Inspection Projects that Utilized UAVs and Robotics

A number of case studies show the effectiveness of UAV and robotics in conducting tools in the inspection of infrastructure. The inspection of the condition of a Golden Gate Bridge in California is one such example as UAVs with the LiDAR and high-definition cameras were used to inspect the condition of that bridge. The UAVs could capture detailed 3D models of the bridge that had structural problems that could have been hard to note by use of traditional methods. This eye check up was done way faster and at an insufficient cost when pitted against conventional manual practices. During another incident, industrial boilers in a coal-fired power plant have been inspected and cleaned using robotic arms. Robots monitored the status of the panel through the help of ultrasonic sensors and thermal cameras to determine the status of the boilers which had been subjected to high temperatures and corrosive gases. Due to the fact that these robotic systems reduced the amount of downtime and overall safety due to the regular ability to perform inspections on the boilers without the need of a human allowing entry into the boilers. Additionally, monitoring of structural integrity of wind turbines in Europe has also been conducted by the use of UAVs. Such UAVs have high-resolution cameras and thermal imaging systems and can be used to carry out a rapid and accurate inspection to the turbine blades and other vital parts. Automating these inspections, operators detect damage at early stages before it causes expensive downtimes, and with new repairs. To sum up, there are significant advantages in the safety, efficiency, and accuracy of integrating UAVs and robotics to perform infrastructure inspection and maintenance. The use of these technologies to inspect and monitor bridges, tunnels, highways, buildings, towers, and industrial plants is on the increase and delivers real-time information about the health of the infrastructure in a dangerous or inaccessible location. The possibilities of UAVs and robotics as a means to revolutionize infrastructure maintenance are also evidenced by case studies and actual use success stories that would promise the sustainability and resilience of critical infrastructure in the long run.

ADVANTAGES OF UAVS AND ROBOTICS IN INSPECTION AND MAINTENANCE

The combination of the UAV and robotics technology in the process of inspecting and assessing the condition of a structure and providing maintenance infrastructures has added a lot of value to the whole process. They have several benefits compared to conventional techniques of inspection such as that they are safe, accurate and efficient, cost effective and accessibility in hard to reach places. The fact that they can remotely manage inspections and do the maintenance work autonomously has not only changed the nature of infrastructure management but has also helped them make the most of the existing resources and create a long-term sustainability of the existing infrastructure.

Improved Safety

The safety is one of the first benefits of conducting an infrastructure inspection with UAVs and robotics, since there will be a significant increase in safety levels. Conventional inspection techniques usually depend on human employees entering hazardous locations physically, like climbing tall structures, operating at altitude, or down tiny spaces, including tunnel system or holds. Such activities subject the employees to high risk of falling, exposure to dangerous chemicals of accidents with heavy machinery. The high-risk actions can also be done with the help of UAVS and robotics by not placing a human at risk. By the use of a camera and sensors, UAVs can also fly structures like bridges, power lines, or tall buildings and take detailed pictures/videos about the state of the assets, without human interaction needed in risky areas. Robotics, especially climber robots/ ground robots, can tunnel into narrow openings, or even into dangerous terrain, such as underground systems or offshore platforms, where otherwise human workers would not be able to attend to. Such robots have the possibility to install sensors that analyze the possible concerns such as cracks, corrosion or leakages of chemicals, which minimizes the risk of exposure to human inspectors to a great extent. As the UAVs and robots are remotely controlled, workers do not have to work in dangerous conditions; hence, there are fewer accidents, injuries, and deaths; therefore, the safety levels within infrastructure management will be increased as well.

Increased Accuracy and Efficiency

Infrastructure inspection Vehicle automation comes with an impressive rise in precision and productivity. The conventional approaches are based on either human visual inspection or simple inspection devices that may be inaccurate either because of the subjectivity of human judgment when looking at objects visually or due to the shortcomings of human senses that are not sensitive to some flaws that are quite minimal. Most UAVs and robotics are also furnished with the latest sensors and imaging systems (LiDAR, thermal cameras and ultrasonic sensors), which can deliver accurate and detailed information consistently at a level that has never been attained by any human evaluation. Such as, the LiDAR-based technology would enable the design of 3D models of infrastructure and, therefore, it would make it easier to detect even the slightest deformations or cracks in the structure. Warm-up scanners are able to detect temperature changes which signal a problem like water leak or electrical problems or even an energy leak which would not have been easily identified. In cases where cracks or rotting in the metals and concrete are not noticeable in the surface, an ultrasonic sensor will detect such flaws, and these cannot be detected in the traditional inspection method. The obtained information by UAVs and robots can be processed in real-time and it will be possible to determine a potential problem immediately, which is why the process of inspection will take less time and inspection can be carried out more often. This degree of elaboration and accuracy facilitates the coming up with timely repairs interventions, something that will improve on efficiency in management of the infrastructure. Another benefit of UAVs and robotics is its capability to carry out live surveillance. The technologies allow constant or frequent checks without the need to have human labour on field, and give constant information of the status of the infrastructure. Real-time data collected makes it possible to make fast decisions since the engineers and the maintenance teams have access to current information regarding the status of infrastructure and can respond fast and make decisions.

Cost-Effectiveness

The other major benefit of applying UAVs and robotics is that they are cost-effective as opposed to the conventional infrastructure inspection and maintenance techniques. Conventional means sometimes involving manual work, scaffoldings, crane and specialized machines may prove to be costly since it is subject to costs of manpower, rent of equipment and idles during the checks. To give one example, inspection of tall buildings, wind turbines or communication towers not only needs a costly apparatus like aerial lifts or scaffolding but trained staff to control those machines. UAVs and robots, in their turn, are a less costly solution. The UAV technology is able to perform aerial inspections of their buildings, bridges or pipelines far faster than it took the human eye to set up scaffolding or aerial lifts to do inspection work. UAVs are also usually cheap in terms of deployment compared to the amount spent on doing the inspection using the traditional inspection equipment. Likewise, robotics systems can be used in dangerous or restricted areas without involving human staff to get into hostile working areas, thereby avoiding risks and expenses. Also, the possibility of carrying out regular inspection using both UAVs and robots implies that problems may be detected before they turn into big unpleasant scenarios and they can be later fixed or substituted within the framework of repair or replacement. The lack of specialized staff and large training programs also help to reduce the cost since it will be achieved using UAVs and robots which do not require extensive training programs or specific staff. Certain parts of the task included in the inspection process, e.g., visual evaluation or data gathering can also be automated, and it is possible to devise a more efficient process that saves on manual work. In the long term, use of UAVs and robotics in to perform maintenance of infrastructure can yield dramatic savings in labour costs, equipment costs, and increased downtime, so using them becomes a more economical way of doing things.

Ability to Access Hard-to-Reach Areas

The presence of areas that are hard to reach and may be out of access by human inspectors is perhaps one of the most convincing cases to adopt UAVs and robotics in infrastructure inspection and maintenance due to the ease by which the inaccessible areas may be reached using UAVs and robotics. The technologies can be used to access high or otherwise inaccessible regions including the roofs of tall buildings, bridges, communication towers, or the oil and gas platforms beyond the shores. High-rise buildings or tall buildings are better to be inspected with the help of UAVs and in particular multirotor drones, because they are capable to hover and maneuver around obstacles easily and produce detailed images of the high-rise building without any scaffolding or lifts. Other than tall buildings, UAVs and robots are able to access places where humans cannot reach because of inaccessibility in deeper water, within tunnels, or remote location. As an example, remotely operated vehicles (ROVs) can conduct inspection of underwater pipelines, oil rigs or submerged infrastructure where human access by divers is dangerous or hard-to-reach in underwater conditions. Likewise, to reach atop the wind turbine or bridges, a climbing robot would be used where manpower would be unacceptably costly by attaching suction cups or magnets to the robot. These are robots that are able to traverse on vertical surfaces with details containing information on the hard to reach inspections using UAVs only.

It is especially beneficial because it can access hard or hazardous locations without putting workers in dangerous situations, the setup does not have to be complicated, human exposure to risky environments is decreased, and the process of inspection could be much faster. This will enable the UAV and robotics technologies to be very useful in the management of modern infrastructure, as even the most difficult and dangerous locations are checked and checked occasionally. To summarize, UAVs, and robotics are both significant when it comes to inspections and maintenance of infrastructure. The technologies enhance the safety, accuracy, efficiency, and cost-effectiveness of performance so that they are suitable in the implementation of various infrastructure applications. Their capability to inspect hard-to-reach and high-risk places, coupled with the sophisticated data gathering powers, allows more regular and comprehensive inspections, which eventually leads to a better long-term resilience and sustainability of the critical infrastructures.

INTEGRATION APPROACHES

The use of UAVs and robotics in combination with the existing process of maintenance repairs of infrastructure has introduced enormous improvement in terms of content, safety, and accuracy. These technologies are still advancing and are being adapted into workflow systems to make the maintenance of infrastructure more automatic, specific and economical. The part examines the topic of the integration of UAVs and robotics into the existing system of infrastructure maintenance, the impact of autonomous systems and human-robot interaction, data management systems integration with UAVs and robotics, including cloud computing, AI-related analysis, and predictive maintenance.

Workflow Integration of UAVs and Robotics with Existing Infrastructure Maintenance Processes

A continuous merger of UAVs and robotics into the current perseveration of infrastructure purposes is important in fully optimizing the performance of the systems. The existing tradition in the maintenance of the infrastructure is a manual and labor-intensive undertaking often implying serious human labor, complicated arrangements and having a long-term downtime of facilities to be examined and fixed. The incorporation of UAVs and robots into these processes has reduced most of the efforts in terms of data collection, on-going analysis, and automation of repetitive tasks. UAVs and robotics are fitted into the working process mainly due to the replacement or supplement of the traditional inspections and repair practices. Considering bridge inspections, high-resolution camera and LiDAR sensors mounted to drones can help provide the live picture of the surface of the structure, whereas the ground robots may be used in inspecting those spots which are inaccessible to UAVs, like under the bridge or narrow spaces. The technologies brings quicker and detailed inspection, which will lessen the need of manual data collection, hand and will take less time to be climbed or scaffold on the building by the human inspector. Data UAVs and robots capture are readily available to analyze, allowing a more expeditious decision-making process and shortening the gap between the time of inspection and maintenance work. The integration of these technologies in current modes of work also enhances the rate of the inspections. To give but one example, UAVs may be stationed to do periodic aerial checks on large and expensive infrastructure, be it highways, power lines or industrial facilities, on a regular basis. Such continuous monitoring will permit the timely detection of problems and avoid huge damages or expensive repairs that could be caused by late monitoring. Incorporation of these tools in routine maintenance programs also facilitated the increased level of inspection of the infrastructure systems resulting in improved planning in maintenance of the system and life cycle of assets.

Figure 1. Integration approaches in infrastructure maintenance

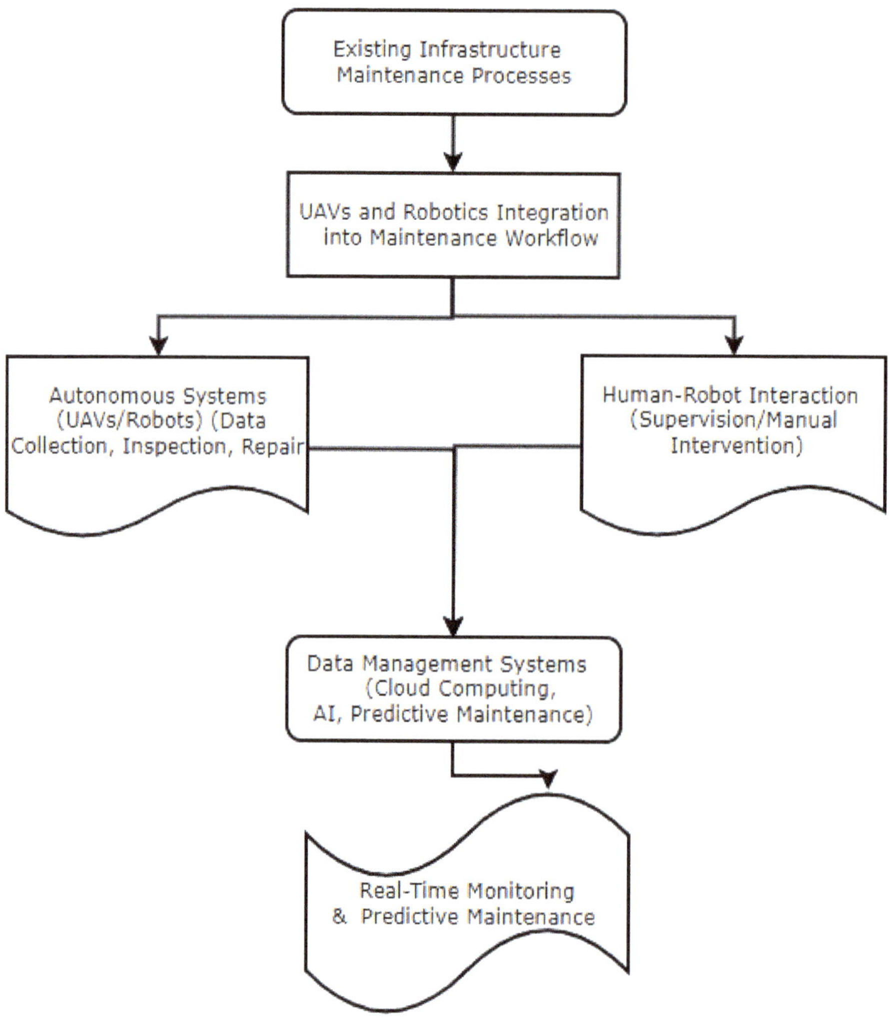

Figure 1 illustrates the, Integration Approaches in Infrastructure Maintenance, depicts how it is contemplated to integrate UAVs and robotics into the current maintenance processes of infrastructure. The starting point of the diagram is traditional maintenance methods of infrastructure which, more frequently than not, require a lot of human interaction and work force. They become integrated with UAVs and robots, which allows making the workflow more efficient as new processes (data collection, inspections, minimizing repair works, etc.) are automated. The major work of such technologies is carried out by means of autonomous systems, when a specific task can be performed by UAVs and robots with limited human supervision during inspection and repair. The human workers can be involved in control of operations, interpretation of data, and intervention in the case of necessity until a decision is taken or a complicated task is fulfilled despite its autonomous capabilities. UAVs and robots capture data which gets incorporated into sophisticated data management platforms such as cloud computing and information analysis inspired by artificial intelligence, which allows effective storage, processing, and real-time access of the retrieved information. Such data management infrastructure is a key component in predictive maintenance, since it enables any early signs of a possible mishap, to be realized before further complications can occur. Consequently, the introduction of UAVs and robotics will allow increasing the efficiency of maintenance operations and minimizing downtime and resource distribution. The ultimate outcome of this integration is that there will be better real-time monitoring and predictive maintenance which will guarantee that the infrastructure systems take even a longer time before failing and this is done by cutting down the chances of failure and a decrease in the total maintenance costs.

Autonomous Systems and Human-Robot Interaction

Autonomous systems are increasingly being used in infrastructure inspection and maintenance with drones and robots being able to complete increasingly complicated tasks with less human supervision. The autonomous systems are designed to work on their own, therefore there is less need of human involvement, thus making the processes of maintaining the infrastructure much more efficient. These systems can conduct various activities with no continuous oversight including data collection and investigation, minor repairs, and maintenance activities. In other words, an autonomous UAV can be used to make the regular inspection of, say, a communication tower, capture high-resolution images and data, and transmit such data to a central system where it is analyzed. In the same manner, one can utilize autonomous robotic arms in industrial manufacturing facilities to implement routine maintenance activities that would otherwise require human labor, such as tightening bolt, cleaning machines, etc. Nevertheless, when it is necessary to engage in decision-making, judgment, or a complex manual operation, the interaction between man and a robot is a necessary element despite the significant role of autonomous systems. Workers of human origin more often will probably be engaged in the management of the functioning of autonomous systems and the interpretation of data obtained due to UAVs and robots and the making of critical decisions guided by them. A supervisory role of human-robot collaboration is common too because humans are in charge and may consider the performance of autonomous systems, fix them in some cases, and give commands when the succeeding task is more precise or adaptable. Practically, such integration leads to a hybrid working process in which the autonomy systems take up routine or dangerous tasks, and a person oversees the process, interprets the information and makes strategic management decisions. As an example, when tunnels or high-rise structures are to be inspected, the robots might be able to autonomously navigate narrow spaces, scan concrete surfaces and areas of structural damage, and transmit them back but it is human labor to interpret the data, prioritize the work to be carried out, and carry out the repair work that is too difficult to do.

Integration with Data Management Systems (Cloud Computing, AI-Driven Analysis, Predictive Maintenance)

nother essential part of the process of increasing the efficiency of infrastructure maintenance is the recreation of UAVs and other types of robotics with computer systems of data management. These technologies collect large data volumes i.e., pictures, sensors signals, 3D maps, and other performance parameters and processing, analyzing, and storing these amounts of data needs optimization to access them in future. By using cloud computing and AI-based processing, the UAV and robotic data can be easily integrated into central systems, which in turn can analyze the data in real-time, archive it to develop long-term monitoring, and generate predictive maintenance policies. Integration of the system of managing data with UAVs and robotics focuses on cloud computing. Since the data UAVs and robots gather are stored in the cloud, infrastructure managers may virtually use the data, work and communicate with other teams and make decisions relying on current information. It is also possible to use cloud-based systems that will easily enable scalable data storage such that when significant sizes of data are produced by UAVs and robots, data can be organized, kept and retrieved as well. The cloud platforms can also be connected to the rest of enterprise systems, like maintenance management software, thus simplifying the process of matching UAV and robot data to a larger-scale infrastructure management process.

AI coupled analysis also increases the worth of data obtained by UAVs and robotics. Auto analysis of images, videos and sensor data can be analyzed to identify patterns, anomalies and predict potential problems with machine learning methodologies, and computer vision techniques. To illustrate, an AI can be used to detect contributions on the surface of the infrastructure or on corrosion of the infrastructure, detect anomalous variations in temperature that might indicate failure in the equipment. Such AI algorithms can handle huge volumes of data in a much shorter period and with increased accuracy as compared to human labor thus allowing more efficient inspections and timely detection of possible maintenance problems. Alongside enhancing the data analysis process, AI-driven systems can also help predictive maintenance, predicting failures through forecasting trends or variations in the infrastructure performance that signify a greater risk of failure. To give an example, in case a UAV finds that a bridge has a minor crack, AI models can interpret past data to determine the possibility of it becoming worse or whole bridge becoming at risk of collapsing and in this case would be able to take preventive action before the situation escalates. The predictive maintenance promotes unnecessary reduction of downtime, maintenance cost and increases the projected life of infrastructure assets by minimizing or eliminating the risk of premature deterioration of the infrastructure asset. Moreover, AI analysis and predictive maintenance can also be used in the most efficient arrangement of inspection and repairs to be carried out so that, resources are now more efficiently directed, and infrastructure systems are not maintained according to some random time schedule but according to real system performance data. This results in improved resource utilisation, enhanced planning and less unimportant maintenance.

The incorporation of UAVs and robotics in the maintenance arrangement of infrastructure is a massive cutting edge in the progress of overseeing, examining, and managing infrastructures. Integration of autonomous systems, data management and interaction between human beings and robots enables maintenance that is efficient, accurate and cost-effective. With cloud computing and analysis based on AI, it is possible to process enormous amounts of data provided by UAVs and robots, and analyze it to carry out predictive maintenance and ensure the correct maintenance of infrastructure objects in a timely and efficient manner. With further developments in technology, the connection of UAVs, robotics with the current infrastructure management systems will be enhanced even further, allowing yet more automation, safety, and a long-term sustainability.

RESULTS AND DISCUSSIONS

The inclusion of UAVs and robotics in the process of infrastructure inspection has brought considerable upgrading of multiple essential parameters such as time savings, the accuracy of results, and reduction of costs. The effectiveness of these technologies can be well illustrated with case studies and pilot programs that show the feasibility of using these technologies in the comparison to traditional methods. These developments have rendered inspection and maintenance of infrastructure more effective, safer and cheaper. Figure 2 is an illustration of the large amount of time that has been saved using UAVs and robotics. Based on the results of case studies carried out on bridge inspection, highway inspections and building inspections, it is evident that UAVs can cut down inspection periods by as much as 50% with robotics cutting down inspection periods by about the same percentage in relation to the traditional methods that still involves human beings having to crawl or use scaffolding in order to carry out inspections. To take an example, the process involved in an ordinary bridge inspection may be up to 120 minutes, and UAVs would cut the time to approximately 60 minutes, and a robotic system can be applied to accomplish an inspection within approximately 90 minutes. Inspections conducted in shorter periods not only end up boosting the productivity, but they also enable more frequent inspections to come into effect and hence a better perception of the infrastructure conditions and early maintenance interventions.

Figure 2. Time savings comparison between UAVs, robotics, and traditional inspection methods

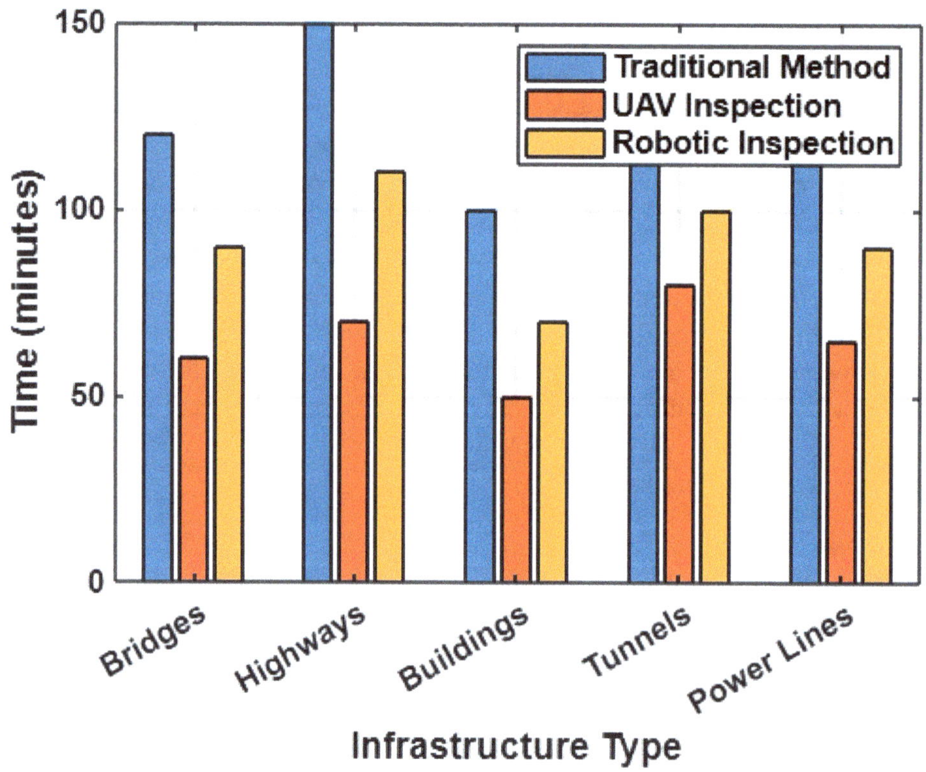

Figure 3 presents the comparisons of accuracy (the percentage of correctly identified cracks, corrosion, or any other defects) between the ideas of traditional methods, UAVs, and robotics to search cracks, corrosion or any other damage in infrastructure. The statistics prove considerable enhancements of crack, corrosion, and wear detection with the help of UAVs and robots. When the UAV is equipped with the improved LiDAR and thermal cameras, structural problems can be identified by the equipment at a rate of around 90 percent, as opposed to the approximately 55 percent rate that the traditional models have been catching. Robotics also act as a good alternative with sensors such as ultrasonic testing, 3D imaging with 85 percent of structural problems uncovered. This increased precision in detecting the faults is very important in avoiding further degradation of infrastructure and also making sure that there is no wastage of resources in the maintenance process where it would not be required.

Figure 3. Accuracy improvements using UAVs and robotics in infrastructure inspections

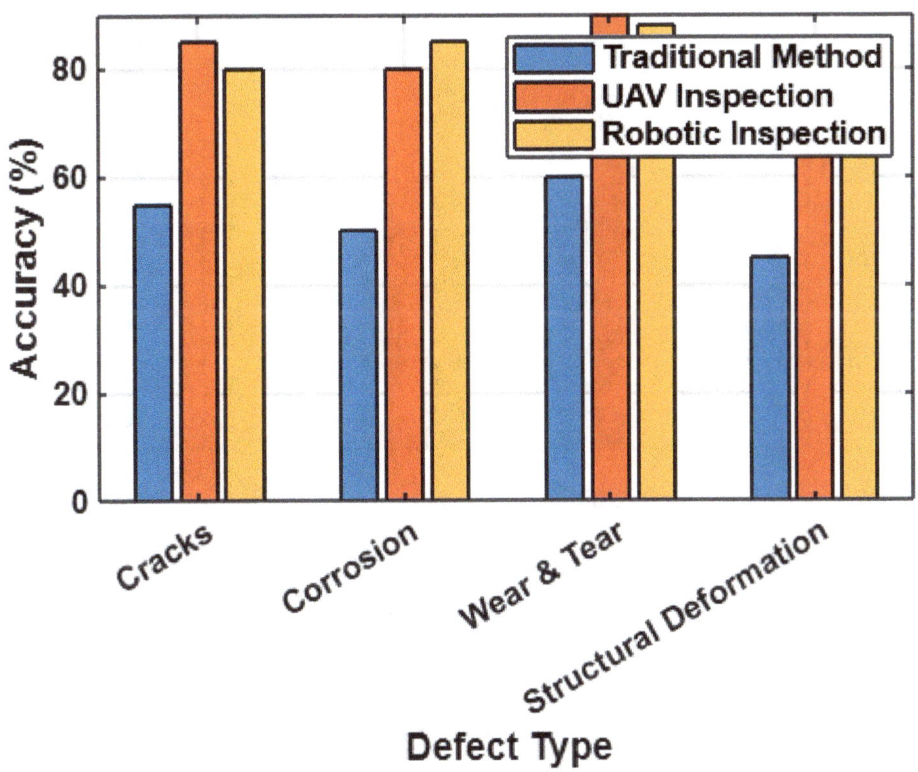

As Figure 4 provides, the cost savings represented by the use of UAVs and robotics during the inspection and maintenance process are gained because these processes are much easier. An example is that conventional inspection of a highway or an industrial plant may cost above 5,000 dollars; this is taking into consideration the labor, the equipment used and the downtime. Comparatively, inspection with UAVs lowers the expenses to nearly 2,000 dollars whilst the robotic checks are normally valued at 3,000 dollars. The main saving will be the savings on the costs of labor and equipment rent as UAVs and robots do not require large crews and such expensive scaffolding and equipment, but they can inspect the site independently. Moreover, the downtimes and disturbances in operations are reduced to minimum due to the use of UAVs and robots and costs of maintenance are further decreased.

Figure 4. Cost reduction comparison in infrastructure maintenance (traditional vs. UAV & robotics)

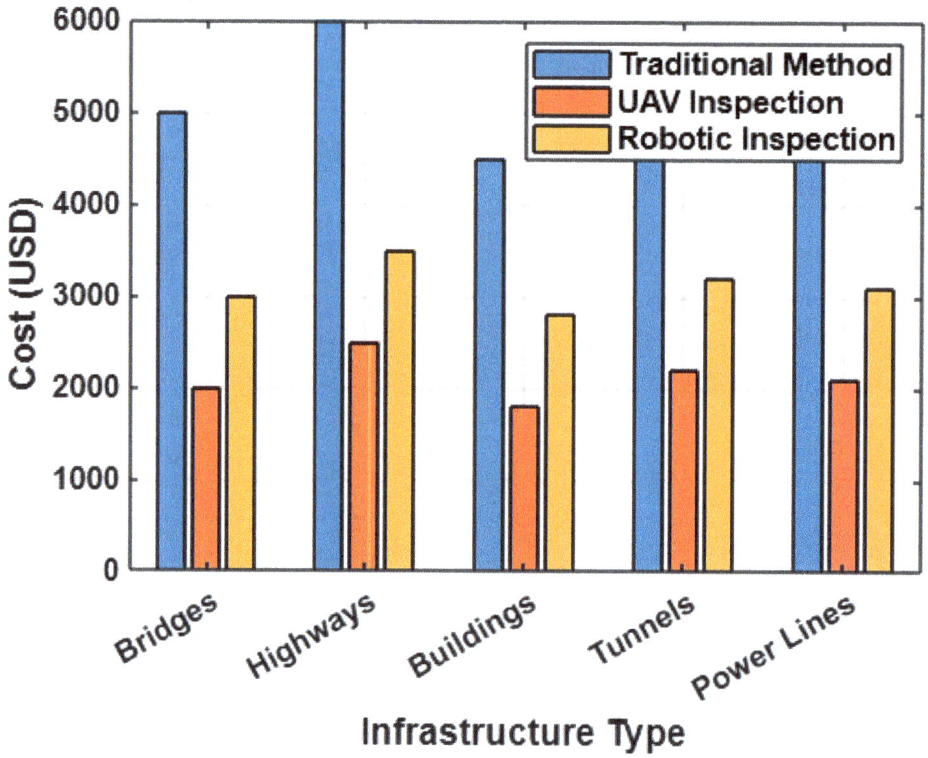

Figure 5 gives an assessment of the actual-time performance of UAVs and robots in the fight of structural infirmities, including its cracks, corrosion, and other forms of weariness. The analysis reveals that UAVs can be especially useful during the process of inspections carried out in a large area, e.g. inspecting bridges or power lines, where the UAV would notice and report the problems almost in real-time. Robotics, however, prove extremely useful in closed or dangerous environments, in other words, spaces where people cannot go to check, including tunnels or industrial facilities. Such technologies help to identify the problems at an early stage, minimizing the chances of their further degradation or critical collapse.

Figure 5. Performance evaluation of UAVs and robotics in detecting structural issues (real-time data analysis)

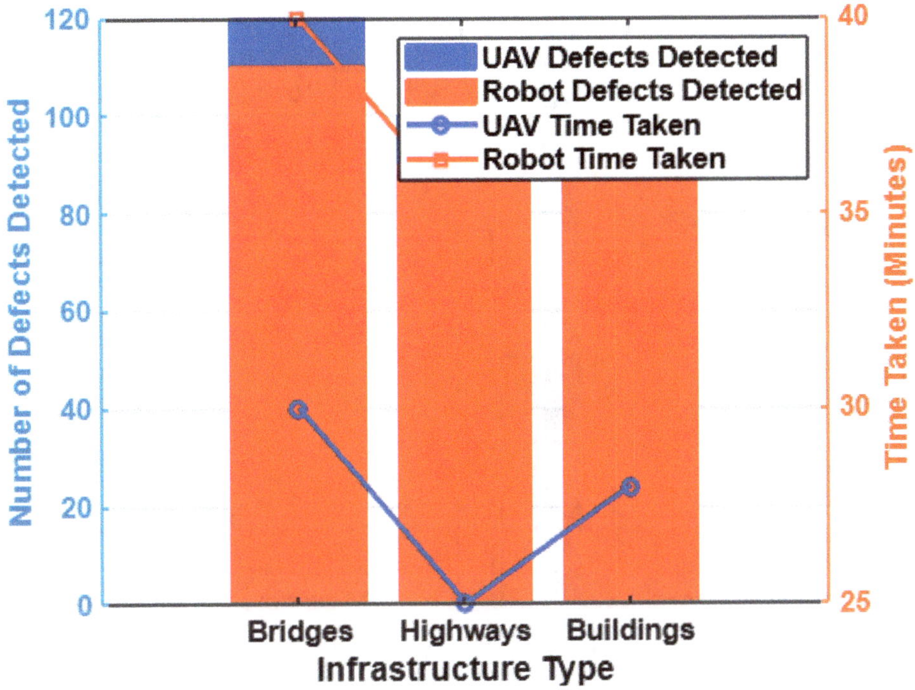

Figure 6 provides a detailed comparison in various parameters, i.e. accuracy, cost, time and safety, between conventional methods of inspection and UAV/robot-based methods of inspection of various types of inspection. The statistics make it clear that UAVs and robotics are much better in almost every aspect compared to the traditional ones. There is a huge improvement in accuracy where UAVs and robots notice more structural problems as compared to their human counterparts. The cost involved is cut down by about 40-50 percent and the time taken during inspection is halved. Safety is increased significantly, especially since UAVs and robots help to avoid sending the workers into risky locations, like working in heights or inside the closed rooms. This data allows concluding with strong signs that the application of UAVs and robotics could make significant contributions to conventional inspection techniques. These technologies make inspections more effective and efficient, cost-efficient, and safer, so the maintenance of infrastructure becomes more proactive and timely.

Figure 6. Comparative analysis of traditional vs. UAV and robotics methods for various inspection metrics

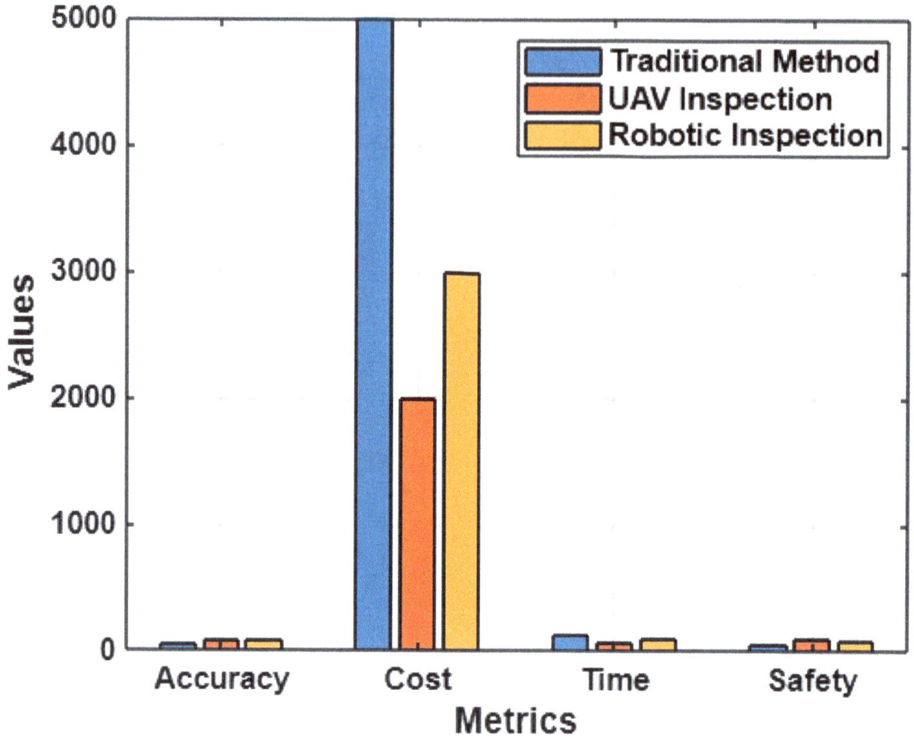

The affordability of UAV and robotics systems to the maintenance of the infrastructure is becoming a feasible option because of technological solutions that keep on cutting down the cost and enhancing functionality. The increasing availability of such systems has facilitated the use of such systems in large-scale infrastructure monitoring like the professional monitoring of the bridges in an entire country or constant surveillance on a highway. Scalability of UAVs and Robotics can also be seen through their capability to be used in different environments including rural roads and urban areas thereby contributing to their extensive usage. Moving ahead, there are many possibilities of the breakthrough that could emerge in the field of UAV and robotics. Amalgamation of 5G networks with artificial intelligence (AI) and machine learning will contribute towards the better functioning of such systems. UAVs and robots will be able to improve their capability to evaluate the collected data, detect the slight global patterns of infrastructure deterioration, and increase the defect recognition accuracy in the long term due to the use of AI-based systems. More specifically, as an example, one may refer to a case where AI would help detect minor patterns denoting the gradual deterioration of the infrastructure, which would not be identified by human inspectors. Machine learning algorithms have a capability to foretell the probable spots of failure of infrastructure and hence, predictive maintenance works, which saves maintenance expenses and avoids catastrophic failure. Moreover, 5G networks will enable more rapid and stable transmission between UAVs, robots, and central information networks and make it possible to conduct real-time analysis and make instant conclusions, further streamlining the process of maintaining infrastructure.

Another major consideration is the effect it can have on the industry and the workforce. Although the UAVs and robotics integration is likely to decrease the demand of human labor in hazardous surveying operations, the trend will still pose opportunities in the employment sectors like operation of robots, artificial intelligence, as well as data mining. Increasingly the number of workers who have the expertise to operate and maintain such systems will be needed with the increase in the demand of such technologies. In addition, standardisation of the UAV and robot inspection procedures will probably see the formulation of new industry requirements and standards of operation, which will majorly dominate the future of infrastructure management. Ethical aspects also have to be elaborated upon especially when it comes to data privacy and safety. Through inspections carried out by UAVs and robots, large amounts of data have been gathered in form of high resolution images as well as sensors. It will be of essence that this data is safely stored, shared, and utilized and in line with privacy laws. As well, the safety of these technologies operating should also be given priority. With the increased use of UAV and robots in infrastructure management, there will be the need to have an intense safety standards to make sure the UAV and robot works safely particularly in a busy city or around areas sensitive infrastructure.

Lastly, the application of UAVs and robotics in maintenance of the infrastructure will require cross-disciplinary study. Engineers, computer scientists, urban planners, and policymakers will need to collaborate to assist in the development of such technologies and advance how they can be implemented into existing infrastructure management systems without a hitch. The long-term sustainability, environmental and cost-effectiveness of the technologies will have to be researched in order to get optimal use of the technologies and achieve more sustainable and efficient infrastructure systems. To sum up, UAVs and robotics have already demonstrated their usefulness with respect to the increasing accuracy, speed, safety, and cost-efficiency of exams of infrastructures. With such technologies continuing to advance, they will be an even more significant support in the maintenance of infrastructure and deliver tremendous benefits to the industry itself and the society on the whole. The capabilities of these systems are already field-tested with continuous AI, machine learning and 5G connectivity development, which will inevitably lead to even more automated, data-based future in infrastructure management.

CONCLUSION

The use of UAVs and robotics in inspection and maintenance of infrastructure has greatly enhanced the operational effectiveness and accuracy and most importantly safety. These technologies are obviously much better in comparison with traditional methods that suggest significant savings in terms of time, better accuracy in detection of defects and, finally, cost-reduction. With robotics and UAVs, inspections can be conducted more quickly and comprehensively and see problems that may remain invisible using manual means. The result is improved maintenance planning, which results in the prolonged life of infrastructure assets and stops failures that cost escalating amounts of money. Additionally, the deployment of UAVs and robots in hazardous areas improves safety because people do not have to endure dangerous situations, like working at a high position or enter a small space. The technologies also reduce the use of big teams of inspectors and costly equipments thus making the process less expensive. UAVs and robotics will grow in autonomy and become increasingly data-driven and with the ability to perform predictive maintenance further streamlining the management of infrastructure as future AI, machine learning and 5G networks gain capability. Nonetheless, these technologies should be used in good ethics that include data privacy and safety standards. Also, the transition towards UAVs and robotics will lead to the appearance of new occupations in such areas as robotics operation and data analysis. To sum it all up, UAVs and robotics will transform infrastructure maintenance with more time, money, and life-saving applications. The future of infrastructure management depends on how they will continue to develop and be integrated into the practices of the industries.

REFERENCES

Addula, S. R., Mamodiya, U., Jiang, W., & Almaiah, M. A. (2025). Generative AI-Enhanced Intrusion Detection Framework for Secure Healthcare Networks in MANETs. *SHIFRA, 2025*, 62–68. DOI: 10.70470/SHIFRA/2025/003

Ali, A., & Ghanem, M. C. (2025). Beyond Detection: Large Language Models and Next-Generation Cybersecurity. *SHIFRA, 2025*, 81–97. DOI: 10.70470/SHIFRA/2025/005

Bruggemann, T. (2022, April). Automated Feature-Driven Flight Planning for Airborne Inspection of Large Linear Infrastructure Assets. *IEEE Transactions on Automation Science and Engineering, 19*(2), 804–817. DOI: 10.1109/TASE.2021.3062154

Car, M., Markovic, L., Ivanovic, A., Orsag, M., & Bogdan, S. (2020). Autonomous Wind-Turbine Blade Inspection Using LiDAR-Equipped Unmanned Aerial Vehicle. *IEEE Access : Practical Innovations, Open Solutions, 8*, 131380–131387. DOI: 10.1109/ACCESS.2020.3009738

Chand, N., Mahalleh, V., Aziz, T., Rahman, A., Mohammed, A., & Zeid, W. (2021) "Small Scale Localized Maintenance of Industrial Infrastructure using Autonomous UAVs," *2021 IEEE International Conference on Advanced Robotics and Its Social Impacts (ARSO)*, Tokoname, Japan, 2021, pp. 168-175. DOI: 10.1109/ARSO51874.2021.9542838

Du, B., Borhan, U. M., Chen, T., Chen, J., Li, J., & Chen, J. (2024) "Robust UAV Policy Learning for Urban Infrastructure Surface Screening," *2024 International Conference on Advanced Robotics and Mechatronics (ICARM)*, Tokyo, Japan, 2024, pp. 1-8. DOI: 10.1109/ICARM62033.2024.10715841

Gil-Castilla, M., Poma, A. R., Caballero, A., & Ollero, A. (2025) "A Cooperative Multi-UAV Framework for Bridge Inspection," *2025 International Conference on Unmanned Aircraft Systems (ICUAS)*, Charlotte, NC, USA, 2025, pp. 408-415. DOI: 10.1109/ICUAS65942.2025.11007792

Hament, B., & Oh, P. (2018) "Unmanned Aerial and Ground Vehicle (UAV-UGV) System Prototype for Civil Infrastructure Missions," *2018 IEEE International Conference on Consumer Electronics (ICCE)*, Las Vegas, NV, USA, 2018, pp. 1-4. DOI: 10.1109/ICCE.2018.8326346

Hammad, O. M., Smaoui, I., Fakhfakh, A., & Hashim, M. M. (2024). Recent Advances in Digital Image Masking Techniques: Future Challenges and Trends—A Review. *SHIFRA, 2024*, 67–73. DOI: 10.70470/SHIFRA/2024/008

Ibrahim, A. A. A. (2024). Features and Functional Components in the Use of Artificial Intelligence in Human Capital Development (Higher Education as a Model). *ESTIDAMAA, 2024*, 1–6. DOI: 10.70470/ESTIDAMAA/2024/001

Liu, K., & Chen, B. M. (2023, September). Industrial UAV-Based Unsupervised Domain Adaptive Crack Recognitions: From Database Towards Real-Site Infrastructural Inspections. *IEEE Transactions on Industrial Electronics*, *70*(9), 9410–9420. DOI: 10.1109/TIE.2022.3204953

Lynch, C., Lyu, L., Buzzatto, J., Heshmati-Alamdari, S., & Liarokapis, M. (2024) "A Control Barrier Function Based Visual Servoing Framework for Safe Operation of UAVs in GNSS Denied Settings using Fiducial Markers," *2024 IEEE International Symposium on Safety Security Rescue Robotics (SSRR)*, New York, NY, USA, 2024, pp. 1-6. DOI: 10.1109/SSRR62954.2024.10770045

Montambault, S., Beaudry, J., Toussaint, K., & Pouliot, N. (2010) "On the Application of VTOL UAVs to the Inspection of Power Utility Assets," 2010 1st International Conference on Applied Robotics for the Power Industry, Montreal, QC, Canada, 2010, pp. 1-7. DOI: 10.1109/CARPI.2010.5624443

Nyboe, F. F.. (2023) "Towards Autonomous UAV Railway DC Line Recharging: Design and Simulation," *2023 IEEE International Conference on Robotics and Automation (ICRA)*, London, United Kingdom, 2023, pp. 3310-3316. DOI: 10.1109/ICRA48891.2023.10161506

Patsiouras, E., Mygdalis, V., & Pitas, I. (2022) "A UAV Object Detection Benchmark for Vision-Assisted Powerline Element Inspection," 2022 29th International Conference on Systems, Signals and Image Processing (IWSSIP), Sofia, Bulgaria, 2022, pp. 1-4. DOI: 10.1109/IWSSIP55020.2022.9854489

Paul, H., Martinez, R. R., Sumetheeprasit, B., & Shimonomura, K. (2023) "A Tiltable Airframe Multirotor UAV Designed for Omnidirectional Aerial Manipulation," *2023 IEEE/ASME International Conference on Advanced Intelligent Mechatronics (AIM)*, Seattle, WA, USA, 2023, pp. 460-465. DOI: 10.1109/AIM46323.2023.10196207

Perez-Jimenez, M., Montes-Grova, M. A., Ramon-Soria, P., Arrue, B. C., & Ollero, A. (2020) "POSITRON: Lightweight Active Positioning Compliant Joints Robotic Arm in Power Lines Inspection," *2020 International Conference on Unmanned Aircraft Systems (ICUAS)*, Athens, Greece, 2020, pp. 729-736. DOI: 10.1109/ICUAS48674.2020.9214022

Petit, L., & Desbiens, A. L. (2024) "MOAR Planner: Multi-Objective and Adaptive Risk-Aware Path Planning for Infrastructure Inspection with a UAV," *2024 IEEE International Conference on Robotics and Automation (ICRA)*, Yokohama, Japan, 2024, pp. 8422-8428. DOI: 10.1109/ICRA57147.2024.10610907

Ramon-Soria, P., Perez-Jimenez, M., Arrue, B. C., & Ollero, A. (2019) "Planning System for Integrated Autonomous Infrastructure Inspection using UAVs," *2019 International Conference on Unmanned Aircraft Systems (ICUAS)*, Atlanta, GA, USA, 2019, pp. 313-320. DOI: 10.1109/ICUAS.2019.8797874

Sadik Hassan Hussein, S. Y., Dziyauddin, R. A., Muhtazaruddin, M. N., Yaghoobi, Y. J., & Din, N. M. (2024) "Development of AI Vision Inspection System for UAV Imagery Surveillance of Transmission Towers," 2024 5th International Conference on Smart Sensors and Application (ICSSA), Penang, Malaysia, 2024, pp. 1-5. DOI: 10.1109/ICSSA62312.2024.10788575

Wu, J., & Qin, H. (2008) "The Design of Wireless Intelligent Home System Base on ZigBee," 2008 11th IEEE International Conference on Communication Technology, Hangzhou, 2008, pp. 73-76.

Yang, G., Zhao, B., Zhang, J., Wen, J., Li, Q., Lei, L., Chen, X., & Chen, B. M. (2025). Det-Recon-Reg: An Intelligent Framework Toward Automated UAV-Based Large-Scale Infrastructure Inspection. *IEEE Transactions on Instrumentation and Measurement*, *74*, 1–16. DOI: 10.1109/TIM.2025.3571118

Chapter 10
Internet of Things Applications in Civil Engineering

Damodharan Palaniappan
https://orcid.org/0009-0003-0721-3068
Marwadi University, India

T. Premavathi
https://orcid.org/0009-0003-0172-2021
Marwadi University, India

ABSTRACT

The IoT is transforming civil engineering into a new real-time monitoring, automated control, and data-based decision-making in infrastructure. In this field, the presence of embedded sensors in structures such as buildings, bridges and dams has increased structural health monitoring, smooth traffic flows, improved construction procedures and management of water resources. These sensors gather information about stress, strain, temperature, and vibration helping to implement predictive maintenance and fault detection. By making traffic, routing, parking more efficient, the IoT can be used in the transportation industry to decrease congestion and emissions. Wearables and sensors enhance safety in construction sites, track the site equipment and facilitate the process of managing materials. The existing problems in its expansion are data security, interoperability and sensor durability. It is expected that future augmented intelligence will include AI, machine learning, digital twins, and blockchain to lead to fully autonomous and predictive infrastructure systems.

DOI: 10.4018/979-8-3373-5248-0.ch010

INTRODUCTION

The concept of the Internet of Things (IoT) has been rising in popularity in the sphere of civil engineering, which is becoming one of the instruments of change that drive the transformation of many fields. IoT represents an interconnected system of devices that have sensors, software, and the ability to interact on the networking level and, hence, enable collecting and sharing data (Choudhary et al., 2025). It can be traced back to the late twentieth century when the initial deployment included toaster control by an Internet-available toaster. Kevin Ashton initially applied the term Internet of Things when he was talking about radio frequency identification technology in 1999. Further than the same wireless communication, data analysis and cloud computing have sparked explosive growth (Parihar et al., 2024). In civil engineering, the IoT presents an unequalled capacity of design, development, and management of infrastructure frameworks. The traditional norm is characterised by manual acquisition of data, intermittent assessment, and responsive maintenance measures. Incorporating IoT technologies, real-time control, automation of the processes, and data-driven decisions are possible thus helping to transform efficiencies, boost safety, and sustainability (Liu et al., 2025). There are a number of backgrounds behind embracement of IoT solutions in civil engineering. To start with, the need to have smart and resilient infrastructure systems is big. An increase in urbanisation and the population pressure on the available networks necessitates the need to establish mechanisms to enhance performance, reduce expenditure, and strengthen safety. Second, IoT solutions have become more affordable due to the availability and popularity of cheap sensors, wireless communications, and cloud computing systems. Lastly, the rising interest toward data-driven decision making is another factor that drives the use of the IoT in civil engineering. The main outcomes of this part are to provide a detailed description of the IoT in civil engineering. The discussion is therefore on the definition and the history of the IoT, its relevance in civil engineering, the reasons that would encourage the use, and what it is meant to cover in other sections.

Defined Problems, Variables, and Contributions of This Chapter

Although the Internet of Things (IoT) has garnered significant attention in the civil engineering field, several practical issues remain underrepresented. These include:

Problem 1: Fragmented integration IoT solutions tend to remain in silos (e.g., traffic/structural monitoring/water management) with no single architecture.

Problem 2: Complexity of data management: The management, analysis, and utilization of massive data generated by civil IoT applications are difficult.

Problem 3: Reliability and interoperability: Making sensors highly performant (and resistant to harsh conditions) and devices able to communicate with each other flawlessly remains an important challenge.

Problem 4: Inability to predict – Most existing systems are monitoring-oriented rather than predictive analysis or preventive maintenance.

Problem 5: Ethical and governance gaps: Policy and ethical ownership, consent, and long-term impact frameworks are absent.

The key Variables explored in this chapter include the following:

- Types of IoT architecture (cloud vs. edge)
- Sensor types and communication protocols
- Integration with AI/ML for prediction
- Use case performance indicators (e.g., efficiency, safety, sustainability)
- Security and governance considerations

Contributions of the Chapter

- An extensive theoretical framework to depict how the IoT is used in different sub-areas within civil engineering, such as structural health, transportation, water, and construction.
- Technology mapping can be applied to elements such as sensors, communication standards, data platforms, and decision-making layers.
- Analysis of practical applications and new developments, such as digital twins, predictive analytics, and the incorporation of AI.
- Determination of policy, regulatory, and ethical issues that are of interest in long-term sustainable deployment.
- A roadmap towards multidisciplinary collaboration, smart cities, and autonomous infrastructure systems in the future.

Thus, the chapter will act as both a technical guide and a strategizing tool for engineers, researchers, and policymakers involved in the implementation of IoT in civil engineering settings.

Objectives and Purpose of the Chapter

This chapter provides a systematic analysis of how the Internet of Things (IoT) is transforming civil engineering. This is not just a descriptive intention but rather an analytical and strategic intention.

The main points of this chapter are as follows:

- To describe the basic ideas and historical development of the IoT in civil engineering settings.
- To classify and specify the fundamental technologies of IoT (sensors, gateways, platforms, and protocols) and their use in sub-areas of structural health monitoring, transportation systems, water management, and construction.
- The arrival of new technologies such as Artificial Intelligence (AI), Digital Twins, and Blockchain, along with IoT, in the development of resilient and intelligent infrastructure is emphasized.
- To evaluate the practical applications, performance effects, and socio-technical consequences of deploying IoT in infrastructure projects.
- To demonstrate the priority problems and hindrances to implementation, such as interoperability, safety, cost, and deficiencies in human capital.
- To describe the emerging trends, research, and policy directions required to achieve a sustainable civilization that embraces the IoT in civil engineering.

This chapter serves to empower civil engineering practitioners, researchers, and policymakers with a unified view of how IoT technologies may contribute to the improvement of the performance, safety, and sustainability of infrastructure systems.

BUILDING AND CONSTRUCTIONS OF IOT SYSTEMS ON CIVIL INFRASTRUCTURES

The structure of an IoT system in civil engineering involves multiple interconnected elements that work together in order to consume, transform and execute data (Alghamedy et al., 2024). These parts are sensors, actuators, gateways, and cloud platforms (Palaniappan et al., 2024).

Fundamental Elements of IoT Architecture for Smart Infrastructure

Please refer to Figure 1. containing IoT Architecture Elements of the smart infrastructure. Sensors are among the main data collecting devices and they play the role of sensing and measuring physical characteristics, e.g., strain, temperature, pressure, displacement (Masmali & Miah, 2021). Such devices can incorporate strain gauges to monitor structural health, accelerometers to measure vibration, environmental sensors to measure air and water quality, amongst others depending on the application. Actuators are working entities that translate processed data into physical response thus implementing feedback-derived control measures. Gateways create the interface between the edge and cloud platforms allowing the exchange of data.

To facilitate real-time processing of data and historical data, cloud platforms can be used to scale both analytical response and the computing resources of a business.

Figure 1. IoT architecture elements

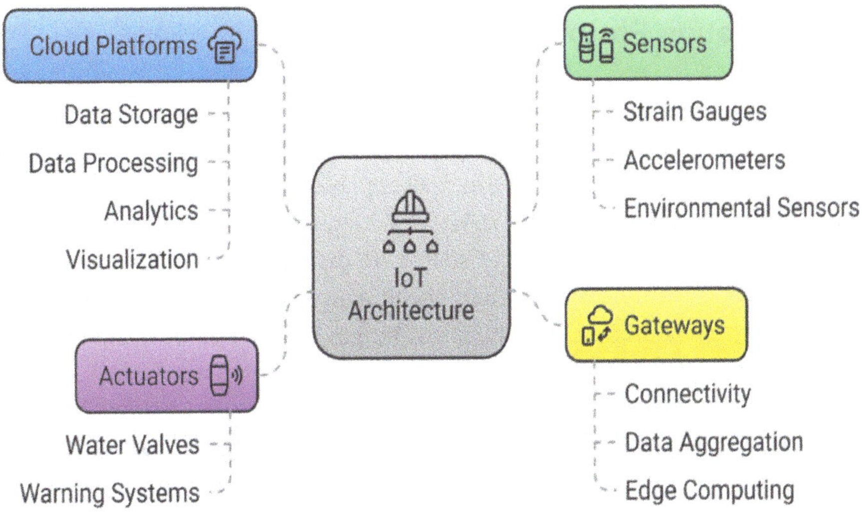

Actuators are gadgets that perform the physical processes on the information data processed. They are used in the management of systems including the regulation of water taps in a drinking water network or triggering of alert systems due to an event of urgency. Gateways act as links between sensors and actuators and cloud platforms, and hence offer connectivity, data collection and edge-computing functionalities (Tariq et al., 2023). The former are often used to do localized data processing, thus decreasing the amount of data sent to the cloud, increasing efficiency and responsiveness of the system. Cloud platforms provide data storage infrastructure and data processing services, information processing, analytics, and data visualization services (Masmali & Miah, 2021). Such services allow remote monitoring, data driven decision-making, as well as integration with other enterprise systems.

The localized processing strategy enables a scenario in which the data can be filtered, aggregated, or analyzed close to the source data, allowing the system to respond to events rapidly without the need to refer to cloud-based calculations. In structural health monitoring or water distribution systems, real-time actions, such as issuing alerts, setting valve positions, or shutting down elements, can be immediately executed at the edge. This not only reduces bandwidth across the network by avoiding unnecessary data transmission, but also guarantees responsiveness in low-latency scenarios, where a safety measure or operation can be considered by the network. In addition, edge data can be coordinated with centralized cloud provisioning, which can interface with enterprise resource planning (ERP) systems, preventive maintenance routines, and automated reporting tools. This smart interconnection of local-scope IoT and civil infrastructure management will increase the efficiency, scalability, and accuracy of decision-making.

Communication Protocols

Civil-engineering IoT systems have various communication protocols, which may demonstrate particular advantages and limitations (Alghamedy et al., 2024). MQTT is a publication/ subscription level protocol capable of working in high latency and low-bandwidth conditions. LoRa WAN is a long-range and Low-Power wireless communication that is best suited to span long distances among the devices. NB-IoT is a low-power wide-area IoT-friendly cellular communication protocol. Choice of these protocols is dictated by aspects of range, bandwidth, power requirements and security aspects.

The Battle Between Edge Computing and Cloud Computing

The general idea of edge computing involves calculating data as close as possible to the sources: a gateway or an edge device instead of sending all the information to the cloud to have it processed (Khalil et al., 2022). Such an approach is capable of reducing latency, increasing safety and providing the capacity to make real-time decisions. Cloud computing, in its turn, enables the centralized storage, processing, and analytics of the data. The best approach toward mixing edge with cloud computing is application specific.

Civil-Engineering Assets Integration

The fields of implementation of IoT with current civil-engineering facilities demand a very careful organization and implementation, incorporating the choice of suitable sensors and communication objectives, effective creation of data-acquisition and process frameworks, and maintaining compatibility with current infrastructure (Palaniappan et al., 2024). To fight the threat of cyber-attacks and preserve the integrity of the critical infrastructure, security denizens become a primary priority (Dirin et al., 2023).

IOT STRUCTURAL HEALTH MONITORING

Structural Health Monitoring (SHM) is the systematization of damage-detection system to aerospace, civil, and mechanical engineering infrastructures. The procedure involves the installation of sensors that constantly monitor the structural performance and use data analytics to identify potential damages or degradation that could develop with time (Choi et al., 2024; Ye et al., 2019). SHM is deemed necessary due to its ability to guarantee safety and reliability of infrastructures, to minimize the maintenance costs, and increase the life spans of the structures. This is assisted by constant surveillance; potential problems can be detected early and addressed on time to prevent critical breakdowns.

Type of Sensors

SHM systems use different types of sensors that are used to measure various parameters related to structural health (Muttillo et al., 2020). The strain gauge measures the strain thus providing data regarding stress and deformation of the structure. Accelerometers allow measuring structural acceleration and identifying vibration and related anomalies. Temperature probes measure the temperature of the structure, which could be used to point to thermal stress and possible damage. Displacement, tilt and corrosion sensors are other types, and each is assigned a different purpose and therefore can be explained together in Figure 2. Sensor Types in Structural Health Monitoring.

Figure 2. Sensor types in structural health monitoring

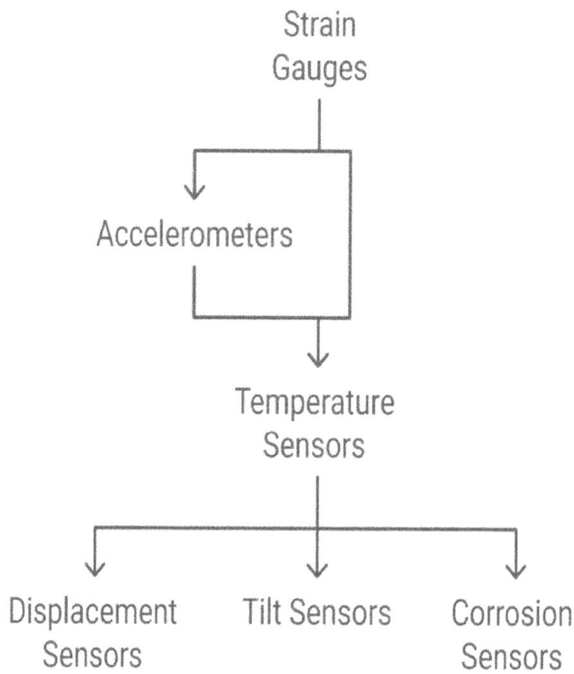

Wireless communication technologies are a basic aspect of a structural health monitoring (SHM) system and they assist in transferring data on the sensors to central analysis nodes. Various wireless methods have been integrated all of which have their own advantages and disadvantages. The Wireless Sensor Networks (WSNs) category continues to be one of the top contenders since it possesses an acceptable power draw and is very scaly. In the same measure, LoRaWAN has incorporated itself into large scale industrial settings due to the long-range and low energy demands on its use. The cellular networks have extended coverage and immense bandwidths making them especially suitable in applications that require real time feedback. Connecting the Internet-of-Things (IoT) systems also enhance SHM, as it allows one to monitor data constantly and analyze the information automatically, which are the essential aspects to proactive safety. A strong monitoring plan should thus contain careful planning of the architect of the sensor network, together with a thorough evaluation of the detection techniques (Tan et al, 2025; Wang et al, 2025).

The main advantage of IoT-enabled SHM is that it allows collecting and interpreting data in real-time. A series of WSN sensors placed in a building can generate never-ending streams of data that can be directed into a central data repository to be analysed in real-time. Existing algorithms are developed to recognize the anomalies or degradation early enough initiating timely action and preventing further damages. Evidence of case studies demonstrates the importance of IoT-based SHM - Bridge integrity monitoring: IoT sensors are able to sense the cracks as well as corrosion as well as other anomalies on the bridge allowing the engineers to create real-time models allowing them to identify the latent defects before the bridges experience critical failure. Detecting sway in skyscrapers: With the help of IoT, engineers can monitor sway so as to understand when the structure is becoming unstable and then adjust the damping system to ease structural oscillations and decrease their effect on tenants and the risk. Dam deformation warning: IoTs will help in detecting minute deformation of dams and trigger warnings that will cause timely evacuation, thus preventing disastrous floods. Altogether, the IoT implementation in SHM brings great benefits: the sooner deficiencies are discovered, the more the preventative approaches can be facilitated, the lower the cost of repair, the safer citizens are, and the better the significant infrastructure can be taught to depend on.

IOT IN TRANSPORTATION ENGINEERING

The Internet of Things (IoT) converges with transportation engineering, which has essentially changed how the transportation network is governed and how people interact with the system (Garg et al., 2021). IoT can incorporate communication systems, traffic-based sensory solutions, and analytical data-driven applications into shared traffic places in order to create an intelligent transport system that focuses more on safety, efficiency and sustainability. This combination creates an inter-linked ecosystem that has the potential of solving numerous transport issues (Abbasi et al., 2021).

Intelligence Traffic Monitoring and Congestion Control

The transportation area is also one where traffic flow will be controlled, and congestion mitigated, which is to be considered one of the biggest applications of IoT. Conventional traffic systems rely on stagnant information and manual controls, which in many cases contribute to poor performance and time wastages. Instead, the IoT-facilitated systems can utilize real-time information provided by cameras, sensors, and automated vehicles to track traffic in real-time and identify conjecture hotspots. The ensuing data is used to re-calculate traffic lights, optimize paths, distribute real time guidance to drivers, and minimise congestion, thereby enhancing performance of the entire network (Palaniappan et al., 2024).

Intelligent Transportation Systems

IoT as a building block of Intelligent Transportation Systems (ITSs) refer in Figure 3. is aimed at increasing the levels of transportation safety, efficiency, and sustainability. ITS is composed of a number of different systems, such as Advanced Traffic Management Systems (ATMSs), Advanced Traveler Information Systems (ATISs) and Connected Vehicle Systems (CVSs). ATMS is based on real-time data gathered in order to control the patterns of traffic and reduce congestion, whereas ATIS informs citizens about the current traffic situation, street closures, and alternative routes. Two-way communications between vehicles and infrastructure are facilitated by CVS, making all-new safety features and emerging applications feasible, like platoons and collaborative driving (Palaniappan et al., 2024).

Figure 3. IoT in transportation systems

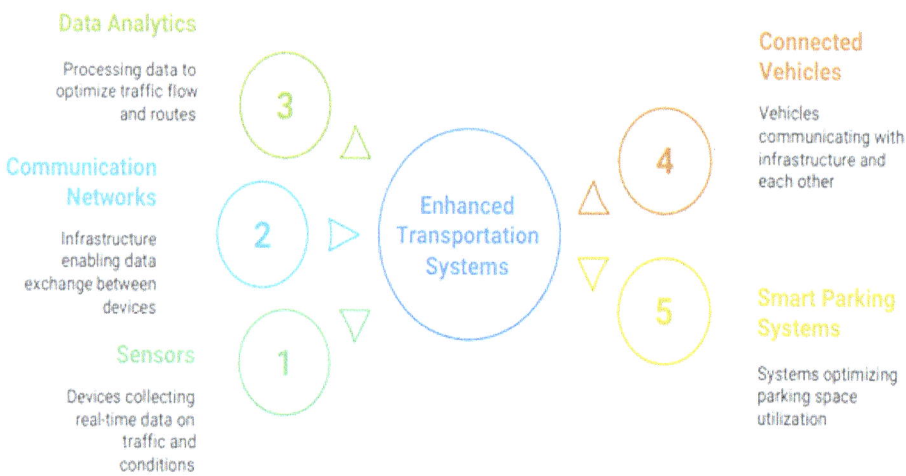

The Internet of Things (IoT) currently assists car users with dynamic route planning and on-demand navigation so that they can plan their actions according to the current situation in the traffic (Murroni et al., 2023). Combined systems receive and analyze live data provided by traffic sensors, on-road cameras, and connected cars and thus provide the drivers with up-to-date information on traffic jams, road closures, and accidents occurred (Palaniappan et al., 2024). With this kind of information, it is possible to make dynamic route changes, eliminate over-crowded corridors and minimize traveling times.

The IoT also has substantial impact on emergency response within the transportation sector as it provides real-time insights on accidents, road blockages, and any other events (Kumar et al., 2023). Connected vehicle sensors automatically send important information regarding the site of incidents and level of injuries to emergency workers. The IoT-based sensors are able to monitor the road conditions and report about any dangerous changes, e.g., floods or landslides, so the necessary efforts could be fulfilled by the emergency agencies on the basis of fast and effective responses (Kumar et al., 2023).

Moreover, smart parking systems use IoT sensors and data-analytics solutions to optimise parking space provision to the user experience. Occupancy sensors installed within each parking space tell whether the parking space is available or otherwise and relay the message to a central system (Kumar et al., 2023). To find parking space, driver can gain access to real-time information through their mobile devices or websites to locate the available parking spaces, book in advance, and pay parking fees electronically. Intelligent parking also allows constant observation of parking occupancy, parking hotspots detection, and parking rates adaptation on the fly to best utilize the available space.

Many cities across the world are adopting transportation systems that use IoT in order to deal with certain challenges. In an example of this, Barcelona has established a smart parking system where sensors are used to determine the available parking spaces and direct the drivers using a mobile application (Palaniappan et al., 2024). Singapore has already introduced an integrated ITS that uses real-time information to optimize the flow patterns, congestion relief, and provide a driver with updated information on traffic-flow situations. These grounded applications are a sign of the potential that IoT solutions can bring to change the transportation system and enhance the everyday lives of citizens (Murroni et al., 2023).

IOT IN CONSTRUCTION MANAGEMENT

Incorporation of the Internet of Things is transforming the construction management in creating smart construction sites by implementing connected machinery to create smart construction sites, enhancing worker safety with wearables and geofencing, maximising equipment and material stewardship (5 Unique Applications of IoT in Construction, 2024). At the same time, the use of AI and predictive analytics in the process of project planning has resulted in better productivity, efficiency, and meeting safety requirements (Richard, 2023).

Intelligent Construction sites and IoT machines

Sensors that are connected to the Internet of Things allow monitoring and controlling construction machinery in real-time which helps to improve efficiency and minimize downtime (Richard, 2023). The sensors mounted on equipment provide performance, location and maintenance need data, allowing pre-emptive maintenance before the costly breakdowns. At the same time, remote surveillance ensures that the project managers have a real-time reference of the site and condition as well as safety and security of resources which is done to ensure the resources are effectively utilized and schedules are met.

The sensors placed at the locations of construction can provide real time data about the environment, ensure compliance of regulations and ensure safety. Air quality, noise, and temperature measurements allow recognizing the hazard early and taking measures to prevent it. Combined with Building Information Modeling, this integration provides a thoroughly developed digital image of the undertaking itself, thus enhancing the level of collaboration across several teams as well as decision-making operations. All the site-level variables become integrated to promote more sustainable operations that are safer and more efficient with the help of the IoT.

The use of IoT to develop intelligent construction sites is a revolutionary move towards the development of construction operations. Stakeholders can track and monitor performance parameters, such as the location of construction machinery and other tools, operating hours, fuel consumption, and mechanical health, in real time by installing sensors in construction hardware and tools. This type of proactive monitoring makes predictive maintenance possible, which lowers downtime and increases the efficiency of all site equipment. One of them is the implementation by Komatsu in cooperation with NVIDIA, where IoT systems with AI power were employed to monitor heavy machinery at various locations. These systems examined the use of machines, fuel consumption, and idling times, causing a 30 percent increase in the level of operations. Likewise, Turner Construction is another large construction firm in the U.S. that has incorporated the use of IoT sensors and wearables throughout its sites to manage the location of equipment, monitor the safety of workers, and maintain compliance with regulations in real time.

In addition to operational advantages, integration via the IoT enables the monitoring of remote sites. Digital dashboards allow project managers to see the progress of a site and coordinate it better, make decisions quickly, and save travel expenses. This can be particularly useful for large-scale or regionally spread construction projects. According to Richard (2023), such intelligent systems not only increase productivity and maintenance efficiencies but also allow project management to be data-driven owing to their ability to analyze data and report in real time. The source describes several application scenarios, such as automated failure alerts, geofencing of the unauthorized use of equipment, and connection with Building Information Modeling (BIM) to achieve unified supervision.

Geofencing and Wearables to be Used to Promote Worker Safety

Smart helmets and vests, which follow the health and position of the workers, and will share alerts in case of an accident or a threatening situation, are all examples of wearable technologies. Geofencing enables drawing invisible fences around construction areas, which trap the human staff inside the safe zone and keep intruders outside the hazardous areas. Fatigue, stress and environment data real time information enable timely action to prevent accidents and protect occupational health. The vitals like the heart rate, body temperature, and fatigue recorded should enable monitoring of the health at all times hence able to aid in evidence-based safety management, especially in locations that are highly prone to risks.

The synergy between wearables and real-time analytics is more effective in controlling individual safety and detecting trends toward risk-proactive mitigation.

Material Management and Tools Tracking

Monitoring systems based on IoT will allow seeing the equipment and materials in real-time thus reduce the threat of theft and loss. Sensors document temperature and humidity to make sure that the storage conditions are maintained, avoiding damage, and preserving quality. Uninterrupted supply checks promote regularity and see to it that no waste or loss occurs due to continuous checking of materials, avoiding shortage of the materials or overstock.

Material management in the form of IoT optimises the supply chain activities and general project performance. Quality control with the use of sensors identifies faults as they appear and allow instant response and thus become more resourceful in utilisation and sustainable.

Building Information Modeling integration

Combining IoTs data with Building Information Modeling provides project managers with an integrated, data-rich view of what is happening with construction matters, enhancing coordination, communication, and decision-making. Instant sensor data can be directly consolidated into the BIM model, which makes the visualization of site developments possible and supports an active resolution of issues. Issues can be detected in real-time so further steps can be taken timely and expensive delays can be avoided, and, finally, activities can be better optimised, ensuring safer working conditions and improved waste.

Predictive Analytics and AI in Project Planning

Algorithms to perform machine learning take in data exemplified by sensors and BIM models, historical records of the project to predict the issues that might arise and optimizing the project plans. Predictive analytics are used to pre-empt delays, cost overruns and safety risks to give the managers time to enact proactive remedies. The resource allocation, budgeting and financial performance are further refined through integration with enterprise resource-planning systems. AI-based models of predictive maintenance allow repairing infrastructure objects and prolonging their service life in a timely manner.

Advantages: Compliance, Efficiency and productivity

Adoption of IoT in construction management provides significant productivity and efficiency, conformity benefits, and real-time data and analytics can help with data-driven decision-making that reduces the cost and enhance the project results.

Table 1. IoT applications in construction management

IoT Application	Description	Benefits
Real-time Monitoring of Equipment	Sensors track location, performance, and maintenance needs of construction equipment.	Improved efficiency, reduced downtime, and proactive maintenance.
Worker Safety Wearables	Smart helmets and vests monitor worker health and location.	Alerts in case of accidents or hazardous conditions, ensuring worker safety.
Geofencing Technology	Virtual boundaries around construction zones ensure workers stay within safe areas.	Prevents unauthorized access to dangerous locations and enhances safety.
Equipment and Material Tracking	IoT-enabled tracking systems monitor equipment and materials in real-time.	Reduces risk of theft and loss, ensures proper storage, and prevents damage.
Integration with Building Information Modeling	Real-time data from sensors overlaid on BIM model.	Provides a visual representation of the construction site, enabling better coordination and communication.
AI and Predictive Analytics	AI algorithms analyze data from sensors, BIM models, and historical data to predict potential problems.	Identifies potential delays, cost overruns, and safety hazards, allowing for proactive mitigation.

Table 1. has a systematized pattern that shows the key Internet of Things (IoT) applications that are used in construction management and its related benefit.

WATER RESOURCES AND ENVIRON WATER RESOURCES AND ENVIRONMENTAL ENGINEERING

Moreover, the Internet of Things presents a revolutionary prospect to environmental engineering and the management of water resources, as it could allow establishing smarter and more sustainable management approaches (Bouziane et al., 2021). The IoT use in this sphere covers multiple aspects, such as smart distribution and leakage, flood prediction and real-time warnings, wastewater plants and pollution control, and sustainable environment and compliance indicators (Alzahrani et al., 2023).

Intelligent Water Distribution and Water Leakage Management

The conventional water delivery networks have been characterized by inefficiencies such as loss of water as a result of leakages in the distribution network, and inabilities to monitor the water quality in real-time. Solutions that incorporate IoT will reduce these challenges due to the implementation of sensor networks across the water distribution system (Devasena et al., 2019). Such sensors measure some of the most important parameters: pressure, water flow, and quality to gather the data that can be used to detect the leaks and improve the water flow (Devasena et al., 2019; Kadar et al., 2018). Such systems are able to save water losses, cause less destruction to the infrastructure by finding and repairing the problem leaks as fast as possible, and resulted in greater efficiency in the entire water distribution system. In addition, predictive maintenance is a feature that can be implemented with the intelligent assistance of big-data analytics, permitting water utilities to forecast possible failures and the maintenance schedule in advance, thus prolonging the structural life of elements of the infrastructure (Bouziane et al., 2021).

IoT makes precision irrigation a futuristic tool, which enables farmers to conserve water consumption in farming, and at the same time maintain or even increase crop yield (Sharifnasab et al., 2023). The systems rely upon sensors to collect data regarding soil moisture, weather patterns, and plant health in order to facilitate controlled reservoir irrigation with all the time hence avoiding extensive water wastage and increasing harvest yield (Palaniappan et al., 2024). With IoT smart water management, high-resolution monitoring of the water level is available along the distribution networks with continuous monitoring, automated detection algorithm of the leak, quickening the response process and preventing high losses.

Forecast of Floods and Real-Time Warning

The flooding related hazards are severe to the people and infrastructures and therefore modelling of accurate forecast and pre-warning systems have to be developed. The IoT technologies are capable of improving flood management, using real-time data collected by a variety of sources, such as water-level sensors and weather stations, as well as rainfall gauges (Kamaruidzaman & Rahmat, 2020). The gadgets acquire information about water levels in rivers, streams, and reservoirs, allowing to detect possible floods and alert the communities at risk in time (Kamaruidzaman & Rahmat, 2020). Through real time warning, early interventions can be made against floods like evacuation and save structures, thus heralding limited effects of floods on the lives and property of people. This can be done through combining IoT data and advanced hydrological models to come up with realistic and dependable flood predictions thus enhancing the success of disaster response and mitigation.

Figure 4. Enhancing flood management with IoT

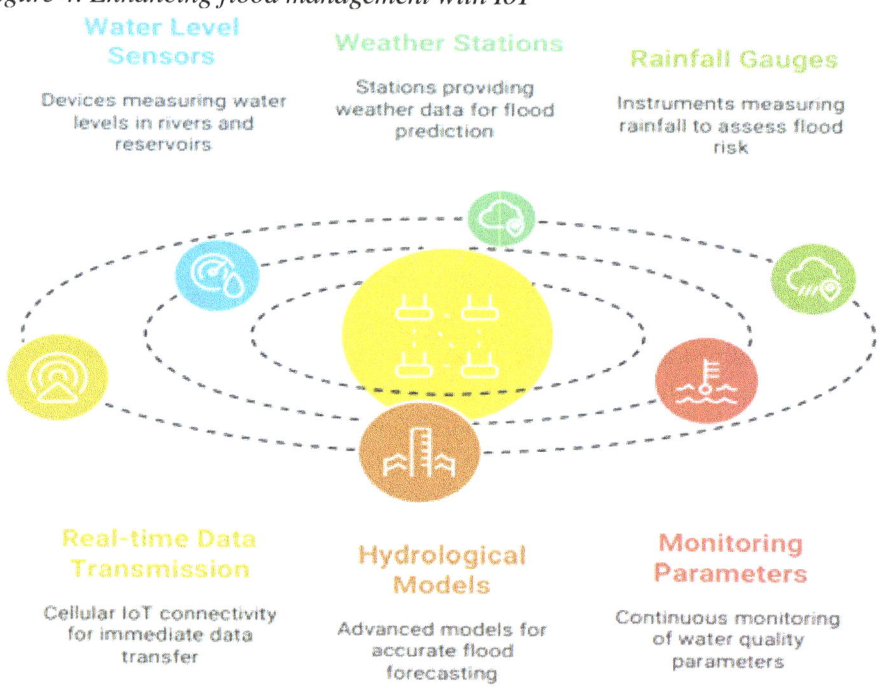

Implementation of cellular Internet of Things (IoT) connectivity in flood monitoring systems adds to the accuracy and stability of real-time information transferring and, thus, enhances flood prediction, along with warning issues broadcasting. Devices should be strategically deployed and installed at the rivers, lakes, and other water bodies that will constantly monitor the parameters of pH, temperature, dissolved oxygen, existence of chemicals and microorganisms (Miller et al., 2023). This functionality can be applied through Enhancing Flood Management with IoT refer Figure 4.

Pollution Control and Monitoring of Water

Wastewater treatment plants (WWTPs) are an epitome in the protection of the water sources against pollution. lOT-based systems facilitate more efficient involved in wastewater treatment by measuring the parameter of wastewater such as the PH, temperature, and substance levels (Alzahrani et al., 2023). Controlling, environment regulation compliance, and efficiency of the performance are achieved through real-time data to achieve accuracy. Simultaneously, the IoT sensors can keep an eye on pollution of rivers, lakes, and reservoirs, so that pollution incidents can be detected in time and mitigated fast to avoid pollution of water and destruction of the aquatic life. Smart water sensors activated via IoT also report about water quality, pressure and temperature, providing an LED array of abilities of the water utility provider to control the capture, treat and remove water (Alzahrani et al., 2023).

Sustainability and Environmental Performance Indicator

As an organization, there is growing consideration of the use of IoT technologies in the monitoring and control of the environmental performance in regards to the increased awareness and responsibility over environmental stewardship and scrutiny brought about by environmental regulations. Sensors will help store the data on air and water quality, energy use, and amount of generated trash, which, in its turn, will support sustainability indicators and indicate the areas in which one can improve. Data based approach focuses on decision makers to consider practices that are sustainable, fulfil regulatory requirements and reduce impact on the environment (Palaniappan et al., 2024).

Use Cases: The Urban Flood Warning System, the Water Quality Monitoring

Some case studies can serve as an example of IoT in the field of water resources and environmental engineering effectiveness. The infrastructure of smart cities implies such IoTs as sensors that control the state of the environment (air quality, temperature, humidity), consume water, and identify water leakage in distribution lines (Palaniappan et al., 2024). Moreover, IoT tools installed in households and shops can identify leaks and the possible issue of loss of water quality, enhancing public health management and water management (Miller et al., 2023).

ISSUES AND CONSTRAINTS OF IOT IN CIVIL ENGINEERING

Although the benefits of IoT to civil engineering are expected, many of the challenges and limitations have to be met in order to ensure successful implementation and the sustainability of its use. Such challenges cut across cybersecurity and data privacy, sensor reliability and robustness, data management, interoperability, cost, and human capital (Mohammad et al., 2024).

Concerns of Cybersecurity and Data Privacy

IoT interconnectedness creates vulnerability that increases the risk of data secrecy and infrastructure protection. Sensitive data must be defended against loss and malicious manipulation, and so strong encryption, authentication and access controls are needed (Palaniappan et al., 2024). The open architecture feature of IoT increases the risks of getting hit by malicious activity, and an infested device may face huge financial and physical losses (Kumar et al., 2023).

Robustness and Sensor Reliability in Harsh Conditions

Such situations are common in civil engineering environments where sensors are subjected to unpleasant environmental conditions of high temperatures, temperature changes, humidity, or high vibration levels among others. To guarantee the accuracy of data, it becomes important to have durable sensors and high-frequency calibration and strict methods of data validation to overcome environmental interference. Limitation of energy limits advanced security procedures and wireless communication exposes a network to eavesdropping. The IoT devices suffer time-based connectivity, resulting in a data loss or interference, which can be attributed to overlapping transmissions and wireless brownouts (Oshan, 2025).

Data Overload and Management

Due to increased IoT devices, it has created huge amount of data leading to issues of data storage, processing and analysis. There is the need to have effective strategies of managing the flow of data, deriving meaningful insight, and preventing data overload. This will include adopting scalable data storage applications, effective data processing algorithms, and data visualization instruments to allow engineers to make useful decisions based on the utilized available information. Attacks may also be predicted in advance with the help of machine learning (Alghamedy et al., 2024). Special significance of big data has in smart cities because they draw information pertinent to different sources (Palaniappan et al., 2024).

Problems of Interoperability and Standardization

IoT protocols and devices are not standardized, and this factor may slow interoperability, which means that it is challenging to combine various systems and exchange information on them (Palaniappan et al., 2024). The solution to this challenge has to come in the form of development of open standards which facilitate interoperability and communications among different systems and devices. Cross-layer communication can be regulated by adopting a de facto standard protocol, using secure APIs (Kumar et al., 2023). Engineering problems are represented by the demand of heterogeneous devices and the related wireless protocols these can support (Bagchi et al., 2020). The communication protocol, data format, and security specifications of the different IoT devices are not homogeneous, and hence, it becomes difficult to unify them into one system.

Cost, Scalability and Infrastructure Preparedness

Since constructing an IoT infrastructure including sensors, communications platforms, and complex data-processing systems can be expensive, this aspect presents a serious financial challenge to many ventures (Palaniappan et al., 2024). Moreover, expanding IoT implementations in terms of large geographic regions or complex infrastructure is another complexity added to the mix, which in many cases results in a price escalation. It is so because there will be a necessity of increased number of devices, greater capacity of the network and more sophisticated features of data analytics. Addressing these obstacles would have to take into consideration careful planning, a thorough cost-benefit analysis, and the investigation of new funding mechanisms to make sure that the IoT solutions would be economically feasible and extendable. Civil engineering initiatives will be important to consider cost-benefit ratio of IoT use before utilizing it in their initiatives (Palaniappan et al., 2024).

Skill Shortages and Skill Gaps

Proper utilization of IoT in the civil engineering needs proper human resource that have the appropriate skills and knowledge to design, deploy and maintain such systems. Nevertheless, the number of skilled specialists in such sphere as data analytics, cybersecurity, and integration of IoT systems is in a critical shortage (Almotairi, 2022). Such a skill gap should be rectified through investment in education and training programs that can provide any engineer and technician with both the knowledge and skills to succeed in the age of IoT.

FUTURE DIRECTIONS AND INTEGRATIONS OF TECHNOLOGIES FUTURE TRENDS AND INTEGRATIONS OF TECHNOLOGIES

Focus on civil engineering future is taken place in a number of enthusiastic trends and technological convergences, which are advanced to transform the way infrastructure is designed, built, and maintained. These are a combination of Artificial Intelligence and Machine Learning, predictive maintenance using digital twins, blockchain as a method of securely logging data, building of autonomous infrastructure and self-healing systems with the proliferation of smart city technologies, and an increased desire to have cross-disciplinary collaboration (Sargiotis, 2024).

Crossing Over with Artificial Intelligence and Machine Learning

AI and ML will revolutionize civil engineering because they allow making decisions using data, automate operations, and increase predictability (Rane, 2023). AI algorithms can process big data generated by sensors and other systems to discover patterns and hence predict and prevent possible failures as well as achieve optimal performance of the infrastructure (Acemoğlu et al., 2022). The models that can be trained using ML techniques can forecast the traffic flows, maximize energy use in buildings and identify structural health anomalies (Alghamedy et al., 2024). AI can lead to knowledge about which populations are in danger of being left behind by utilizing information (demographics and preferences) to make better policies and solutions to the unequal access to AI and its answer to all walks of life. Artificial intelligence and smart construction technologies are causing a profound shift in infrastructure delivery and are having a far-reaching influence on everything, including design and operation (Liu et al., 2025). The combination of the AI with finite element analysis can be used to improve the accuracy of simulation and strengthen the infrastructure resilience. With the help of AI algorithms, one will be able to predict the outcomes of various planning strategies, suggest the most optimal land use strategies, identify particular locations, where infrastructural improvement is essential, and service expansion must be made (Acemoğlu et al., 2022).

The Role of Digital Twins in Predictive Maintenance

A virtual representation of physical assets, digital twins are becoming very strong predictive maintenance and infrastructure management tools. These digital twins dynamically and comprehensively exhibit the detailed representation of the real infrastructure in the world including information on the design, materials, and operating parameters. Through constant monitoring of the physical asset with the help of sensors and real-time data feed and comparing it in relation to the digital twin, engineers will be able to detect the associated issues early on and intervene by taking preventive measures to avoid failures, (Machalski et al., 2025; Yang et al., 2025). Digital twins may also support varying variables, including extreme weather situations or higher traffic volumes and can help optimize maintenance timelines, boost the durability of infrastructure assets, and salvage on maintenance expenses. Moreover, with the assistance of digital twins, collaboration between stakeholders is improved by centralizing the access to infrastructure data, the possibility of its visualization, the decision-making process, and fostering transparency across the asset lifecycle.

Blockchain-Secure Infrastructure Data Logging

Blockchain technology provides the capacity of secure and transparent data management of infrastructure data, data integrity and tampering prevention (Rahman et al., 2022). That is done through the establishment of a decentralized, un-edit-able ledger that logs transactions in a verifiable and auditable way. It is possible to track the lineage of construction materials, track performance of infrastructure assets and manage contracts and payments using a blockchain. The blockchain allows reducing disagreements, streamlining operations, and increasing the trust of stakeholders by offering a single source of truth of all information related to the infrastructure. The decentralized structure reduces any chances of cyberattacks and guarantees that information is always present and accessible.

Self-Healing and Autonomous Infrastructure

Creation of autonomous infrastructure and self-healing systems is long term and would ensure that a good part of maintenance costs could be saved and the infrastructure can be much more resilient. Robots and drones can be used in autonomous systems to perform a full check and make the required adjustments and solutions so that the threats are detected on time. On one hand, the existing self-healing systems could exploit the use of more advanced materials capable of repairing any type of damage automatically, reducing downtimes, and maximizing the life of any infrastructure assets. Such innovations are predictable to transform the civil engineering profession, improving the structural strength, sustainability and the general performance of the projects (Liu et al., 2025; Rojas et al., 2025).

Applications of Smart City

Innovation In civil engineering, the smart city applications implement the principles of the IoT-based devices and data analytics in making urban infrastructure more efficient and sustainable (Palaniappan et al., 2024). Intelligent streetlights can ensure they become brighter in traffic and dim during the evenings; intelligent system of water can pick up the leaks and optimize water distribution; intelligent transportation can transport people more efficiently and hence reduce traffic jams and their air-degrading effects. The joint development of AI through IoT may be seen in how interconnected objects can collect data and process them to optimize the processes (Acemoğlu et al., 2022). These applications get translated into different areas of the urban living such as management of waste, energy distribution, safety of people and citizen participation. As an example of smart waste management systems, one can prescribe the maximum efficiency of routes used in waste collection and reducing the amount of waste that is delivered to landfills, and smart grids may enhance energy efficiency and lower carbon emissions. The data on IoT devices can also be utilized to enhance the safety of the population by identifying and reacting on an emergency in real-time. Cities should use data analytics and AI to make informed decisions, enhance their living standards, and develop sustainability and resilient cities (Abdullahi et al., 2025).

Demand of Inter- Disciplinary Partnerships

The multidimensional nature of problems that civil engineering has to deal with demands cross-disciplinary cooperation between engineers, computer scientists, data scientists, and other professionals. The commonality of these experts has the ability to design novel solutions, which may take advantage of the most up-to-date technologies and recognize the social, economic, and environmental outcomes of infrastructure construction. Moreover, these partnerships also lead to a integrated way of doing things, whereby stakeholders see to it that projects done in infrastructure have a technical aspect, but a social aspect as well as aspect of sustainability. Such collaborations may result in establishing additional educational courses and training systems that will raise the new generation of civil engineers who will be able to deal with the future issues.

POLICY, GOVERNANCE, AND ETHICS

Including the IoT in civil infrastructure will require taking the policy, governance, and ethical aspects carefully into consideration, so that a deployment is done responsibly and advantageously. The main parent aspects are regulatory standards, ethical use of data, use of public-private collaboration, the role of the academia, industry, and government (Kumar et al., 2023).

IoT Regulation Standards of Civil Infrastructure

This includes the need to establish clear regulatory standards in the design and deployment of the IoT in the civil infrastructure to facilitate the development of trust and adoption of the IoT in civil infrastructures. Such standards need to focus on such important issues as data privacy, cybersecurity, interoperability, and safety (Kumar et al., 2023). However, governments are very instrumental in setting these standards as indicated in the paper, A framework of Internet of Things-enabled smart government: A case of IoT cybersecurity policies and use cases in U.S. federal government (Babiceanu, 2019). There should be standards put in place which can conform to the technological advancements and one that is ethically sound. Similarly, legislative systems must support the provisions of data protection regulations and local regulations regarding blockchain administration, as in the case of convergence of the IoT, blockchain, and computational intelligence in smart cities (Kumar et al., 2023). Lack of such standards may create certain uncertainties and even risks in deploying IoT.

Concerns about Ethical Data Usage and Surveillance

This vast amount of data that is stored through IoT devices brings about serious ethical issues in regards to privacy invasion, surveillance and abuse of the data. To secure the rights to individual privacy and avoid discriminating or unethical behavior, it is important to define clear policies on the data collection, storage, and usage (Jacobs et al., 2020). Trust and ethical use of IoT data depend on the importance of transparency and user consent. With the spread of AI, a concern is emerging to solve possible inequities by exposing demographics and preferences of the population to data as the basis of policies and resolutions to equitability distributions (Dhirani et al., 2023). Strict data regulations and data transparency frameworks should be established in cities to ensure data is not used out of context and to meet the confidence of the people (Palaniappan et al., 2024).

PPPs

The use of the public-private partnership can encourage the implementation and operation of IoT infrastructure because it helps to utilize the experience and resources of the civilian firms without undermining the interests of the government (Kelly et al., 2020). Academic institutions can assist governments in conducting research to contribute to the development of infrastructure by working with technology suppliers in such domains as AI, IoT, and cybersecurity (Palaniappan et al., 2024). This type of collaboration may contribute to IoT solution faster implementation and innovation. SUCCPI PPP needs the partnership between the government officials and the technological providers so that funding and advancement of the infrastructures related to it could be achieved (Palaniappan et al., 2024).

Academia, Industry and Government Role

The development and deployment of IoT in civil engineering are vital roles of academia, industry, and government (Chatterjee, 2020). The academia can investigate possible risks and advantages, create ethical frameworks and educate future workers in the sphere of Internet of Things. The industry is able to create and adopt safe, trustworthy and ethical solutions of IoT. Governments are able to build regulatory standards, foster innovations, and make sure that the interest of the population is not violated. Moreover, governments must take into consideration pro-innovation solutions in regulation of the processes and focus on the need to protect citizens, such as privacy laws and online security policies (Palaniappan et al., 2024).

The advent of the AI of enterprise resource-planning system is a revolutionary change that shows potential in streamlining different business processes as well as becoming effective in making general business operations more optimized.

CONCLUSION

To sum up, the application of IoT to civil engineering is the game changer, which is expected to transform the infrastructure design, construction, and operation. IoT technologies can provide unequaled possibilities in terms of creating greener, smarter, and more resilient infrastructure solutions, as they improve structural health monitoring and make predictive maintenance possible, to optimize smart city applications. Enhancing cross-disciplinary collaborations, policy, governance, and ethical issues could allow engineers and planners to fully utilize the power of the IoT to enhance the efficiency, safety, and sustainability of all civil infrastructure. The coincidence of AI and IoT, as mentioned by the editor, is the most optimistic one, as it allows making data-based decisions and automating several dimensions of urban life.

Its vision of the future of civil engineering is that the civil engineering infrastructure system would be resilient, intelligent and sustainable, which is smoothly intertwined with urban environment. The vision will need the whole package that would take into account the social, economic and environmental effects of infrastructure development, as well as use the recent technologies in dealing with the multidimensional challenges facing cities. Other important suggestions to engineers and planners have to do with the adoption of cross-disciplinary collaborations, the first-place usage of data and privacy, and direct involvements in the formulation of regulatory guidelines of the IoT within civil infrastructures. Through the adoption of such principles, we will be laying the foundation toward a future where infrastructure systems may be not only cutting edge technologically, but also socially responsible and sustainable. With the development of smart cities, the safe, efficient, and modular use of technologies, including blockchain, IoT, and 6G, should improve city life.

REFERENCES

Abbasi, S., Rahmani, A. M., Balador, A., & Sahafi, A. (2021). Internet of Vehicles: Architecture, services, and applications. *International Journal of Communication Systems*, *34*(10), e4793. Advance online publication. DOI: 10.1002/dac.4793

Abdullahi, U., Zhang, W., Cao, Y., & Irankunda, G. (2025). Integrating IoT Technology for Fire Risk Monitoring and Assessment in Residential Building Design. *Buildings (Basel, Switzerland)*, *15*(8), 1346. DOI: 10.3390/buildings15081346

Acemoğlu, D., Autor, D., Hazell, J., & Restrepo, P. (2022). Artificial Intelligence and Jobs: Evidence from Online Vacancies. *Journal of Labor Economics*, *40*(S1), S293–S340. Advance online publication. DOI: 10.1086/718327

Alghamedy, F., El-Haggar, N., Alsumayt, A., Alfawaer, Z. M., Alshammari, M., Amouri, L., Aljameel, S. S., & Al-Bassam, S. (2024). Unlocking a Promising Future: Integrating Blockchain Technology and FL-IoT in the Journey to 6G. *IEEE Access : Practical Innovations, Open Solutions*, *1*, 115411–115447. Advance online publication. DOI: 10.1109/ACCESS.2024.3435968

Almotairi, K. H. (2022). Application of internet of things in healthcare domain.

Alzahrani, A. I. A., Chauhdary, S. H., & Alshdadi, A. A. (2023). Internet of Things (IoT)-Based Wastewater Management in Smart Cities. *Electronics (Basel)*, *12*(12), 2590. DOI: 10.3390/electronics12122590

Babiceanu, R. F. (2019). A framework for Internet of Things-enabled smart government: A case of IoT cybersecurity policies and use cases in U.S. federal government. https://www.sciencedirect.com/science/article/pii/S0740624X17304847

Bagchi, S., Abdelzaher, T., Govindan, R., Shenoy, P., Atrey, A., Ghosh, P., & Xu, R. (2020). New Frontiers in IoT: Networking, Systems, Reliability, and Security Challenges. arXiv (Cornell University). https://doi.org//arxiv.2005.07338DOI: 10.48550

Bouziane, S., Aghoutane, B., Moumen, A., Sahlaoui, A., & Ouali, A. E. (2021). Proposal of a Big data System for an Intelligent Management of Water Resources. E3S Web of Conferences, 314, 2002. DOI: 10.1051/e3sconf/202131402002

Chatterjee, S. (2020). The safety of IoT-enabled system in smart cities of India: Do ethics matter? *International Journal of Ethics and Systems*, *36*(4), 601–618. DOI: 10.1108/IJOES-05-2019-0085

Choi, Y. J., Park, H. W., Mi, Y., & Song, S. (2024). Crack Detection and Analysis of Concrete Structures Based on Neural Network and Clustering. *Sensors (Basel), 24*(6), 1725. DOI: 10.3390/s24061725

Choudhary, V., Guha, P., Pau, G., & Mishra, S. (2025). An Overview of Smart Agriculture Using Internet of Things (IoT) and Web Services. *Environmental and Sustainability Indicators, 100607*, 100607. Advance online publication. DOI: 10.1016/j.indic.2025.100607

Devasena, D., Ramya, R., Dharshan, Y., Vivek, S. S., & Darshan, T. (2019). IoT based water distribution system. *International Journal of Engineering and Advanced Technology, 8*(6), 132–135. DOI: 10.35940/ijeat.D6295.088619

Dhirani, L. L., Mukhtiar, N., Chowdhry, B. S., & Newe, T. (2023). Ethical Dilemmas and Privacy Issues in Emerging Technologies: A Review, Sensors, 23(3), 1151. Multidisciplinary Digital Publishing Institute. DOI: 10.3390/s23031151

Dirin, A., Oliver, I., & Laine, T. H. (2023). A Security Framework for Increasing Data and Device Integrity in Internet of Things Systems. *Sensors (Basel), 23*(17), 7532. DOI: 10.3390/s23177532

Garg, S., Mehrotra, D., Pandey, H. M., & Pandey, S. (2021). Accessible review of internet of vehicle models for intelligent transportation and research gaps for potential future directions. *Peer-to-Peer Networking and Applications, 14*(2), 978–1005. DOI: 10.1007/s12083-020-01054-6

Jacobs, N., Marković, M., Cottrill, C., Edwards, P., Corsar, D., & Salt, K. (2020). Made-Up Rubbish: Design Fiction as a Tool for Participatory Internet of Things Research. *The Design Journal, 23*(3), 419–440. DOI: 10.1080/14606925.2020.1744259

Kadar, H. H., Sameon, S. S., & Rusli, M. E. (2018). SMART2L: Smart Water Level and Leakage Detection. *IACSIT International Journal of Engineering and Technology, 7*(4.36), 448–452. DOI: 10.14419/ijet.v7i4.36.23914

Kamaruidzaman, N. S., & Rahmat, S. N. (2020). *Water Monitoring System Embedded with Internet of Things (IoT) Device: A Review IOP Conference Series Earth and Environmental Science, 498(1), 12068.* IOP Publishing., DOI: 10.1088/1755-1315/498/1/012068

Kelly, J. T., Campbell, K. L., Gong, E., & Scuffham, P. (2020). The Internet of Things: Impact and Implications for Health Care Delivery. Journal of Medical Internet Research, 22(11). JMIR Publications. DOI: 10.2196/20135

Khalil, U., Malik, O. A., Uddin, M., & Chen, C. (2022). *A Comparative Analysis on Blockchain versus Centralized Authentication Architectures for IoT-Enabled Smart Devices in Smart Cities: A Comprehensive Review, Recent Advances, and Future Research Directions Sensors, 22(14), 5168.* Multidisciplinary Digital Publishing Institute., DOI: 10.3390/s22145168

Kumar, R., Jain, V., Leong, H. W., & Teyarachakul, S. (2023). Convergence of IoT, Blockchain, and Computational Intelligence in Smart Cities. In *CRC Press eBooks*. Informa., DOI: 10.1201/9781003353034

Liu, K., Wang, T., & Zhang, X. (2025). Intelligent Techniques Applied in Infrastructure, Engineering, and Construction. *Buildings (Basel, Switzerland)*, *15*(11), 1799. DOI: 10.3390/buildings15111799

Machalski, A., Szulc, P., Błoński, D., Nycz, A., Nemś, M., Skrzypacz, J., Janik, P., & Satława, Z. (2025). The Concept of a Digital Twin for the Wały Śląskie Hydroelectric Power Plant: A Case Study in Poland. *Energies*, *18*(8), 2021. DOI: 10.3390/en18082021

Masmali, F. H., & Miah, S. J. (2021). Different Applications and Technologies of Internet of Things (IoT). arXiv (Cornell University). https://doi.org//arXiv.2110.10452DOI: 10.48550

Miller, M., Kisiel, A., Cembrowska-Lech, D., Durlik, I., & Miller, T. (2023). IoT in Water Quality Monitoring—Are We Really Here? *Sensors (Basel)*, *23*(2), 960. DOI: 10.3390/s23020960

Mohammad, N., Khatoon, R., Nilima, S. I., Akter, J., Kamruzzaman, M., & Sozib, H. M. (2024). Ensuring Security and Privacy in the Internet of Things: Challenges and Solutions. *Journal of Computer and Communications*, *12*(8), 257–277. DOI: 10.4236/jcc.2024.128016

Murroni, M., Anedda, M., Fadda, M., Ruiu, P., Popescu, V., Zaharia, C., & Giusto, D. (2023). 6G—Enabling the New Smart City: A Survey. Sensors, 23(17), 7528. Multidisciplinary Digital Publishing Institute. DOI: 10.3390/s23177528

Muttillo, M., Stornelli, V., Alaggio, R., Paolucci, R., Battista, L. D., de Rubeis, T., & Ferri, G. (2020). Structural Health Monitoring: An IoT Sensor System for Structural Damage Indicator Evaluation. *Sensors (Basel)*, *20*(17), 4908. DOI: 10.3390/s20174908

Oshan, P. (2025). Ensuring Reliability and Accuracy in IoT-Generated Data: Challenges, Solutions, and Future Trends. https://www.encardio.com/blog/ensuring-reliable-accurate-iot-data-challenges-solutions

Palaniappan, D., Premavathi, T., Jain, R., Parmar, K., & Jhanvi, M. (2024). Blockchain-Based IoT-Enabled secure 6G smart City applications. In *Advances in civil and industrial engineering book series* (pp. 335–364). DOI: 10.4018/979-8-3693-8029-1.ch012

Parihar, A., Prajapati, J. B., Prajapati, B. G., Trambadiya, B., Thakkar, A., & Engineer, P. (2024). Role of IOT in healthcare: Applications, security & privacy concerns.

Rahman, Z., Yi, X., & Mehedi, Sk. T., Islam, M., & Kelarev, A. (2022). Blockchain Applicability for the Internet of Things: Performance and Scalability Challenges and Solutions. arXiv (Cornell University). https://doi.org//arXiv.2205.00384DOI: 10.48550

Rane, N. (2023). Integrating Leading-Edge Artificial Intelligence (AI), Internet of Things (IoT), and Big Data Technologies for Smart and Sustainable Architecture, Engineering and Construction (AEC) Industry: Challenges and Future Directions. SSRN Electronic Journal. DOI: 10.2139/ssrn.4640926

Richard, (2023). IoT in Construction: Top Benefits, Use-Cases & Application. https://toolsense.io/equipment-management/iot-in-construction-top-benefits-use-cases-application/

Rojas, L., Peña, Á., & García, J. (2025). AI-Driven Predictive Maintenance in Mining: A Systematic Literature Review on Fault Detection, Digital Twins, and Intelligent Asset Management. *Applied Sciences (Basel, Switzerland)*, *15*(6), 3337. DOI: 10.3390/app15063337

Sargiotis, D. (2024). Advancing Civil Engineering with AI and Machine Learning: From Structural Health to Sustainable Development. DOI: 10.51483/IJARP.4.2.2024.54-81

Sharifnasab, H., Mahrokh, A., Dehghanisanij, H., Łazuka, E., Łagód, G., & Karami, H. (2023). Evaluating the Use of Intelligent Irrigation Systems Based on the IoT in Grain Corn Irrigation. *Water (Basel)*, *15*(7), 1394. DOI: 10.3390/w15071394

Tan, P., Wang, Z., & Chang, X. (2025). Road Obstacle Detection Method Based on Improved YOLOv5. *Algorithms*, *18*(6), 300. DOI: 10.3390/a18060300

Tariq, U., Ahmed, I., Bashir, A. K., & Shaukat, K. (2023). *A Critical Cybersecurity Analysis and Future Research Directions for the Internet of Things: A Comprehensive Review Sensors, 23(8), 4117*. Multidisciplinary Digital Publishing Institute., DOI: 10.3390/s23084117

Wang, T., Cui, T., Qi, X., & Huan, X. (2025). *Urban Viaduct Structural Health Monitoring: A Review of Wireless Sensor Approaches Buildings, 15(10), 1619.* Multidisciplinary Digital Publishing Institute., DOI: 10.3390/buildings15101619

Yang, S.-W., Lee, Y. Z., & Kim, S.-A. (2025). Design and Validation of a Real-Time Maintenance Monitoring System Using BIM and Digital Twin Integration. *Buildings (Basel, Switzerland)*, *15*(8), 1312. DOI: 10.3390/buildings15081312

Ye, X., Jin, T., & Yun, C. (2019). A review on deep learning-based structural health monitoring of civil infrastructures. Smart Structures and Systems, 24(5), 567. Technopress. DOI: 10.12989/sss.2019.24.5.567

Chapter 11
Lifecycle Assessment and Structural Degradation

Mohammed Ameen Haitham Ahmed Alhababi
College of Administration and Economics, Samarra University, Iraq

Hameed Hasan Khalaf
College of Administration and Economics, Samarra University, Iraq

Moqdad Ibrahim Jassem
College of Administration and Economics, Samarra University, Iraq

Shihab A. Shawkat
https://orcid.org/0000-0002-9529-2151
University of Samarra, Iraq

ABSTRACT

This research presents an integrated analysis that combines Life Cycle Assessment (LCA) and structural degradation of structural materials and components, with the aim of understanding the relationship between long-term performance and the environmental impact of buildings. LCA is an important tool for analyzing environmental impacts from the extraction of raw materials to the end of their service life. Structural degradation represents a direct challenge to the sustainability of buildings and includes mechanisms such as corrosion, fatigue, and creep. The study highlights the importance of shifting from static models to dynamic models based on real data, using artificial intelligence and real-time monitoring techniques, as illustrated in Figure (6). It recommends applying this methodology in the design, maintenance, and environmental assessment phases to enhance the sustainability and operational efficiency of facilities.

DOI: 10.4018/979-8-3373-5248-0.ch011

INTRODUCTION

Modern societies face increasing challenges related to environmental sustainability, structural safety, and the economic efficiency of resources used in infrastructure and industrial projects. In this context, two supplements have emerged to solve these challenges using accurate analytical devices: life cycle assessment (LCA) and structural decline (Klüppel 2005). Each forms an independent framework to analyze and understand the performance of products and features in its lifetime, but their integration opens widespread possibilities to develop smart design and maintenance models based on accurate data and predictions, contributing to reducing environmental impacts and improving the longevity of features. Life Cycle Assessment (LCA) is a systematic and standardized tool for estimating the total environmental impact of any product, process, or system over its entire life cycle (Guinée 2001).

This cycle involves a sequential chain of steps, which begins with the extraction of raw materials, through manufacturing, transportation and use, and ends with final disposal or recycling. Therefore, the life cycle assessment depends on the "Cradle-to-Grave" principle, which enables analysis of each phase cumulative environmental effects, whether greenhouse gas emissions, energy and water consumption, or solid and liquid waste production (Ding 2008). Structural deterioration refers to a set of physical and chemical processes that lead to a decline in the mechanical and functional properties of structures and materials over time. These processes include various types of deterioration, such as creep, fatigue, corrosion, and cracking resulting from environmental interactions or overloads (Yang *et al.* 2024). Understanding these mechanisms is critical in determining the "service life" of structural components and making early decisions regarding maintenance, replacement, or reinforcement (Liu, F. *et al.* 2017).

Both approaches converge at a central point: managing the service life of structures to meet environmental and economic stability requirements. While the life cycle assessment provides a composite approach to the environmental impact of a material or product, the structural fall analysis (SDA) provides quantitative tools to identify when and how the material loses its ability to perform effectively (Brischke & Rapp, 2008). By combining these two approaches, smart, more realistic assessment systems can be developed, support decision-makers during the initial design phase and subsequently reduce the need for emergency or expensive intervention (Asdrubali *et al.* 2013). In this context, recent years have witnessed a growing trend toward integrating LCA with structural degradation models into a unified framework, supported by the use of artificial intelligence, numerical simulation tools, and live facility monitoring systems (such as the Internet of Things (IoT) (Buyle *et al.* 2013).

This integration has become a key research focus in civil engineering, materials engineering, and industrial design, and has extended to the performance assessment of critical infrastructure such as bridges, smart buildings, sewage systems, and energy networks, sewage and energy networks, and others (Gervasio *et al 2018*). Based on this, this research examines and analyzes the complex relationship between life cycle assessment and structural deterioration, with the aim of building a comprehensive knowledge model that contributes to enhancing sustainability in the design and maintenance of facilities, and achieving a balance between technical performance and environmental and economic efficiency (Bertolini *et al.* 2013).

Despite the growing literature on LCA and structural degradation, most studies have treated them as separate frameworks, limiting their capacity to address long-term sustainability challenges (Russo, G. 2004). Existing research often emphasizes environmental impacts without accounting for real-time deterioration, or conversely, focuses on degradation mechanisms without linking them to environmental performance (Hamad *et al.* 2024). This chapter addresses this gap by proposing an integrated framework that operationalizes the synergy between LCA and structural degradation, thus providing both a conceptual and practical contribution to sustainable infrastructure research.

IMPORTANCE OF THE TOPIC

Studying the relationship between life cycle assessment and structural deterioration is of paramount importance in light of the rapid changes the world is witnessing in terms of urban growth, increasing demand for resources, and global environmental challenges such as climate change, rising pollution levels, and carbon emissions from the industrial and construction sectors (Niedźwiedzińska, 2021). With the development of concepts related to sustainable development, engineering decisions are now not only based on only efficiency and ideas of performance, but a long -term perspective that takes into account environment, social and economic impacts over time (Abdulbaqi. *et al.* 2023).

The subject is particularly important in the civil engineering sector, where features and infrastructure are among the largest contributors of carbon emissions and natural resources. Several studies have shown that a significant proportion of the environmental impact of buildings, roads and bridges, especially during operation, maintenance and repair, results from post -construction stages (Abdullah Hamad, *et al.* 2025). Therefore, integrating the concepts of structural deterioration with life cycle assessment enables a better understanding of how a facility's performance evolves over time, contributing to reducing life cycle costs and achieving environmental efficiency.

Furthermore, unexpected structural deterioration poses a significant risk to public safety and community security, especially in vital facilities such as airports, dams, hospitals, and schools. Sudden structural failure in these facilities, due to deterioration factors not considered during design or maintenance, can lead to significant human and economic losses. This is the place where environmental impact assessment is the right value to integrate a decline analysis in functioning (Hussein *et al.* 2024). This integration enables us to predict future problems and plan active maintenance interventions that ensure stability and safety of convenience.

The importance of this integration extends beyond environmental considerations to encompass economic dimensions. By understanding the relationship between deterioration and environmental impact, designers, engineers, and planners can make smart economic decisions, such as selecting more durable or less environmentally impactful materials, modifying designs to match actual operating conditions, or improving maintenance programs to reduce resource waste (Hameed *et al.* 2020). In this way, sustainability becomes not only an environmental requirement but also an effective tool for achieving economic efficiency and reducing government and private spending in the long term.

The topic is becoming increasingly important in light of the rapid development of measurement and monitoring technologies. Using technologies such as the Internet of Things, smart sensors, and non-destructive monitoring, it is now possible to track the condition of buildings in real time and update LCA and deterioration models based on real-time, continuous data (Kang & Lee, 2018). Artificial intelligence and advanced data processing technologies also enable the development of accurate predictive systems that can support design and operational decisions with unprecedented efficiency.

Linking life cycle assessment and structural deterioration is not an academic luxury or scientific theory; it is a realistic and necessary response to the challenges facing the construction and materials sector in the modern era. As the world moves towards a circular economy, clean energy, and a reduced carbon footprint, this multidimensional approach becomes a vital entry point for building smarter, longer-lasting, and less environmentally impactful buildings (Aborujilah *et al.* 2023).

Beyond technical considerations, the research problem also reflects deeper structural and epistemological shifts in global knowledge systems. Integrating LCA with structural degradation illustrates how engineering research is moving from isolated, discipline-specific approaches toward systemic, foresight-driven methodologies that bridge environmental science, material durability, and policy design. This epistemological shift redefines not only how sustainability is measured but also how science contributes to societal resilience and governance.

THE LIFE CYCLE ASSESSMENT (LCA) FRAMEWORK

Life Cycle Assessment (LCA) is one of the most prominent and internationally recognized methodological tools for analyzing the environmental impact of products, processes, or systems over their entire life cycle. This device was developed as the standardization (ISO) standards for the International Organization, especially as ISO 14040 and ISO 14044, which aims to provide a standardized approach to identifying, analyzes and assessing overall environmental impacts. This approach supports strategic decisions in industry, engineering, and public policy.

General Concept

LCA is based on the principle of a comprehensive, multi-stage assessment, whereby each stage of the life cycle is viewed as an additional element to the "total environmental footprint" of a product or system. This cycle begins with extraction of raw materials (such as mining, agriculture, or chemical extraction), then manufacturing and processing, followed by transport and distribution, then use and maintenance, and ending with final disposal or recycling. This full path is often referred to as the concept "cradle-to-grav", but it can sometimes be brief for "cradle-to-gate" if the analysis is limited to pre-utilization stages.

Life Cycle Assessment Phases

According to international standards, a life cycle assessment goes through four main, integrated phases:

a. Goal and Scope Definition:

This phase includes defining the primary objective of the study (for example, comparing design alternatives, or identifying a phase with a high environmental impact) and identifying the "functional unit" against which environmental performance is measured. This phase also defines the system boundaries and clarifies the assumptions used.

b. Life Cycle Inventory (LCI):

This is the quantitative phase in which all data related to the inputs and outputs for each phase of the life cycle are collected. This data includes energy use, material consumption, air emissions, effluents, and other resources and impacts.

c. Life Cycle Impact Assessment (LCIA):

In this phase, inventory data is converted into environmental impact assessment indicators such as climate change (Global Warming Potential), resource depletion, acidification, biodegradation, and others. Advanced models (such as CML, TRACI, ReCiPe) are used to analyze multidimensional environmental impacts.

d. Interpretation:

This phase includes analyzing results, identifying sources of uncertainty, comparing scenarios, and providing clear recommendations for informed decision-making. This phase ensures that the data and assumptions are consistent with the initial objective of the study.

Applications of LCA in Engineering and Construction

LCA is widely used to assess the environmental sustainability of building materials, insulation systems, and design alternatives in buildings, bridges, and transportation networks. For example, the environmental impact of conventional concrete versus high-performance concrete or composite materials can be compared not only in terms of direct emissions, but also across their entire life cycle, including maintenance and end-of-service requirements.

This tool is also used to determine the environmental feasibility of initial design options (such as steel columns versus concrete), and to analyze the impacts of material selection on the thermal performance of buildings and the carbon emissions resulting from their life cycle and maintenance. LCA enables decision makers, designers, and engineers to make choices based on accurate quantitative analysis rather than relying on intuition or preconceived preferences. It also enhances the transparency of industrial and engineering processes, contributes to compliance with environmental legislation, and improves companies' image in the eyes of society and investors. Because LCA is not limited to environmental impact alone, but can be expanded to include economic and social analysis (known as LCSA – Life Cycle Sustainability Assessment), it has become a central tool in managing major construction projects, planning maintenance strategies, and estimating life cycle costs (Life Cycle Costing – LCC).

STRUCTURAL DEGRADATION CONCEPTS

Structural degradation is a fundamental phenomenon that directly impacts the efficiency and safety of engineering structures, particularly as materials age and are exposed to various environmental and mechanical factors. Structural degraded deterioration refers to the cumulative process that leads to a gradual or sudden decline in the physical and mechanical properties of a material or structural element, which may result in impaired functional performance, reduced service life, or even catastrophic collapse. From this perspective, understanding degraded structures in its various types and mechanisms has become a critical factor in the design of structures, assessing their in-service condition, and planning preventive and corrective maintenance operations.

1. Classification of Structural Degradation: Structural degrading can be classified into several main types, varying depending on the underlying cause and operating environment. The most important of these are:
   a. Mechanical Degradation: This occurs as a result of repeated or excessive loads. Its most prominent manifestations are:
      - Fatigue: This occurs as a result of cyclic or reciprocating loading, leading to the appearance of microcracks that gradually develop until failure.
      - Creep: This occurs in materials under the influence of constant, long-term loads, especially at high temperatures, and causes permanent deformations.
      - Strain Cracking: This occurs as a result of uneven expansion or contraction between different parts of the structure or material.
   b. Chemical and Environmental Degradation: This includes interactions between materials and their surrounding environment, such as:
      - Corrosion: This occurs particularly in metal structures as a result of exposure to moisture, salts, or chemicals.
      - Chemical decomposition: As in polymeric or organic materials, under the influence of environmental factors or ultraviolet radiation.
      - Carbonation: A process that affects concrete through the reaction of carbon dioxide with its components, reducing the alkalinity of the medium and increasing the likelihood of iron rusting.
   c. Deterioration resulting from harsh environmental conditions: Such as repeated exposure to freeze-thaw cycles, extreme humidity, or thermal expansion resulting from temperature changes, all of which affect the cohesion and durability of materials.

2. Biological degradation: This occurs as a result of the growth of microorganisms, fungi, or algae, especially in humid or marine environments, leading to the decomposition of materials or increased permeability.
   a. Deterioration indicators and their impact on functional performance:

Structural deterioration is measured through multiple indicators, such as: a decrease in the modulus of elasticity, a change in tensile or compressive strength, a crack growth rate, a change in the shape of the structural element, or a change in dynamic properties such as natural frequency. Non-destructive testing (NDT) technology, such as thermography, ultrasound, X-rays and vibration monitoring, can be used for monitoring these indicators. The fall, if not detected early or properly addressed, can give rise to the cumulative effects that endangered the security of the convenience, increase the cost of maintenance, and reduces long -term economic viability. For this reason, linking deterioration models to design planning and life cycle assessment is an essential step toward developing smart and proactive maintenance systems.

b. The impact of deterioration on the sustainability of facilities:
   - Deterioration negatively impacts sustainability in three ways:
   - Environmentally: by increasing the need for materials and corrective interventions that consume energy and release emissions.
   - Economically: by increasing the costs of repair or premature replacement of materials.
   - Socially: by disrupting facilities or compromising the safety of their users.

Therefore, a precise scientific understanding of deterioration mechanisms forms the basis for developing design and operation plans that preserve the life of facilities and minimize their negative impacts.

THE INTEGRATION BETWEEN LCA AND STRUCTURAL DEGRADATION

The integration of Life Cycle Assessment (LCA) and structural degradation analysis represents the intersection of environmental sustainability and mechanical and material engineering for structures. It provides a comprehensive framework for scientific and environmentally sound design and maintenance decisions. Although both approaches are used separately in design and evaluation practices, their integration provides many benefits, from better environmental efficiency to precise service life prediction and reduces operating risks and costs.

1. Practical Pathways for Integration:

While the rationale for integrating LCA with structural degradation models is evident, a practical roadmap is necessary to guide implementation. This integration can be achieved through three key steps:

a. Data Coupling: Employ real-time monitoring tools (IoT sensors, NDT methods) to feed degradation indicators directly into LCA databases, ensuring dynamic updates.
b. Predictive Modeling: Use AI-based simulations and scenario modeling to translate degradation patterns into quantifiable environmental impacts, allowing planners to forecast life cycle outcomes more accurately.
c. Decision Support Systems: Develop digital platforms that combine LCA outputs with degradation forecasts to assist policymakers, designers, and engineers in choosing optimal materials, maintenance strategies, and regulatory frameworks. These mechanisms transform the conceptual integration of LCA and structural degradation into actionable tools for sustainable engineering practice.

RESULT

The Relationship Between Long-Term Performance and Environmental Impact

Studies show that structural deterioration directly impacts the environmental impact assessment of structures and materials. Accelerated deterioration leads to:

1. Increased resource consumption for maintenance or replacement.
2. Increased emissions from frequent repairs.
3. Reduced actual service life compared to the expected design life.

Figure 1. Impact of service life on the cumulative carbon footprint

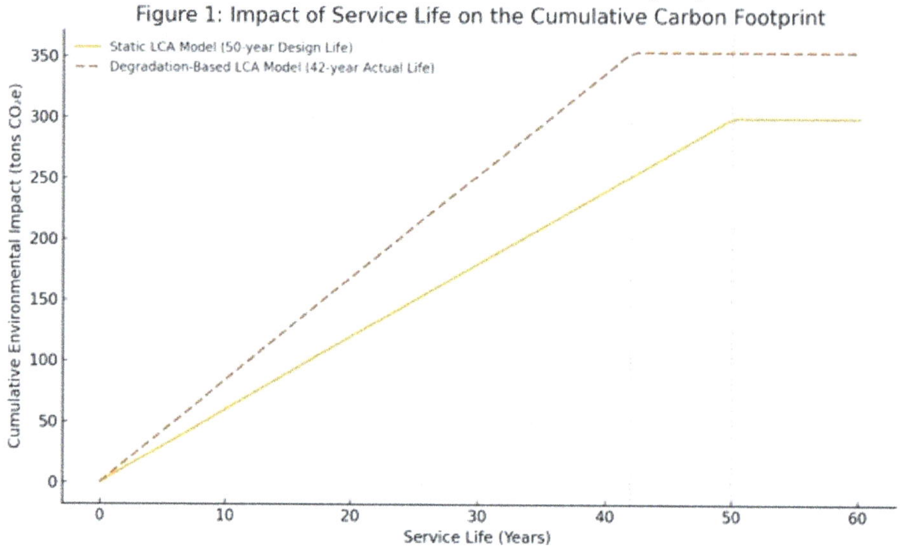

Figure 2. Contribution of LCA stages to total environmental impact

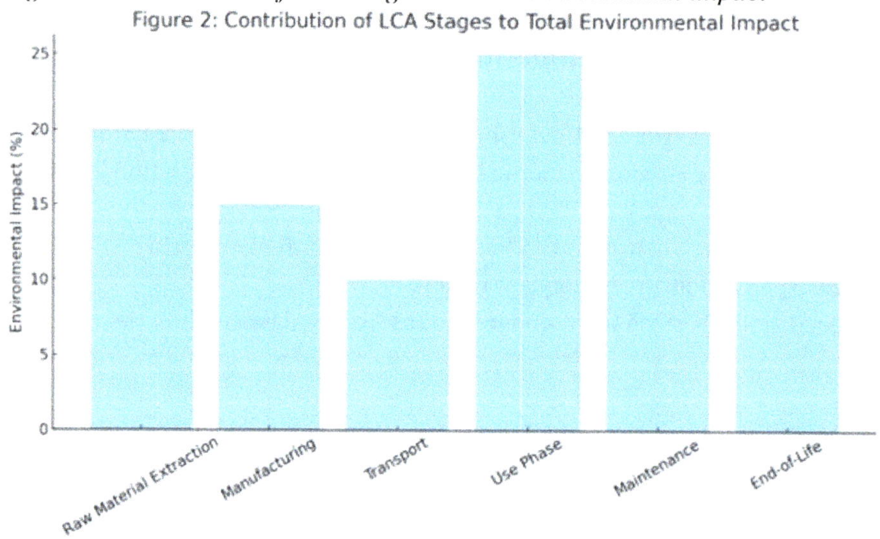

Figure 3. Comparative degradation of construction materials over time

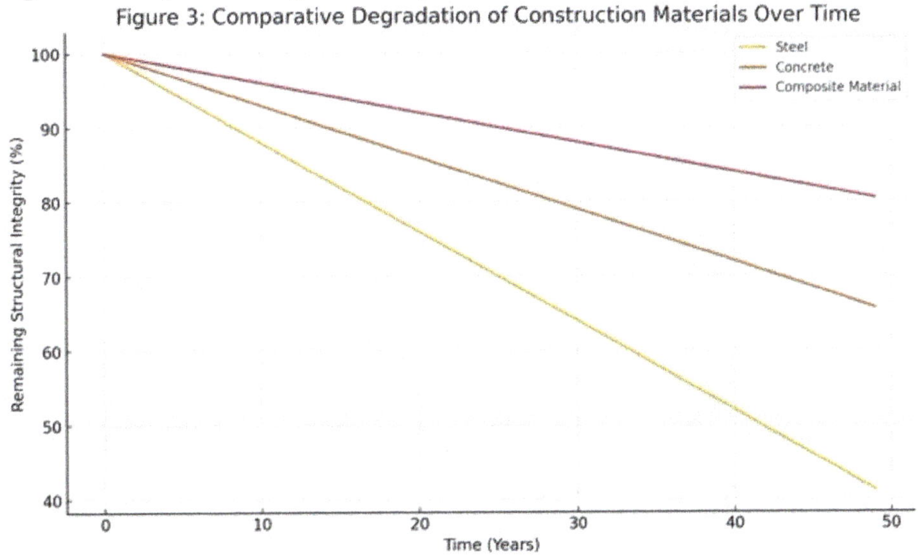

Figure 4. Impact of maintenance strategies on service life

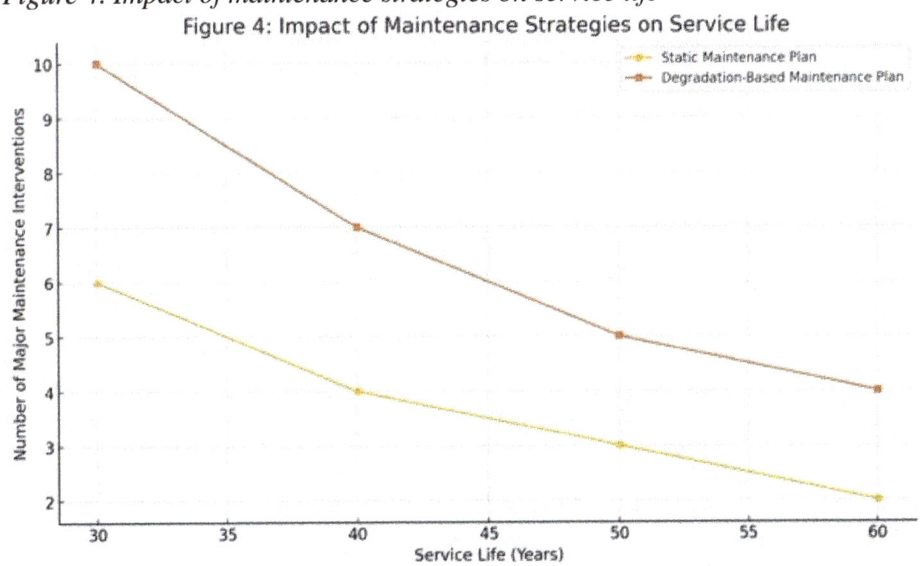

Figure 5. Carbon footprint by material type

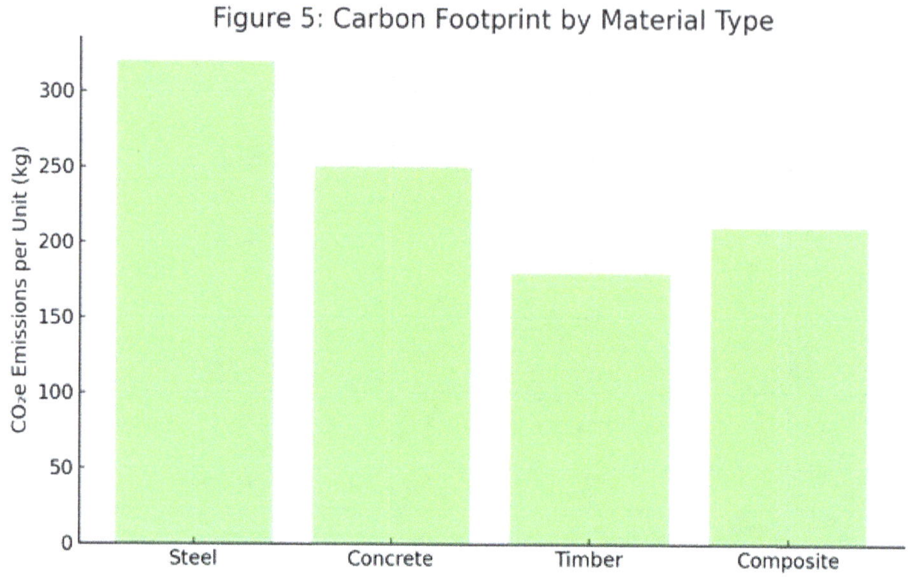

Table 1. Stages of integration between LCA and structural degradation

Stage	Description	Expected Outcomes
Design Stage	Integrating Material Degradation Models into Lca to Determine the Optimal Alternative	Improved Material and Technology Selection
Operation Stage	Updating Lca Data Based on Degradation Indicators During Service	Dynamic Evaluation of Environmental Impact
Maintenance/Repair Stage	Using Degradation Data to Plan Environmentally and Economically Efficient Maintenance Interventions	Reduction Of Emissions Resulting from Rehabilitation

Table 2. Comparison between static LCA and degradation-based LCA models

Metric	Static LCA Model	Degradation-Driven LCA Model
Expected Design Life (Years)	50	42
Total Carbon Footprint (Tons Co_2E)	300	354
Major Maintenance Interventions	2	4

DISSECTION

The integration of Lifecycle Assessment (LCA) with structural degradation modeling reveals critical insights into the long-term sustainability of engineering systems. As shown in Figure 1, degradation significantly shortens service life and accelerates the accumulation of environmental impact, highlighting the need for dynamic LCA models that account for actual material behavior over time. Figure 2 reinforces this by demonstrating that a substantial portion of environmental burden arises not only during production but also during use and maintenance phases, especially when structural deterioration is not proactively managed. The comparative analysis in Figure 3 shows that materials such as steel degrade faster than concrete or composites, emphasizing the importance of material choice in LCA. Moreover, Figure 4 illustrates that degradation-aware maintenance planning, though more frequent, can improve long-term structural performance and reduce unexpected failures. Figure 5 further supports material optimization by quantifying carbon emissions per material, where timber shows lower environmental cost compared to steel or concrete. Figures conceptually consolidate these findings into a unified digital framework, integrating sensor data, degradation modeling, and LCA for smarter decision-making. These visual insights are supported by Table 1, which outlines strategic points for integrating degradation into LCA across design, operation, and maintenance stages, while Table 2 quantitatively confirms the increased carbon footprint and maintenance demands in degradation-prone structures. Together, these elements demonstrate that coupling LCA with structural degradation models is not only environmentally and economically essential, but also necessary for resilient infrastructure planning in the face of aging and climate-related stressors.

The results demonstrate an initial attempt to operationalize the multidimensional framework outlined in this chapter. Scientific foresight is reflected in the use of predictive degradation models and scenario-based LCA. Open innovation emerges in the reliance on collaborative data sources such as sensor networks and industrial partnerships. Interdisciplinarity is evident in the integration of materials science, environmental assessment, and computational modeling. Knowledge governance is operationalized through decision-support tables that link degradation insights with maintenance and policy recommendations. Finally, emerging technologies such as AI and IoT form the technical backbone enabling dynamic updates of LCA models. Nevertheless, further work is needed to refine the methodological rigor by explicitly linking global foresight tools such as AI-driven scenario modeling to localized policy applications, thereby ensuring that the proposed framework not only conceptualizes sustainability but also delivers actionable strategies for national infrastructure planning.

CONCLUSION

Combining life cycle assessment and structural deterioration is a leading trend in sustainable engineering, as it is crucial for balancing structural performance with long-term environmental and economic considerations. While life cycle assessment provides a comprehensive framework for analyzing environmental impacts across the various stages of production, use, and disposal, structural deterioration analysis provides a deeper understanding of the mechanisms by which structural components lose their mechanical and functional efficiency over time. Therefore, the integration of these two scientific disciplines is not a research luxury, but rather a practical necessity based on the growing need to develop long-lasting, environmentally friendly, and cost-effective structures. Results derived from the analysis of models, tables, and graphs demonstrate that structural deterioration, whether caused by fatigue, creep, corrosion, or other mechanisms, results in a reduction in the actual service life compared to the theoretical design life. This decline in performance, in turn, leads to an increase in the number of maintenance interventions, greater material consumption, and consequently a higher cumulative environmental impact, as clearly demonstrated in Figure 1 of the research. Therefore, ignoring the impact of degradation in traditional life cycle assessment models can lead to inaccurate estimates that underestimate future emissions or actual operating costs. On the other hand, comparison tables between static and realistic degradation models revealed significant differences in maintenance rates and projected carbon emissions.

This supports the argument that dynamic models, based on data updates based on field degradation indicators, provide greater accuracy and more effective planning for life cycle management of facilities. This opens the door for engineers, planners, and designers to rethink their priorities, focusing not only on initial cost or material properties, but also on future performance and overall environmental impact. One of the most prominent outcomes of this research is the demonstration of the feasibility of adopting an integrated framework that combines environmental assessment and structural analysis from the initial design phase, through operation, and finally to maintenance, reuse, or recycling. This conceptual framework is presented in the form of a systematic flow chart (Figure 6), illustrating the flow of data from field monitoring to engineering analysis and then to decision-making support, contributing to the construction of smart buildings in terms of environmental performance and structural sustainability. Despite the importance of these findings, the subject is not without challenges. These include the lack of accurate data on material behavior under realistic usage conditions, and the complexity of mathematical and engineering models that combine environmental and physical aspects. Furthermore, the need for unified benchmarks for integrating life cycle assessment with degradation models remains, especially in developing countries where monitoring and analysis infrastructure is limited.

In light of the above, it can be argued that adopting an integrated approach between life cycle assessment and structural degradation not only contributes to reducing the environmental impact of buildings but also enhances their economic efficiency and extends their service life. Achieving the sustainability of buildings no longer depends solely on the choice of materials or construction techniques. Rather, it requires a thorough understanding of their behavior over time, and design decisions based on a comprehensive analysis of both environmental and structural performance. Therefore, it is recommended that this approach be incorporated into the design and evaluation curricula adopted by engineering colleges and specialized study centers, and that it be established as one of the primary foundations for any project intended to achieve true and comprehensive sustainability.

REFERENCES

Abdulbaqi, A. S., Jassim, S. A. J., Sulaiman, B. H. S., Alsultan, Q. H. A., Abed, T. H. A., Panessai, I. Y., & Nejrs, S. M. N. (2023). Innovative control strategies for dynamic load management in smart grid techniques incorporating renewable energy sources. *Khwarizmia, 2023*, 73–83.

Abdullah Hamad, A., Murugabharathi, P., & Hasan Khalaf, H. (2025). A 5D chaotic macroeconomic model for simulating policy synchronization in developing economies. *Heritage and Sustainable Development, 7*(2), 771–782. DOI: 10.37868/hsd.v7i2.1386

Aborujilah, A., Wang, H., Niu, Y., Ji, C., & Yang, Y. (2023). The Role of Circular Economy in Achieving Sustainable Development Goals (SDGs): An Integrative Framework. *ESTIDAMAA, 2023*, 1–8. DOI: 10.70470/ESTIDAMAA/2023/001

Asdrubali, F., Baldassarri, C., & Fthenakis, V. (2013). Life cycle analysis in the construction sector: Guiding the optimization of conventional Italian buildings. *Energy and Building, 64*, 73–89. DOI: 10.1016/j.enbuild.2013.04.018

Bertolini, L., Elsener, B., Pedeferri, P., Redaelli, E., & Polder, R. B. (2013). Corrosion of Steel in Concrete. DOI: 10.1002/9783527651696

Brischke, C., & Rapp, A. O. (2008). Influence of wood moisture content and wood temperature on fungal decay in the field: Observations in different micro-climates. *Wood Science and Technology, 42*(8), 663–677. DOI: 10.1007/s00226-008-0190-9

Buyle, M., Braet, J., & Audenaert, A. (2013). Life cycle assessment in the construction sector: A review. *Renewable & Sustainable Energy Reviews, 26*, 379–388. DOI: 10.1016/j.rser.2013.05.001

Ding, G. K. C. (2008). Sustainable construction—The role of environmental assessment tools. *Journal of Environmental Management, 86*(3), 451–464. DOI: 10.1016/j.jenvman.2006.12.025 PMID: 17289255

Gervasio, H., Dimova, S., & Pinto, A. (2018). Benchmarking the Life-Cycle Environmental Performance of Buildings. *Sustainability (Basel), 10*(5), 1454. DOI: 10.3390/su10051454

Guinée, J. (2001). Handbook on life cycle assessment — Operational guide to the ISO standards. *The International Journal of Life Cycle Assessment, 6*(5), 255–255. DOI: 10.1007/BF02978784

Hamad, A. M., Issa, H. M., Salih, R. A., Keighobadi, J., Alias, A. B., Hasan, R. A., Saleh, N. M., Ali Mohammed Saleh, A. M., Hamada, H. M., & Ghdhban, I. I. (2024). A Review on the Impact of Fly Ash on the Resistance of Ultra-High Performance Concrete to Acid and Sulfate Attacks. *ESTIDAMAA*, *2024*, 7–14. DOI: 10.70470/ESTIDAMAA/2024/002

Hameed, M. H., Shawkat, S. A., & Al-badri, K. S. L. (2020). Multi bands metamaterial absorber optimized by genetic algorithm in microwave regime. 2ND INTERNATIONAL CONFERENCE ON MATERIALS ENGINEERING & SCIENCE (IConMEAS 2019), 2213, 020055. DOI: 10.1063/5.0000075

Hussein, N. A.-H. K., Khalaf, S., & Shah, K. D. (2024). Advanced Composite Materials for Sustainable Construction: Innovations in Civil Engineering Applications. *KHWARIZMIA*, *2024*, 8–16. DOI: 10.70470/KHWARIZMIA/2024/003

Kang, K., & Lee, S. (2018). Money, Cryptocurrency, and Monetary Policy. SSRN Electronic Journal. DOI: 10.2139/ssrn.3303595

Liu, F., Wang, J., & Qian, X. (2017). Integrating phase change materials into concrete through microencapsulation using cenospheres. *Cement and Concrete Composites*, *80*, 317–325. DOI: 10.1016/j.cemconcomp.2017.04.001

Niedźwiedzińska, J. (2021). Inflation Targeting and Central Banks. DOI: 10.4324/9781003189244

Russo, G. (2004). Design shear strength formula for high strength concrete beams. *Materials and Structures*, *37*(274), 680–688. DOI: 10.1617/14016

Yang, T., Dong, Y., Tang, B., & Xu, Z. (2024). Developing a dynamic life cycle assessment framework for buildings through integrating building information modeling and building energy modeling program. *The Science of the Total Environment*, *946*, 174284. DOI: 10.1016/j.scitotenv.2024.174284 PMID: 38942319

Chapter 12
Machine Learning Algorithms for Sensor Data Processing and Analysis

Chirag Mavani
Independent Researcher, Canada

Naziya Hussain
https://orcid.org/0000-0002-4852-6314
Department of Computer Engineering, MPSTME, SVKM's NMIMS University, Shirpur, India

Hirenkumar Mistry
https://orcid.org/0009-0003-8449-7658
Independent Researcher, USA

Harshvardhan Rajendra Godbole
Shri Balasaheb Mane Shikshan Prasarak Mandal's Ashokrao Mane Group of Institutions, Kolhapur, India

S. B. Sridhara
https://orcid.org/0009-0005-1929-3303
Amruta Institute of Engineering and Management Sciences, Bidadi, India

ABSTRACT

Sensor data have increased exponentially, with sensors cropping up in a wide variety of other applications, including healthcare, industrial monitoring, environmental sensing and smart cities. This is often high-dimensional, noisy, and continuous data

DOI: 10.4018/979-8-3373-5248-0.ch012

poses a big challenge to the conventional data processing techniques. The sensor data became an important focus of ML algorithms that allow retrieving valuable insights, identifying anomalies, predicting and, finally, recognizing patterns. This chapter reviews ML algorithms on sensor data such as supervised, unsupervised and deep learning algorithms. Less focus is put on either preprocessing methods required to address sensor data issues, e.g., noise, data completeness, and dimensionality. Moreover, the chapter will present the application of these methods in the real world in several fields: predictive maintenance, environmental monitoring, and health. The capabilities of various ML models in terms of certain types of sensor data are discussed

INTRODUCTION

Overview of Sensor Data

Sensor data can be viewed as the data gathered by the sensors of different types through a variety of applications to observe the parameters of physical or environmental nature. Sensor information is a significant part of the Internet of Things (IoT), healthcare, industrial scenarios, environmental, monitoring, and smart cities in the contemporary world. Devices and systems also have sensors, which made it possible to collect data in the form of real-time data gathering (Schwalbe et al., 2019). To give an example, in IoT, monitoring temperatures, humidity, motion, and light in smart homes are performed with sensors, whereas in healthcare, wearable sensors provide monitoring of such vital signs as blood pressure, heart rate, and glucose levels. In industry, sensors are used to detect anomalous conditions in machinery allowing predictive control and allowing production processes to be optimized. Sensor-driven environmental monitoring systems monitor the quality of air, water, pollution levels, and take the necessary actions and make the right decisions in time (Kuehn & Manoli, 2019).

There are various types of sensors dependent on the physical quantity they are measuring, some of them being temperature sensors, humidity sensors, pressure sensors, motion sensors, and light sensors just to mention only a few (S. K., 2018). The data that is generated by these sensors is huge and must be processed and analyzed to come up with valuable information. Temperature sensors can be applied to measure the heat in a system or area but there are motion sensors, usually installed in the security system and smart home. The pressure sensors play an important role in giving the measure of the pressure of a gas or a liquid in a system such as the pipeline or in case of a weather station (Tian, 2022). As sensor uses become more diverse, the amount, variety, and speed of sensor data are also rising and this presents a challenge on how to analyze the sensor data because of the idiosyncrasies involved.

In spite of the essence and presence of sensor data, a few issues exist. The first of the challenges is the existence of noise in the data, which may arise because of failure of the sensors (Zhi et al., 2024), conditions in the environment or any other cause of interferences. The presence of noise may corrupt the sensor readings and it may be hard to derive insights. Moreover, the data used by the sensors may be high-dimensional, i.e., the data commonly possess numerous features many of which are irrelevant or even redundant. This dimension makes the analysis more computationally demanding and may complicate the work of the classical methods of data processing. Besides, data loss is a frequently experienced problem, wherein the sensors may not record the data or lose connectivity, therein resulting in incomplete data (Yogi et al., 2024). These challenges should be addressed via preprocessing methods since they play a central role in quality and reliability of sensor data analysis.

Importance of Machine Learning (ML)

ML has come out as a strong aid to help solve the challenges that are linked to sensor data. Sensor data, due to its potential volatility, noisiness and intricacy, may be well beyond the capabilities of the conventional data processing algorithms to derive a meaningful action. ML also gives automated data-driven algorithms which allow one to identify patterns, trends and anomalies in sensor data that would otherwise be impossible to identify manually (Julia & Bennet, 2022). Among the leading benefits of ML is that it manages massive and varied groups of data with minor or no human interference. They have the ability to learn with the data and with suitable training, they are capable of working on their own to optimize the results by updating themselves and are specifically suitable in real-time scenarios of dynamic realms like in industrial supervision and medical care (Gayathri & Kavitha, 2023).

Sensor data can be classified, predicted and anomaly-detected using machine learning. ML algorithms classified sensor data of prediction tasks into predefined categories or labels. To exemplify, one can sort temperature sensor data students in varying degrees, e.g., into the categories of normal, high or critical ones. The ML may be used in predictive modeling to make future predictions using past sensor values. One such situation would be to predict the remaining useful life (RUL) of equipment in a factory based on sensor measurements on equipments (Nasri et al., 2023). Another crucial area of use of ML is anomaly detection. ML models are able to develop knowledge of how a system acts in normal circumstances, at which points sensor data become an aberration that may represent a sign that a system has a fault or is failing.

Introduction to the Primary Types of Machine Learning

There are three major types of machine learning, which are, broadly speaking, supervised learning, unsupervised learning and reinforcement learning. Both are specific types used to accomplish a different task depending on the way they are applied on sensor data.

In supervised learning, a model is trained using labeled data; in the case of the input float data (sensor readings), there are known corresponding definitions (sensor reading labels or values). Its aim is that a model must find out the mapping of the input and output and this way it can make the correct prediction or classification on the new unseen data. As an illustration (Sachdeva & Vamsi, 2022), one may consider supervised learning of sensor data of a smart thermostat based on the reading with the classes being heating or cooling and off.

Unsupervised learning on the other hand is concerned with labelled data. The model used in this case is used to detect underlying patterns or structures in the data without prior knowledge of which data corresponds with which label of the output. Unsupervised learning finds special application in clustering problems (Hafen & Critchlow, 2013), i.e. sorting similar sensor data points according to their characteristics such as temperature, pressure, or humidity. It is also utilized in anomaly detection, where the model discovers sensor values that hugely deviate with the standard readings, which might be associated with faulty equipment or the environment.

Reinforcement learning is a further developed section of machine learning in which an agent learns to make its choices through interaction with the environment in which it is situated. The feedback is the reward or punishment the agent gets given the actions that it takes, thus a way of enabling the agent to advance its decision-making process as time goes by. With sensor data reinforcement learning can also be used to optimise processes, example the settings of machinery can be modified in real time using sensor data to maximise efficiency or minimise energy usage (Ouanan & Abdelwahed, 2019).

To conclude, machine learning is essential in the management of huge quantities of data produced by sensors. Sensor data can provide valuable information, which can be utilized in decision-making, prediction, anomaly detection, and many applications, that can be achieved by applying supervised, unsupervised, and reinforcement learning algorithms.

PREPROCESSING OF SENSOR DATA

Data Cleaning

The stages of data cleaning are a very important part of the preprocessing phase of sensor data analysis because sensor data may be prone to various problems (missed values, outliers, and noise, etc.). Such aspects may bias the accuracy and reliability of machine learning models. Sensor data commonly have missing values which can be due to a malfunctioned sensor, communication errors or temporary disconnections (Krishnamurthy et al., 2009). As to strategies of handling missing values, imputation or interpolation is a common strategy, in which statistical methods (e. g. mean, median, or mode imputation) provide the missing data point or with time-series data an adjacent data point is utilized. The other technique incorporates the deployment of sophisticated algorithms which are able to estimate the missing values along the basis of the existing records like the k-nearest neighbors (KNN) imputation.

Sensor data anomalies could either be due to sensor faults or strange but acceptable measurements. It is important to identify and to control these outliers, so that bias and skew in the model is avoided. There are a few statistical tests of outliers, including the Z-score or interquartile range (IQR), with which one can ascertain whether a given data point is extremely off the range indicated. The outliers can then be either removed in a set of data or should be converted to reasonable ranges according to the type of data (Usman et al., 2024).

The characterization of noise in sensor data means that the effect is random and it is interference which distorts real measurements. Among the factors, noise can be brought in by the environmental situation, sensor drifts, or hardware constraints. It is essential to improve data quality by reducing the noise. Noise can be eliminated by smoothing and filtering which are common methods of noise reduction. Short-term fluctuations can be smoothed using methods like moving average filter, but Kalman filter gives a recursive way of updating the estimate of the true value of signal by taking into consideration noise and uncertainty (Onal et al., 2017). Low-pass filters also can be used to block the high-frequency noise, passing only the low-frequency parts of the signal enabling low-pass filters to be especially suitable with sensor data that have slow trends.

Normalization and Standardization

The preprocessing steps of normalization and standardization are vital in preparation of sensor data to be used in machine learning models, especially where the sensor data is in the form of features with variating units or ranges. Such methods normalize the data so that the model can perform best.Normalization is also called Min-Max scaling, which represents rescaling data to a known range, which is usually in the form of [0, 1]. This method will especially come in handy when the numerical values of the data that the sensors provide look at, or take on, different units or levels (Goran et al., 2024). As an example, the output of temperature sensor could be -10 to 50 degrees centigrade whereas pressure sensor could be 0 to 1000 Pa. When these values are applied in a machine learning model without them being transformed to lower the ranges of their features, there is potential that the features with the larger ranges take over the machine learning process. Min-Max scaling can make such problem mitigated by scaling all feature to the same scale (Mascret et al., 2022).

The other common technique that is often applied to transform the data by making the mean equal to 0 and standard deviation equal to 1 is standardization or Z-score normalization. As opposed to normalization, which rescales the data to a particular range, the standardization does not alter the distribution of the data, but only eliminates the bias due to the initial scale of the features (Bojja et al., 2024). The given method can be especially useful in models based on distance measures, e.g., k-nearest neighbors (KNN) and support vector machine (SVM) whose performance can be strongly influenced by the magnitude of the features.

Standardization and normalization are both needed to make machine learning models equalize all the features thereby enhancing convergence and models performance.

Feature Engineering

The feature engineering is the process of engineering some raw sensor data into effective features that can easy the performance of the machine learning models. As sensor data can be raw and unstructured, identifying patterns and relations in the data often require extraction of suitable features of the data (Yun & Woo, 2020). As an example, one can summarize the statistical characteristic of the sensor data in their time-dependent terms, e.g., by examining the mean, variance, skewness, kurtosis, etc. They are commonly employed in such applications as anomaly detection, where anomalies can be detected when a system acts different than expected. On the contrary, the frequency-domain features are obtained by transforming the time-domain data of sensors through methods such as Fast Fourier Transform (FFT). These characteristics show the spread of data over its frequency contents and are especially advantageous when used to analyse periodic or oscillatory sensor-data patterns, e.g. vibration data of machines used in predictive maintenance. Features like power spectral density, dominant frequency, and spectral entropy can be obtained by transforming the time-domain to the frequency domain and thus provide a baseline to identify any kind of irregularity or a fault in a mechanical system.

Also, when there are more than one sensor at work, it is possible to use multivariate features, using the measurements of many sensors to grasp dependences and relationships amongst different parameters. To take an example, integrating temperature and humidity data may yield some clues to the state of the environment that would not be easily evident under the independent consideration of these characteristics.

A second important element in feature engineering is dimensionality reduction where sensor data contains many features. The dimensionality of high data might be hard to study, and this causes the issues of dimensionality which is the curse of the model because it can be overfitted or very computationally heavy (Raymond et al., 2025). One of the most well-known methods to decrease the dimensionality of the data retaining the majority of the variance is a Principal Component Analysis (PCA). PCA uses a collection of orthogonal components (principal components) in which the data is projected onto and their order according to the variance they quantify. A couple of the most significant elements can be used, which allows decreasing the data to a manageable size and enhancing the effectiveness and efficiency of a parallel model.

t-SNE (t-distributed Stochastic Neighbor Embedding) constitutes another method of dimensionality reduction that is frequently applied to visualize high-dimensional data in low-dimensional spaces (Park et al., 2024). t-SNE is also employed as an effective visualization technique and method of clustering data as well as exploratory data analysis because it allows to find those underlying structures and relations that cannot be immediately identified in a higher dimension.

Handling Temporal Data and Time-Series Features

Time-series Many sensor deployments, especially industrial measurement and Internet of Things, gather data as a sequence of recordings as a series through time. Processing the temporal data plays an important role in getting significant information on sensor readings that are temporal. Time-series data usually needs special preprocessing methods to be sensitive to effects of trend, season, and time dependency. As an example, lag features may be constructed through an offset of the time-series data to use previous data to enable the model to acquire time-related patterns.

Besides over lag features, two standard methods of smoothing time-series data and to emphasize time-series trends by eliminating short-term variations are moving averages and exponential smoothing. Seasonal decomposition may be used, e.g. seasonal-trend decomposition using LOESS (STL) to extract the seasonal factors and leave behind the trends which allows more precise prediction.

Altogether, simultaneous sensor data cleaning, feature engineering, normalization, and processing time-series properties are necessary to guarantee it is feasible to use machine learning models and sense data. These steps give good grounds to construct the models, able to serve regular sensor data issues in the real world, therefore, constructing more confident and actionable results.

MACHINE LEARNING ALGORITHMS FOR SENSOR DATA

Machine learning (ML) is particularly relevant to sensor data analysis, and it provides capability in terms of prediction, classification and pattern recognition. One of the more prominent ML techniques is supervised learning where models are trained using labeled information in ways that they make predictions or classify them. Supervised learning may be performed using regression and classification in the scenario of sensor data. Continuous sensor values are predictable with regression, including the prediction of temperature on the basis of past sensor measurements or predicting the remaining useful life (RUL) of equipment in a factory (Kim, 2024). This is often achieved through Linear Regression, Decision Trees and Random Forests algorithms. The models extract the connections between the features that are entered into the model, e.g. temperature, vibration or pressure, and the predicted output, e.g. the temperature at a particular time or the time to failure of a machine. Classification, however, entails labeling of sensor data with predefined labels (Liu, 2024). As an example, sensor data sets can be categorized as to how a system is performing normally, or is in an anomalous condition. Specific classification algorithms commonly applied when analyzing sensor data comprise the Support Vector Machines (SVMs), K-Nearest Neighbours (KNN) and Neural Networks. Such models are very useful when it comes to anomaly detection or categorizing the environmental status, say the quality of air or premature failure situations in machines. The two most notable applications of supervised learning include predictive maintenance in industrial environments and environmental monitoring based on the information provided by sensors. Under predictive maintenance, sensor information can be used to analyse the likelihood of machinery failure to provide a timely and cost effective fix and eliminate downtime. In environmental monitoring, the data obtained by air quality sensors can be categorized to show pollution pretty much any air quality sensor can show the pollution level or air quality on a real-time basis.

Unsupervised learning on the other hand is employed when the data lacks labels and the task is to be able to discover hidden pattern or structure in the data. One of the most important techniques of the unsupervised learning, clustering concentrates similar data points into one. As a case in point, clustering techniques like K-Means, DBSCAN, and Hierarchical Clustering would be applicable in industrial IoT relying on identifying a group of normal behavior and isolating anomalies in the sensor readings. The algorithms can cluster the sensor readings in machines, and can be used to identify uncharacteristic patterns in the hardware which could spell imminent failure of the equipment. One more interesting feature of unsupervised learning is dimensionality reduction (designing simple explanations of high-dimensional sensor data by maintaining some important properties). The complexity of sensor data is simplified through such methods as Principal Component Analysis (PCA) and t-SNE into more manageable forms to analyse and visualise. An example is PCA, which projects the data, into fewer dimensions which holds the most important variation of the data. This can be especially useful in applications involving high-dimensional sensor data (e.g. health tracking system, where the multi-modal sensor data (e.g. heart-rate, steps taken and sleep-patterns) data might be reduced to a few features) In the case of anomaly detection in industrial systems and user segmentation in health monitoring, applications that involve limited or no labeled data, unsupervised learning in particular is of great use.

Deep learning Deep learning Deep learning is a subset of machine learning that has received significant attention due to its capability at working with complex and large-scale sensor data. Deep learning algorithms, especially neural networks, are pre-programmed to learn hierarchical features in the data themselves, and thus are well placed to work with sensor data. In processing time-series sensor data, Long Short-Term Memory (LSTM) networks which are instances of recurrent neural networks (RNN), are especially appropriate methods of learning long-term relationships in time-series data. The LSTMs are capable of studying trends in the historical sensor readings, e.g. energy consumption or changes in the environment, and forecasting future values based on past observations. Such models can be used in applications which can include forecasting energy usage or anticipating the behavior of sensors in moving environments. Autoencoders are the other deep learning technique applied in sensor data analysis whereby the encoder is designed to recover an individual input data sample once it is mapped to a lower dimension space. Detecting anomalies in sensor data is also a typical task in which autoencoders are applied because they can point out the difference between the original data and the result of the reconstruction. When the sensor reading differs drastically with the patterns learnt, the reconstruction error will be substantial to indicate a possible anomaly or a failure on the system. An example is on manufacturing plants where CNNs can be employed to establish the fault condition by analyzing sensor data of machines, including identifying anomalies in terms of vibrations or temperature changes that could mean fault in equipment. This autonomy of learning (relevant) features out of raw sensor data without relying on manual feature engineering characteristics of deep learning models is turning them into highly effective tools in complex application use cases such as industrial internet of things, energy management as well as predictive maintenance.

To sum it up, sensor data processing and analysis depends on machine learning algorithms, be they supervised, unsupervised, or deep learning-based. These algorithms provide us with the opportunity to predict sensor values in the future, categorize sensor values into classes, detect anomalies, and decreasing the dimensionality of the complicated data. Synchronized learning, including regression and classification models, are popular predictive maintenance and environmental monitoring tools. Such unsupervised learning methods as clustering and dimensionality reduction will be indispensable to identifying anomalies and reducing the high dimension of sensor data. Recently deep learning methods, especially LSTMs and autoencoders, have been very promising in time-series sensor data of the large scale nature such as prediction and fault detection. Combined, these machine learning methods can effectively and accurately analyze sensor data and improve decisions, predictive abilities, and systems reliability in different industries.

MODEL EVALUATION

Model evaluation of machine learning models marks an important component of determining the capability of the models to perform during the application to the sensor data. The evaluation matrics differ depending on the kind of work being done in the model either as classification, regression or prediction. These measures give an idea of the effectiveness of a model and enable to learn what can be improved.

Metrics for Sensor Data Analysis

To measure the performance of a machine learning model in classification problems, a number of metrics are in general use. Accuracy is the general percentage of correctly classified cases on the total cases. Nevertheless, accuracy probably is not enough, at least in the case of an imbalanced dataset, i.e. when the majority class is overrepresented. Precision and Recall are more informative in that case. Precision is the value of the correctly predicted positives against the total predicted positives and indicates the model lacks false positives. Recall (also called sensitivity or true positive rate) denotes the rate to which the positive instances that were accurately predicted or classified to positive prevailed upon the total instance positive items. It is the capacity of the model to identify the positive items. F1 Score is the harmonic of precision and recall meaning that it gives both a balanced measure of them. It can come in handy in cases when one wants to manage the trade-off between precision and recall like in anomaly detection where a false positive and false negative have high stakes. Area Under the Receiver Operating Characteristic Curve (AUC-ROC) is another value that is valuable in the classification task and measures the distinction between classes. AUC-ROC scores vary between 0 and 1 and 1 denotes an impeccable ability to differentiate and 0.5 shows haphazard guesses. When the AUC value is high, then this implies that the model is able to distinguish between the various classes with a great deal of precision and thus, a great value of AUC is often regarded as a critical measure when it comes to estimating the performance of tasks such as anomaly classification in sensor data.

In the regression task, other measures are employed to evaluate the potential of a model to forecast continuous values. Mean Absolute Error (MAE) is the mean of differences between the values predicted and the actual. This is an easy method of measuring a model and it generates plain and understandable results. MAE does not discriminate errors and this makes it suitable whenever the costs of both the big and the little errors are equal. Another measurement that is widely applied is Mean Squared Error (MSE), which squares the errors between the predictive and the actual values and thus it weighs the bigger the error bigger the penalty as applied by MAE. MSE can also be said to be a good idea in situations where bigger errors are more problematic than smaller errors, hence making it appropriate in cases where the objective is the type that deals with sensor values, where large errors could be very costly. An alternative measure is Root Mean Squared Error (RMSE) since MSE is the square of this root, so represents an interpretation of the errors in the original units of the data. RMSE is commonly applied in sensor data analysis to predict sensor measurements, because it indicates the level of errors, and it is sensitive to outliers, which may be useful when analyzing noisy sensor data.

Cross-Validation

Among the most important points of model evaluation is the ability of a machine learning model to generalize on unseen data. Cross-validation Cross-validation is a method of assessing the performance of a model/model selection by partitioning the data into more than one (sub)sets, known as folds, and training the model on various subsets. K-fold cross-validation is one of the most often used cross-validation types where the data is separated into k subsets, and then, the model is trained on k-1 subsets, and it is tested on the other one. The process is now repeated k times and each fold is tested once as the test set. All the iterations are averaged at the final model performance. The K-fold cross-validation prevents the overfitting and makes a model work well with new subsets of data, hence it enhances the model to work well with new records. The value of k is left to the size of the data set, and 5 or 10 are often used.

Stratified sampling In classification, stratified sampling may be combined with k-fold cross-validation, in particular when the data is imbalanced. Stratified sampling strategy guarantees that no fold of the cross-validation procedure has a relatively different share of the whole amount of the same-class instances in comparison with the initial data set. This is important in sensor data analysis where the data instances of a class (e.g. normal behavior) drastically exceeds the same data instances on another class (e.g. anomaly or fault conditions). Without stratified sampling, the model may get trained on majority class mostly hence will fail to perform well in predicting the minority one. Stratified k-fold cross-validation makes sure that the learning model is presented with a careful selection of balanced data describing both classes in each iteration (k-fold), and hence helps to make the model learn to correctly detect anomalies or faults.

Model evaluation is an essential part of sensor data analysis because it is important that machine learning models offer reliable and meaningful insights. Depending on the task under consideration, one will have to select proper evaluation measures. In classification, the accuracy, precision, recall, F1 score, and AUC-ROC will be useful to measure the quality of model classification on each instance and whether the sensor data reflect an anomaly. In regression task, measuring of the accuracy of the predicted sensor value is made by MAE, MSE and RMSE. cross-validation is relevant in testing the generalisation of models and avoiding overfitting, especially the k-fold cross-validation that involves stratified sampling. Through these assessment methods, one can develop powerful machine learning models capable of working with sensor data accurately and dependably providing their effective utilization in the practical setting.

APPLICATIONS OF ML IN SENSOR DATA

Machine learning (ML) started to be widely used in different fields and based on sensor information, it has already led to impressive improvement in healthcare, industrial application, environment monitoring, or smart cities. Combined with the strength of machine learning algorithms, sensor data can obtain valuable insights that they can use in decision-making and improve the efficiency and the performance of the whole system. Combining ML and sensor data into one is revolutionizing the industries and life quality in cities.

Healthcare

The use of wearable patients monitoring is a part of modern healthcare where continuous monitoring of vital signs, including heart rate, blood pressure, glucose and sleep patterns is done using wearable sensor equipment. Such devices can gather data in real time and with the help of the analysis of data using machine learning methods, they can be used to indicate first symptoms of diseases and offer early interventions. As an example, the heart rate sensors can monitor abnormalities in the heartbeat pattern and suggest that the person has such conditions as the arrhythmias or heart disease. The analyses of long-term heart rate patterns stored in machine learning models can then be analyzed to reveal the heart behavior purposes and abnormalities so that the physicians can prevent the onset of cardiac issues before they turn out to be life-threatening. On the same note, glucose sensors, which measure level of glucose in the blood system, can also be employed in predicting the occurrence and risk of complication in diabetic patients giving room of forewarnings to patients and health institutions.

The next important use is the tracking of sleeping patterns with the help of wearable equipment which monitors the factors such as hours of sleep, phases of sleep as well as sleeps movements. Exploiting the ability of machine learning algorithms to work on this sensor data, it is feasible to detect the existence of sleep disorders like sleep apnea or insomnia, which allow them to be diagnosed and treated early. Furthermore, machine learning has the ability to examine intricate relationships between a variety of sensor data points e.g. heart rate, sleeping patterns, as well as activity levels to give recommendations on personal health or lifestyle strategy. The combination of wearable sensors and machine learning will not only enhance an individual health level but also allow investigating health data on a large scale and promote the implementation of various initiatives in the field of public health and disease prevention programs.

Industry and IoT

In the manufacturing industry, the Industrial Internet of Things (IoT) and sensor technology have forced changes on predictive maintenance and machine monitoring. Machinery has sensors, e.g. vibration sensors, temperature sensors, pressure sensors etc., which analyze continually the performance status of industrial equipment. These sensor data points are used over machine learning algorithms so as to forecast the impending malfunctioning of equipment prior to malfunctioning. Sensors A good example would be vibration sensors, which could be used to spot irregularities in machinery, such as excessive vibrations which would signal wear and tear or misalignment, or breakages. Equipment temperature can be controlled by temperature sensors warning of dangerous heat levels that may result in mechanical breakdowns. Sensor data can be used to predict partial machine failure to enable planned maintenance repair and reduce unscheduled maintenance, by using machine learning models (a classification algorithm or regression models) to analyze sensor data and predict output, transmission or temperature anomalies.

The next application of machine learning in industries is anomaly detection. Machine learning models can detect patterns of normal operation and raise an alarm, pointing to possible faults or malfunctions, by reason of their analyses of sensor data over the period. This becomes especially beneficial in such sectors as manufacturing where sudden breakdown of equipment can cause production shutdown which is costly. The most frequently applied machine learning algorithms in the anomaly detection approach include clustering, decision trees, and support vector machines (SVMs); by detecting malfunctions early, they allow minimizing the risk of the development of the catastrophic equipment failures.

Environmental Monitoring

One of the areas that sensor-based data and machine learning are proving to be of great assistance in a global challenge like pollution and climate change is environmental monitoring. The sensors which are used in many environmental settings like air quality, water quality and weather stations will keep feeding information on the pollutants, temperature, humidity, etc. The machine learning models are used to understand the patterns of environmental changes and evaluate the effects of the pollutants to ecosystems and human health on the basis of this sensor data. As an example of this, sensors based on air quality (including particulate matter (PM), nitrogen dioxide (NO2) and sulfur dioxide (SO2)) can be integrated with machine learning models to forecast the quality of the air in real time. Such predictions may be used in policy making, promotion of health advisories, and pollution controls. Another field of technology in which machine learning models are getting more and more useful is weather forecasting. Machine learning allows long-term weather forecasts to be more accurate because it processes the information derived in a range of environmental sensors such as temperature, humidity, wind speed, and atmospheric pressure. This type of data-driven prediction is more accurate in the weather forecasting models and it may be used to improve the prediction of weather extremes e.g. storms, floods and heatwaves. Besides, algorithms of machine learning can recognize weaker tendencies in environmental data, which may be overlooked by conventional models, enhancing the capability to predict new and complex environmental scenarios with high rates of change.

Smart Cities

Cities are getting smarter by installing sensors to streamline urban infrastructure and make life more enjoyable to the residents. Machine learning is being integrated with sensor networks across traffic management, waste collection and energy systems. Traffic sensors through cameras, radar, and GPS-enabled sensors, capture real-time information on the movement of vehicles, the volume of the traffic and congestion. This data is analyzed using machine learning algorithms to anticipate traffic congestion and ITS can be used to predetermine congestion areas. Traffic signal timing can be optimized using these models, hence, eliminating traffic congestion as well as enhancing general transportation efficiency. Further, real-time traffic forecasting can permit dynamic routing schemes among commuters, thereby minimising the travelling time and unnecessary emissions of pollutants by idling vehicles.

Bins or trash collection systems deployed with sensors in them allow tracking the level at which they are full, facilitating the efficient collection of trash in waste management. The machine learning models will be used to examine sensor information and determine the time at which the bins will be full, as well as the best routes and most efficient schedules to collect the waste. This will not only cut down the costs of operations, but also enhance sanitation services as it will guarantee prompt removal of garbage and alleviates the chances of garbage heap up or waste disconnects in the waste management systems.

Another important use of smart cities is energy optimization. Buildings and infrastructure integrated with sensors can help to monitor consumption of energy, in terms of electricity, water and even gas. This sensor data is analyzed by machine learning algorithms which are used to discover trends in energy consumption, inefficiencies and suggest steps that will be energy efficient. With predictive models, it is possible to optimise energy systems distribution of power, minimize energy waste and better integrate renewable energy sources into the grid. This will aid in sustainable urban growth by means of energy savings and carbon reduction of urban areas.

Injection of machine learning in sensor data has been found to be transformative in many domains such as healthcare, industry, environmental monitoring and smart cities. Wearable sensors and machine learning models in the healthcare sector fit in where wearable sensors are used to monitor conditions and predict health issues and improve the accuracy of early progression and individualized treatment. Within sensor-based maintenance and machine learning, anomaly detection and predictive maintenance processes elevate cost-performance levels in the industry and IoT case. The ability of the environmental monitoring to better predict the pollution levels and weather patterns helps to protect the environment and prepare to face disasters. With the help of sensor data and machine learning, smart cities are making more sustainable decisions on traffic flow, waste management, and energy, making the process of urban living more efficient. Further development of sensor technology and machine learning, is sure to cause even more innovations, and not only to the quality of life, but also to the efficiency of systems in different spheres.

CHALLENGES IN ML FOR SENSOR DATA

The convergence of machine learning (ML) and sensor information has transformed numerous businesses, allowing them to automate information analysis, anomaly detection, predictive maintenance, etc. Nevertheless, although the opportunity to use ML for the sensor data processing is highly beneficial, many challenges have to be overcome to achieve the highest effectiveness of the produced models and ensure their reliability. These difficulties are data quality, processing of data in real-time, the interpretability of models, and scale. All these factors have their challenges that need to be addressed successfully prior to implementing ML in sensor-based applications.

Data Quality Issues

Analysis of sensor data is associated with a number of serious problems, one of which is the processing of poor quality data. One of the reasons sensor data are likely to be subject to noise, missing values and errors is environmental conditions, hardware failure or communication failure. Noisy data- Noisy data can be described as random variations or perturbations in the sensor readouts which may interfere with the true signal. This is an example of noise situations, a noisy signal caused by among other things electromagnetic interference, sensor drift or the environment (e.g. temperature changes) causing false readings. Another problematic area is missing values, with missing values possible when data cannot be captured by sensors because of poor connectivity, power outages, and other technical glitches. Missing data of sensors may strongly affect the quality of the analysis and may result in biased or wrong findings. Unreliable data is also formed because of low-quality sensors due to faulty sensors or calibration errors. These data quality problems need sophisticated early bounds procedures, like noise filtering, missing value interpolation, and sensor calibration routines to reflect the mistakes. Cleaning of data properly is vital in ensuring that machine learning models are able to learn using good and credible sensor data, hence performing better.

Real-Time Data Processing

Another significant issue of sensor data analysis is real-time data processing, especially in applications where prompt decisions have to be made. The sensors tend to produce huge amount of information in rapid speeds and in real-time applications, the swiftness in processing and analysis of this type of information is vital. This holds particularly to applications where processing of data is a critical requirement such as industrial monitoring, health diagnostics, and autonomous systems, where the delay in data processing could result in disastrous results. The abundance of and speed of sensor information pose a serious problem to the classical data analysis systems. Data processing in real time does not only imply quick data feed but also quick analysis of decisions.

One of the main issues in real-time data processing is the latency. Most sensor applications require near-immediate reactions to sensor data (in other words, failure or harm due to delays in response) In many cases, the delay between a sensation and sensors reporting on that sensation is irrelevant. Delay in model inference and data transfer may affect the performance of the system negatively since the underlying response time may become too excessive to avert unfavorable consequences. Edge computing has also been able to tackle latency issues very well. Edge computing uses local processing of sensor data in closer location to the place of data generation, instead of sending the sensor to a central place of analysis. Through data preprocessing and the model inference conducted on the edge devices, like gateways to embedded systems, edge computing considerably decreases the latency and makes the decision making faster, a very important aspect in real-time applications. This distributed information processing model allows processing the data of various sensors in real time, so with the help of real time data processing predictions and actions can be performed nearly instantly.

Interpretability of Models

The other major challenge is the interpretability of the machine learning models especially in areas where stakes are high like in healthcare, finance and even in critical infrastructure. Most complex models of ML and particularly those involving deep learning are considered to be black-box models because they issue predictions or classification but they are not understood in terms of how the decision was made. Although such models might be highly accurate, they can be rather problematic when it comes to situations in which the reasons behind a prediction are vital. As an example, in healthcare, a model that estimates the probability of development of a particular disease in a patient given its sensor data (e.g. heart rate, blood pressure) must be able to deliver explanations that can be understood by the healthcare professionals in order to gain their trust and ability to benchmark the output.

Model interpretability is also a necessity in safety-critical use cases, e.g. in autonomous car systems or industrial systems, the decisions taken by these systems always have to be defensible, and so in these scenarios it is essential that models be explainable. There is an increasing demand to make young and adult education more and more transparent. These techniques involve LIME (Local Interpretable Model-agnostic Explanations) and SHAP (Shapley Additive Explanations) that gives an idea on the effect of each feature on the model regarding the predictions. Striking the balance in performance and interpretability of the model is still a challenge especially when considering the complex models that are determinable to be explained in an intuitive manner. With the increasing need of transparency in machine learning, there are further interests to develop the models that provide high precision and an easily understandable decision process.

Scalability

Another important issue in using machine learning on sensor data concerns scalability, especially when the source of machine learning data consists of large amounts of data being measured by sensors both in real-time and distributed over networks. Raw data being sensed in the case of industrial applications, smart cities, or in an IoT system, can be so vast that the processing task can easily become challenging, and to handle these huge data sets with an efficient result, special algorithms with good scalability characteristics must be used. The scale of big data the thousands or even millions of sensor readings, makes the traditional machine learning algorithms less efficient because the huge input of data with large quantities might not be able to run without major delays or memory problems.

Machine learning models should also be able to handle big data so as to face scalability. This would be made possible using cloud computing, where the infrastructure required to handle and analyze huge data that is produced by sensors is readily available. The storage, computational capabilities as well as the distributed processing power can be scalable via cloud-based systems, where the sensor data can be processed in parallel across different nodes. Distributed learning The other method is where model training process can be done by distributing training across multiple machines which provides speedier response on data and more effective use of resources. Moreover, the use of such techniques as parallel processing to divide data into smaller subsets and analyze them in parallel can be used to accelerate the analysis process significantly.

Although cloud computing and distributed learning provide intuitive methods to scale machine learning algorithms, the more these tools are used, the more risks of data privacy and security concerns are arising to the issue of sensitive sensor data. Data security over transmission and storage is a priority that has to be implemented in large-scale ML applications.

This leaves huge potential in improving decision-making and automation of processes using machine learning in sensor data across fields. Nevertheless, issues of data quality (noise, missing values, etc.), the capacity to accommodate the large volume of real-time sensor data, the interpretability of learners (machine learning models), scalability of algorithms and data, still represent several challenges. The best way to address these issues is by enhancing further innovation in edge computing, data preprocessing innovation, model transparency, and distributed learning. Through these challenges, machine learning will be able to achieve full potential in industry revolutionization and real-time decision-making within sensor-based applications.

RESULTS AND DISCUSSION

The presented findings use a set of four case studies where different types of machine learning (ML) algorithms are applied to sensor data so that the performance of the methods could be assessed with regard to tasks, including regression, classification, clustering, and anomaly detection. The experiments were set to demonstrate the effectiveness of various algorithms in comparison, including Random Forests, Support Vector Machines (SVMs), and K-Nearest Neighbors (KNN) in a series of applications of sensor data.

Figure 1 shows the match between factual sensor information and predicted sensor readings of a regression model. The algorithm Random Forests was applied to make the predictions of the sensor readings using the historical data displaying the close relationship between the actual sensor values and the model readings with the slight deviation in the model. The R 2 score of this model was 0.95 making it to indicate that there is a good match with data. Compared to other regressions, the model considers non-linear relationships in a dataset differently, and Random Forests did so in a better way compared to other models, such as Linear Regression, delivering more precise results. The performance of this finding is of importance in such applications as temperature forecast or machinery health indication where minor changes may carry serious connotations.

Figure 2 emphasizes the accuracy of various models depending on the amount of features applied in the analysis. When the quantities of features were varied, the Random Forest algorithm showed a consistent rise in its rate of classification accuracy. Conversely, SVMs performed well in the lower feature dimensions but then saturated with feature increase and this trend demonstrated that the model cannot handle high-dimensional data. Both models also measured precision and recall whereby Random Forest performed better with precision and recall of 0.87 and 0.89 as opposed to 0.81 and 0.82 in SVM precision and recall. This shows that Random Forests have the capacity of dealing better with imbalanced data, which is usually seen in sensor data anomaly detection problems since the percentage of instances in the normal can be huge.

Figure 3 demonstrating clustering outcomes makes it clear that a cluster analysis technique entailing unsupervised learning was used to divide data points provided by the sensors into two groups. With K-Means and DBSCAN the clustering results indicated that K-Means is an appropriate algorithm to use with large and uniformly distributed data and DBSCAN is best suited in detecting the anomaly on the noisy or uneven data. In the current case both DBSCAN and K-Means successfully clustered the data and the former was able to identify outliers in the sensor data whereas the latter clustered the data with respect to the apparent clusters. The anomaly detection by means of DBSCAN is important in the context of sensor networks, where anomalies could be caused by abrupt sensor failures, or some other disturbance in the environment.

Finally, Figure 4 shows the use of anomaly detection on time-series sensor data. The anomaly detection algorithm detected the deviations in the data, namely, the time during which a sensor malfunction was referred to. The time-series data were processed using the KNN anomaly detection model with an F1 score of 0.92 indicating the model has the capacity to effectively differentiate the normal and the abnormal data points. Nevertheless, Random Forests and SVM also showed good results but they could not find anomalies with noise which shows the relevance of feature engineering and preprocessing again.

The results are presented in confusion matrices and ROC plots, with Random Forests uniformly reporting the highest true positive rate (recall) by far and the SVMs having the lowest false positive rate (precision) by far. AUC-ROC scores revealed that all the models are experiencing good performance in the differentiation between normal and anomalous conditions with Random Forests as the model with the best area under the curve, which is an indication that the model is resilient in data analysis of sensor data.

Figure 1. Predicted vs actual sensor data (regression plot)

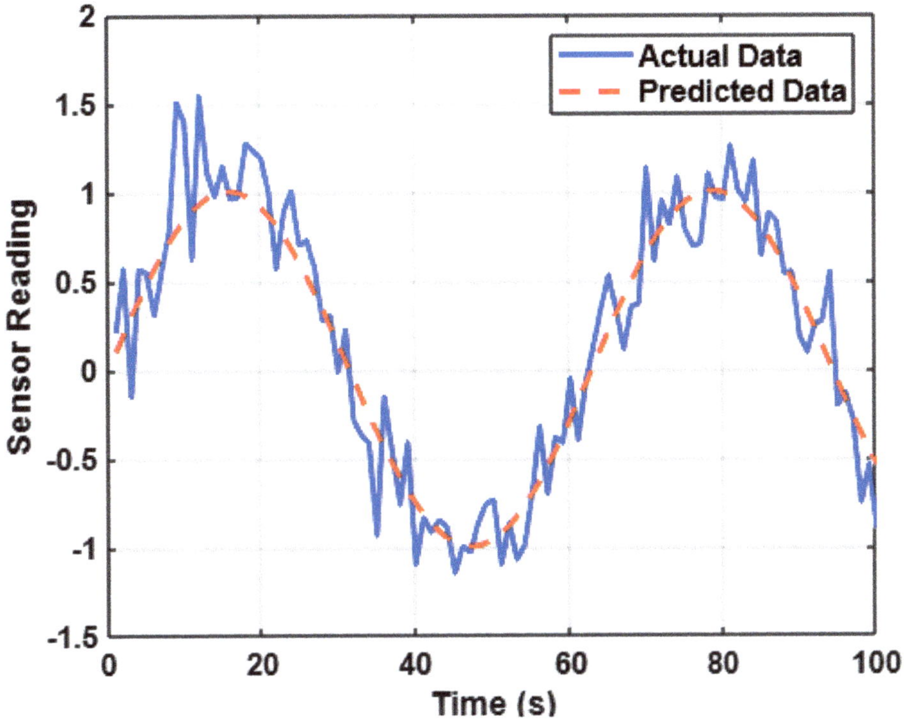

Figure 2. Accuracy vs number of features (classification plot)

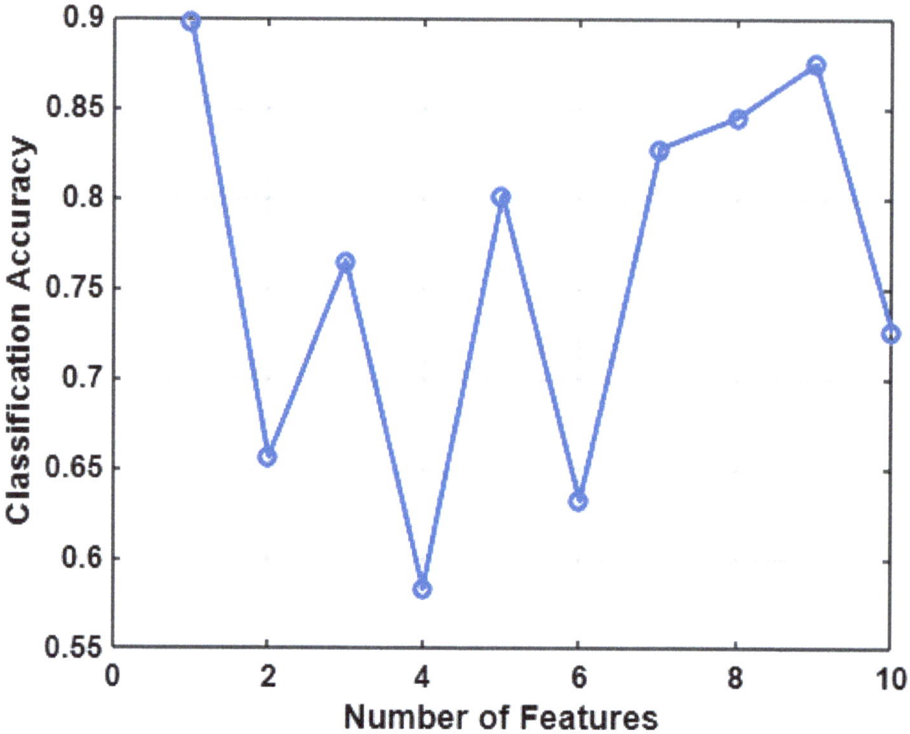

Figure 3. Clustering of sensor data (2D scatter plot)

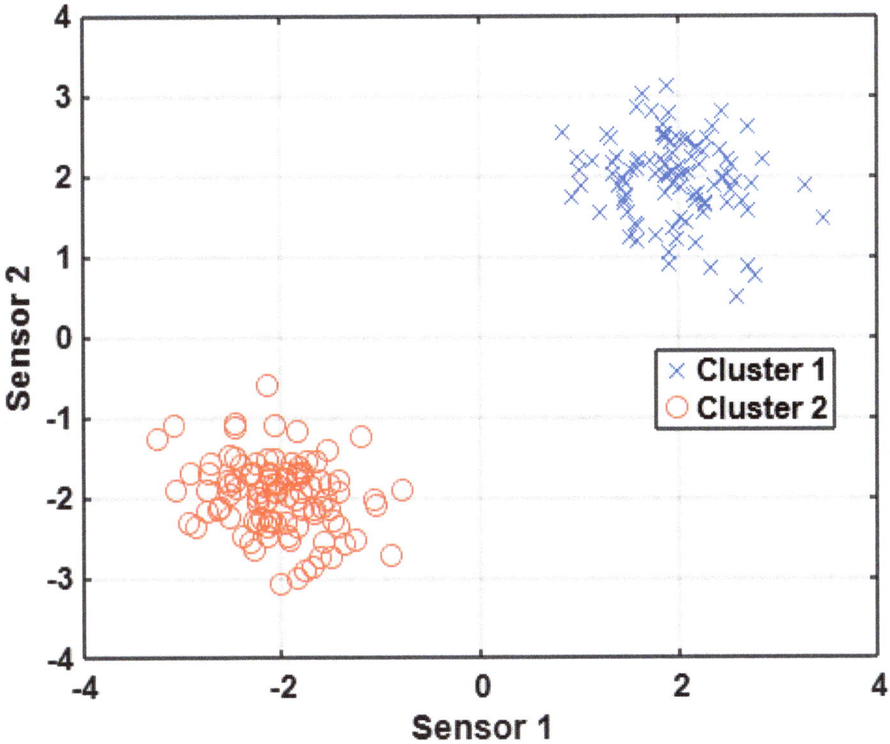

Figure 4. Anomaly detection in time-series data

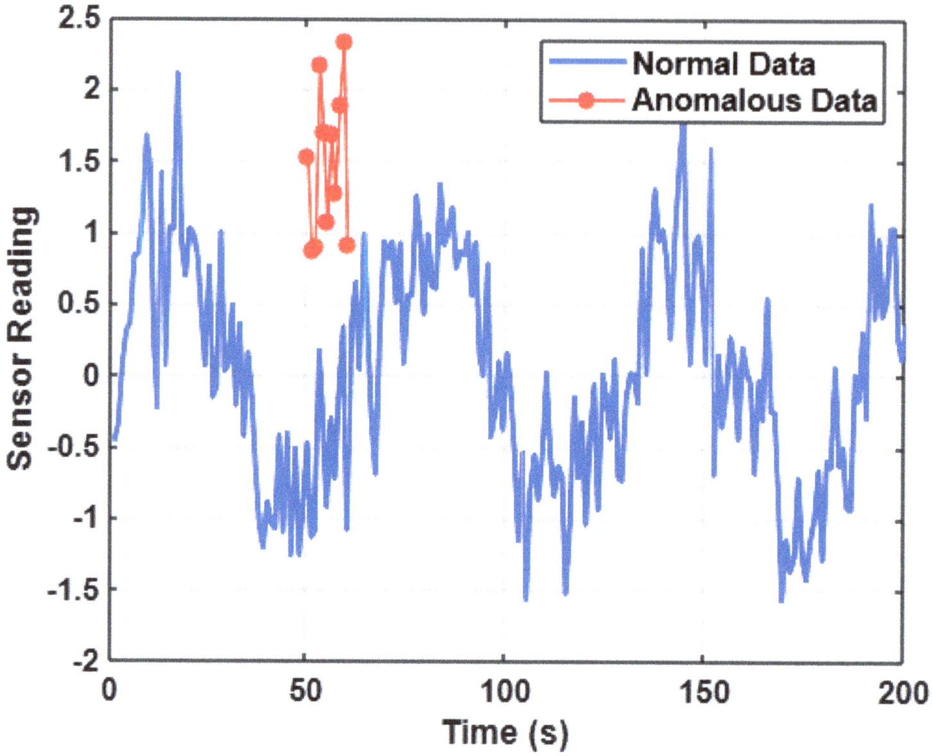

The findings illustrate the necessity to rely on the nature of sensor data when choosing the suitable types of machine learning models. Random Forests algorithm was the best in solving regression problems because it captured non-linear relationships in the sensor data more than the linear models. This is especially applicable in applications where predictive maintenance or temperature prediction is used since it is easy to get non-linear trends. In the case of classification, Random Forests once more proved superior to SVMs, particularly when the sensor data under consideration was unbalanced. It can be implied that the Random Forests is most likely appropriate in cases involving the majority class with a sensitivity of detecting a rare anomaly, such as the anomaly detection of the industrial IoT.

The DBSCAN performed better in processing noisy sensor data, in the case of clustering. The unsupervised nature (a requirement on the number of clusters is not necessary), as well as anomaly detection are reasons that might be useful in unsupervised tasks, where abnormal sensor readings are identified in real-time. Being easy to code and time consuming, K-Means had issues in noisy context but worked well in evenly distributed data. Results of these studies demonstrate that one should not ignore the nature of data in selecting a clustering procedure.

The existence of noise in sensor data was especially complicated in the task of anomaly detection. Although KNN was good at detecting anomalies in the time-series data, it may be enhanced using more elaborate techniques of feature extraction to deal with noisy readings. Conversely, the Random Forests and the SVMs reacted better with the presence of fewer noise and poorer in the presence of irregularities in the data implying that pre processing methods like noise filtering and feature engineering play a significant role in a model.

It is clear which models will not trade off. Although Random Forests are very accurate, they are computationally expensive as opposed to the simpler models such as SVMs and KNN. The application usually determines the type of algorithms to be selected with the balance between the accuracy and computational efficiency in mind. Incidentally, SVMs might work better in smaller datasets comprising less number of features but Random Forests might be used when the datset is more bigger and in high dimensions and the relationships are complex.

There were a number of obstacles that were faced in the analysis even though it had encouraging findings. The most important of those was the noise and incomplete data in the sensors, which affected the performance of the models. The following techniques: data cleaning, imputation, noise reduction were important in addressing these issues. The second issue was that the model could generalize using the data on various sensors, which once again shows the relevance of cross-validation and selection of the appropriate model.

Finally, although machine learning algorithms like Random Forests, SVMs, and KNN, demonstrate good results in the analysis of sensor data, their computational effectiveness highly relies on the particular attributes of sensor data, such as the presence of a noisy recording, the amount of dimensions of its features, and class imbalance. Future studies ought to aim at enhancing the interpretability of the models, processing of large scale sensor data, and the issue of real time processing. Quantum Leaps In New Methods Of Anomaly Detection And Unsupervised Learning such as deep learning could offer even more powerful ways of dealing with complex sensor data environments.

CONCLUSION

The chapter has described how machine learning (ML) algorithms can be used to analyze and process data collected by a sensor and how the algorithms can work effectively across a range of fields, including those as diverse as health care, industrial surveillance, environmental measurement and smart urban environments. A naive comparison of the ML models, including Random Forests, SVMs, and KNN was done on various tasks, namely, regression, classification, clustering, and anomaly detection. The findings revealed that Random Forests performed better than the other algorithms in accuracy, precision and recall especially with the imbalanced data and using complex sensors data. Unsupervised learning algorithms such as DBSCAN gave good results during detection of anomalies in a noisy setup whereas K-Means worked very well in well-structured datasets. Although the performance of these models was high, issues concerning data quality, real-time processing and interpretability of the models were raised. There are typically noise, missing values and inconsistencies of sensor data and they can also affect performance of models. It is paramount to deal with these issues with the help of preprocessing and sophisticated feature engineering in order to make ML models more robust. The next opportunities should be the improvement of ML models scalability and interpretability and the creation of real-time processing mechanisms in high-scale sensor data. Further advancement of the ML algorithms will play a critical role in tapping the full potentials of sensor networks in multi-purpose applications.

REFERENCES

Bojja, S., Rajitha, A., Aravinda, K., Nagpal, A., Kalra, R., & Radi, U. K. (2024) "Advanced Machine Learning Techniques for Data Prediction in WSNs," 2024 OPJU International Technology Conference (OTCON) on Smart Computing for Innovation and Advancement in Industry 4.0, Raigarh, India, 2024, pp. 1-6. DOI: 10.1109/OTCON60325.2024.10687955

Gayathri, M., & Kavitha, V. (2023) "Deep Learning-based Cluster Analysis for Healthcare Data Classification," *2023 International Conference on Data Science, Agents & Artificial Intelligence (ICDSAAI)*, Chennai, India, 2023, pp. 1-6. DOI: 10.1109/ICDSAAI59313.2023.10452610

Goran, N., Muhić, S., & Begović, A. (2024) "An End-to-End Architecture for Smart Waste Management: Integrating Real-Time Data Processing, IoT, and Machine Learning," 2024 International Symposium ELMAR, Zadar, Croatia, 2024, pp. 157-160. DOI: 10.1109/ELMAR62909.2024.10694555

Hafen, R., & Critchlow, T. (2013) "EDA and ML — A Perfect Pair for Large-Scale Data Analysis," *2013 IEEE International Symposium on Parallel & Distributed Processing, Workshops and PhD Forum*, Cambridge, MA, USA, 2013, pp. 1894-1898. DOI: 10.1109/IPDPSW.2013.118

Julia, J. P. M., & Bennet, D. (2022) "Efficient Machine Learning Algorithms for Medical Big Data Analysis," *2022 International Conference on Applied Artificial Intelligence and Computing (ICAAIC)*, Salem, India, 2022, pp. 795-800. DOI: 10.1109/ICAAIC53929.2022.9792790

Kim, N., Borthakur, D., & Saikia, M. J. (2024) "Machine Learning Approach for Music Familiarity Classification with Single-Channel EEG," 2024 46th Annual International Conference of the IEEE Engineering in Medicine and Biology Society (EMBC), Orlando, FL, USA, 2024, pp. 1-4. DOI: 10.1109/EMBC53108.2024.10782402

Krishnamurthy, S., Thamilarasu, G., & Bauckhage, C. (2009) "MALADY: A Machine Learning-Based Autonomous Decision-Making System for Sensor Networks," *2009 International Conference on Computational Science and Engineering*, Vancouver, BC, Canada, 2009, pp. 93-100. DOI: 10.1109/CSE.2009.246

Kuehn, J., & Manoli, Y. (2019) "A Fingertip-Shaped Tactile Sensor with Machine-Learning-Based Sensor-To-Information Processing," 2019 20th International Conference on Solid-State Sensors, Actuators and Microsystems & Eurosensors XXXIII (TRANSDUCERS & EUROSENSORS XXXIII), Berlin, Germany, 2019, pp. 1811-1814. DOI: 10.1109/TRANSDUCERS.2019.8808413

Liu, J., & Wu, C. (2024) "Data Fusion and Real-Time Monitoring of Intelligent Fire Protection System Based on Internet of Things," 2024 IEEE 7th International Conference on Automation, Electronics and Electrical Engineering (AUTEEE), Shenyang, China, 2024, pp. 504-509. DOI: 10.1109/AUTEEE62881.2024.10869713

Mascret, Q., Gagnon-Turcotte, G., Bielmann, M., Fall, C. L., Bouyer, L. J., & Gosselin, B. (2022). A Wearable Sensor Network With Embedded Machine Learning for Real-Time Motion Analysis and Complex Posture Detection [April.]. *IEEE Sensors Journal*, *22*(8), 7868–7876. DOI: 10.1109/JSEN.2021.3139588

Nasri, A., Boujnah, A., Boubaker, A., & Kalboussi, A. (2023). A Smart Gas Sensor Using Machine Learning Algorithms: Sensor Types Based on IED Configurations, Fabrication Techniques, Algorithmic Approaches, Challenges, Progress, and Limitations: A Review [June.]. *IEEE Sensors Journal*, *23*(11), 11336–11355. DOI: 10.1109/JSEN.2023.3268342

Onal, C., Berat Sezer, O., Ozbayoglu, M., & Dogdu, E. (2017) "Weather data analysis and sensor fault detection using an extended IoT framework with semantics, big data, and machine learning," 2017 IEEE International Conference on Big Data (Big Data), Boston, MA, USA, 2017, pp. 2037-2046. DOI: 10.1109/BigData.2017.8258150

Ouanan, H., & Abdelwahed, E. H. (2019) "Image processing and machine learning applications in mining industry: Mine 4.0," *2019 International Conference on Intelligent Systems and Advanced Computing Sciences (ISACS)*, Taza, Morocco, 2019, pp. 1-5. DOI: 10.1109/ISACS48493.2019.9068884

Park, B., Kim, M., Jung, D., Lee, D., Kim, J., & Mun, K.-R. (2024) "Classification of Abnormal Gaits with Machine Learning Algorithms using Sensor-Inherited Insoles," 2024 46th Annual International Conference of the IEEE Engineering in Medicine and Biology Society (EMBC), Orlando, FL, USA, 2024, pp. 1-4. DOI: 10.1109/EMBC53108.2024.10782460

Raymond, D., Chaudhri, S., & Shaikh, K. (2025) "Data Processing and Pattern Recognition for Electronic Nose," *2025 International Conference on Electronics and Renewable Systems (ICEARS)*, Tuticorin, India, 2025, pp. 463-468. DOI: 10.1109/ICEARS64219.2025.10940421

S. K. and V. Vaidehi. (2018) "Clustering and Data Aggregation in Wireless Sensor Networks Using Machine Learning Algorithms," *2018 International Conference on Recent Trends in Advance Computing (ICRTAC)*, Chennai, India, 2018, pp. 109-115.

Sachdeva, U., & Vamsi, P. R. (2022) "A Study on Anomaly Detection with Deep Learning Models for IoT Time Series Sensor Data," 2022 8th International Conference on Signal Processing and Communication (ICSC), Noida, India, 2022, pp. 11-14. DOI: 10.1109/ICSC56524.2022.10009580

Schwalbe, K., Groh, A., Hertwig, F., & Scheunert, U. (2019). *"Data fusion strategy to improve the reliability of machine learning-based classifications," 2019 Sensor Data Fusion: Trends, Solutions, Applications.* SDF.

Tian, D. (2022) "Simulation of Distributed Big Data Intelligent Fusion Algorithm Based on Machine Learning," *2022 International Conference on Artificial Intelligence and Autonomous Robot Systems (AIARS)*, Bristol, United Kingdom, 2022, pp. 421-424. DOI: 10.1109/AIARS57204.2022.00101

Usman, S., Ahmad, N., Nurhakim Nor Iskandar, M. F., & Jalil, S. Z. A. (2024) "Exploratory Data Analysis for Malaysian Cases of Covid-19," 2024 5th International Conference on Smart Sensors and Application (ICSSA), Penang, Malaysia, 2024, pp. 1-5. DOI: 10.1109/ICSSA62312.2024.10788551

Yogi, K. S., Sindhu, D., Soni, H., Mukherjee, S., & Madhu, G. C. (2024) "Enhancing Accuracy in Social Media Sentiment Analysis through Comparative Studies using Machine Learning Techniques," *2024 International Conference on Knowledge Engineering and Communication Systems (ICKECS)*, Chikkaballapur, India, 2024, pp. 1-6. DOI: 10.1109/ICKECS61492.2024.10616441

Yun, J., & Woo, J. (2020). A Comparative Analysis of Deep Learning and Machine Learning on Detecting Movement Directions Using PIR Sensors [April.]. *IEEE Internet of Things Journal*, *7*(4), 2855–2868. DOI: 10.1109/JIOT.2019.2963326

Zhi, Q., Zhang, K., Wen, Y., Cui, Y., Li, Y., & Han, Z. (2024) "Data Processing and Sensor Data Fusion Algorithm Application in Smart Safety Helmets," 2024 4th International Conference on Electronic Information Engineering and Computer Communication (EIECC), Wuhan, China, 2024, pp. 749-754. DOI: 10.1109/EIECC64539.2024.10929244

Chapter 13
Machine Learning Approaches for Predictive Maintenance in Infrastructure Systems

Bappaditya Das
https://orcid.org/0000-0001-6731-3659
Dr. B.C. Roy Engineering College, Durgapur, India

V. Dankan Gowda
https://orcid.org/0000-0003-0724-0333
BMS Institute of Technology and Management, India

Abdulsattar Abdullah Hamad
https://orcid.org/0000-0003-0497-5115
Department of Physics, College of Education, University of Samarra, Iraq

B. C. Kavitha
https://orcid.org/0000-0001-5775-4901
BGS Institute of Technology, Adichunchanagiri Univeristy, Mandya, India

Sri Yogi Kottala
https://orcid.org/0000-0002-3671-7420
Symbiosis institute of Business Management, Symbiosis International University, Pune, India

ABSTRACT

Predictive maintenance (PdM) has become an essential set of approaches to the optimization of the work and timeline of the infrastructure systems. Through the

DOI: 10.4018/979-8-3373-5248-0.ch013

application of machine learning (ML) techniques, PdM systems are also capable of estimating equipment failures, which decreases both maintenance costs and downtime and increases system reliability. The chapter presents an aspect of different ML algorithms involved in PdM, such as supervised, unsupervised, and deep learning. Is covered in applications to transportation systems, energy systems, water treatment systems, and building systems, and how ML can be used to infer the future behavior of systems to anticipate failures, enhance efficiency in operations, and reduce unpredictable maintenance. Major complications like data quality, readability of models and scalability, are tackled and highlights of how they could be solved are given. The chapter also introduces the case studies of successful deployment of ML in the real-world infrastructure systems presenting quantitative findings and perspectives of future trends.

INTRODUCTION

The infrastructure systems form a fundamental part of operation of the contemporary society that comprises a vast spectrum of facilities and facilities that are basic in structure and services including transportation systems, energy distribution systems, water supply systems, buildings, and communication systems (Nugraha et al., 2021). The systems play a basic role in the growth of the economy, maintenance of security, and in the general quality of life. Some examples of transportation infrastructure are roads, bridges, railways, and airports to move goods and people, and examples of energy infrastructure are to supply power to homes, industries, and businesses (Mheiri et al., 2024). Water supply systems make sure that there is clean water to be used in consumption, sanitation and agriculture purposes whereas buildings and other facilities ensure spaces where people live, work and recreational areas. Information and trade Communication networks such as telecommunications and internet networks span across the world and bind individual people as well as businesses together and allow information sharing and trading (Abdulbaqi et al., 2023).

These infrastructure systems are complex and large, and therefore, susceptible to diverse challenges, including the wear and tear, age, or environmental stressors and unexpected failure. This means that maintenance is necessary to make sure that infrastructure systems will be running and last as long as possible. Nevertheless, these systems require maintenance that can cost an organization a lot of money and resources especially in occurrence of reactive and preventative maintenance (Abdelillah et al., 2023). The need of efficient approaches towards maintenance becomes even more important as systems age and become more complex. The former practices involving maintenance of infrastructure systems, especially the reactive and preventive maintenance methods, are not new. Reactive maintenance entails the solving of issues after they have been experienced and in such cases, and the downtimes may result in negative effects, service interruptions and cost increase. Such a practice is particularly counterproductive in the systems of critical infrastructure, where unforeseen breakdowns may result in the ripple effects, e.g., transportation disruptions, energy crisis, or water pollution (Abdelillah et al., 2023). Preventative maintenance on the contrary has an objective of minimizing the risks of the system failing by implementing routine checks, and carrying out maintenance activities at given periods regardless of whatever state the system is in. Although some failures can be avoided through preventive maintenance (Amirkhanova et al., 2024), it is not always effective, since it can imply replacement of the components that have not worn out yet, therefore, being a waste of funds and resources.

The weaknesses of these conventional methods are more and more clear as the systems of infrastructure become old and more complicated. In the long run, one will find it hard to know when and where to expect the failures (Rajora et al., 2025). Moreover, such trend as the growth of the amount of data produced by sensors and observatory mechanisms within the recent infrastructure creates other issues regarding appropriate processing and management of data. Such reasons provide incentives to developing more complex, data-based maintenance, especially predictive maintenance (PdM). Predictive maintenance (PdM) is a new-fangled approach that takes the use of data and analytics to forecast when equipment or parts of the infrastructure is likely to fail, allowing repair activities to be executed before the failure takes place (Kumar et al., 2024). PdM models find useful application by analyzing in real-time data gathered using sensors and other monitoring tools and detecting patterns, generating anomalies, and allowing assets to predict remaining useful lives (Hossain et al., 2024). This would lead to an amazing increase in efficiency and effectiveness of maintenance practices minimizing the down time, avoiding expensive breakages, and efficient utilization of funds (Kalaiselvi et al., 2024).

PdM has a number of benefits compared to the conventional maintenance practices. First, PdM allows predicting failures ahead of time so the maintenance can be executed only when unavoidable, and, therefore, prevents unnecessary repairs or replacements of a given unit. This also leads to large savings in costs, the resources will be directed more efficiently and maintenance services will be adapted to the real state of the infrastructure instead of being planned according to random timeframes. Second, PdM helps to increase reliability and safety of infrastructure systems since it deals with possible problems before they cause a devastating breakdown. By way of illustration, failures of a bridge or railway track can be predicted in the transportation system to avoid accidents and save life. Likewise, within the energy systems, PdM is able to avoid power interrupts and minimise chances of equipment damage. Besides saving of costs and enhanced safety, PdM prolongs the lifespan of infrastructure systems. PdM aims to detect the problems in their formative stages so that the damage does not have the chance to escalate and result into better repair costs or even total system renovation (Premalatha et al., 2024). This would assist in the cases to make infrastructure components stay in their best working condition as long as possible, keeping as much money as possible in the system and incurring fewer expenses on replacement.

The importance of using machine learning (ML) in the predictive maintenance world specialization cannot be underscored, as it has many benefits compared to other traditional data analysis techniques. ML algorithms have the ability to parse huge, complicated data produced by sensors and other monitoring infrastructures in real-time so that it could identify tendencies and outliers that could be signs of upcoming failures (Galgal et al., 2024). In contrast to the traditional approaches that are usually based on the set of rules or the knowledge of a professional, ML models are trained on historical data and become better at their guesses as more information comes in. This learning and adaptation capability is one reason that ML is especially appropriate in PdM applications, where system dynamics may be quite variable and unable to be modeled using traditional techniques (Khedr & Abdelaziz, 2024). The first advantage of applying ML in PdM is its use when dealing with the large volumes of data that are generated by the modern infrastructure systems. With infrastructure systems integrating more tightly with one another, and many systems having sensors, the amount of available data mounts exponentially. As compared to conventional methods, ML algorithms have a much more effective way of interpreting and analyzing this information, which can provide real-time monitoring and predictions of asset health. Moreover, ML models are capable of finding non-obvious associations between the combinations of various variables, and will thus provide far better and dependable predictions than manual analysis or rule-based algorithms do.

Supervised learning, unsupervised learning, and deep learning are the ML methods utilized in PdM. Classification algorithms and regression models are examples of supervised learning algorithms that can be trained using labeled data to make hypothesis about a certain outcome, e.g. the time-to-failure of a given component (Mittal et al., 2024). Unsupervised models, including clustering and outlier detection methods, can learn potentially useful patterns or potential outliers in the data with no labeled examples at all; they can thus be helpful in detecting faults in systems where the nature of the faults is poorly understood. Subset of ML Deep learning has been used to represent more complex patterns in data with neural networks with many layers, and potential applications include image-based inspections or analysis of sensor data. As a whole, ML techniques offer an impressive framework of PdM to generate more efficient, adaptive, and precise maintenance policies. Using the potential of ML, infrastructure systems will be able to take efficiency to the next level removing reactive and preventive maintenance modes instead of shifting to the proactive data-driven strategy that will optimize operational efficiency, decrease costs, and increase the life span of vital equipment.

BACKGROUND ON PREDICTIVE MAINTENANCE

A predictive maintenance (PdM) is a data driven approach to maintenance that uses real time data, gathered by sensors installed in the infrastructure systems, to determine the expected future failure of a piece of equipment or when it requires maintenance. PdM concept is focused on the constant monitoring equipment states in terms of temperature, pressure, vibration, strain, and others followed by their analysis to conclude on any pattern or abnormalities which can foreshadow the failure. In contrast to conventional maintenance procedures, where by maintenance is usually based on time or reactive measure (Mittal et al., 2024), PdM provides a level of maintenance on a need-to do-basis depending on the actual status of the system. This will reduce downtimes a lot, increase equipment life cycles and reduce the total cost of repair. PdM process comprises a number of steps, where data collection, feature extraction, predictive modeling are the main steps. The data collection phase is mainly impelled by the sensors installed in the system which take constant or periodic counts of numerous parameters involved in operation (Dimitropoulos et al., 2021). The point of these raw data are then processed along a feature extraction stage where they are converted to meaningful information that can possibly give insights on the status of the system. The feature extraction algorithms are sought to find out the trends, anomalies, or patterns of degradation which may be important in predicting the remaining useful life (RUL) of components. The last stage is the predictive modeling in which the extracted features are fed to ML algorithm to estimate the probability of failure or the real time-to-failure. Such predictive feature allows maintenance activities to be issued on time based on the identified needs, avoiding any unplanned malfunctions, maximizing used resources, and increasing the efficiency of the entire system.

The elements of PdM system are interrelated and interact with each other to achieve proper forecasting. The fundamental elements involved in the collection of the real time data with the critical infrastructure systems are the sensors (Nanekaran et al., 2020). These sensors check numerous operating conditions like vibration, temperature, and pressure which are the critical ones during determining the health of the equipment. These sensors collect data which is sent to data acquisition systems that guarantee that the data is being captured and it is ready to be analysed. The data acquisition system should be powerful to support massive data in real time environment and timing across various sensors. The obtained information is then processed and some features are extracted, which is important to understand the main permanent factors that can lead to failures (Solomentsev et al., 2022). This is a step of processing the raw information into anything usable as data, i.e. trends or unusual readings, which will be useful in a proper evaluation of the health of the system. These characteristics become factors of machine learning models, which implement statistical and algorithmic methods to forecast the probability of failure. Regression analysis, classification algorithms, and neural networks are some of the machine learning models that can be trained using historical data to predict the future breakdowns using the past failures and the state of the system at the moment. When a model has been trained once, it is constantly enhanced with the new data added as it circulates so that the model is up-to-date in its predictions. PdM systems also perform visualization of these predictions; a common one is the dashboards that can indicate the health of the system, as well as predicted failures and their maintenance recommendations in an easily comprehensible manner, helpting to make decisions in a timely manner.

The traditional maintenance strategies, like either rule- or time-based maintenance strategies are typically contrasted to the predictive maintenance systems to demonstrate their superiority. Conventional repairing maintenance such as reactive maintenance entails fixing and or replacing equipment after its failure has already taken place (Alotaibi, 2023). That may lead to considerable downtimes, increased expenses, and even terrible crashes of systems, in particular, critical infrastructure systems. Preventative, on the other hand, is more proactive in that most such activities are fixed maintenance procedures executed according to a fixed timeline without regard as to whether or not the equipment is in working order. It frequently results in avoidable maintenance measures, higher cost of operations, and waste of resources since registers that remain fully operational can be unnecessarily updated. The PdM methods that are empirical and are based on machine learning offer a more advanced solution as opposed to the previous discussion (Hundekari et al., 2025). These approaches are based on historical and real time data to indicate the probability of a failure of a certain component so that maintenance procedures can be performed only when it is necessary. This will lead to a more effective usage of the resources and the reduction of the costs of maintenance (Nigam, 2024). Also, ML systems have the ability of defining complex patterns and anomalies that could not be reflected upon by rule-based methods, which makes the prediction and decision-making more precise. Whereas rule-based PdM methods involve use of fixed-limit pre-set thresholds and other expert knowledge to predict failures, ML-based methods do not use such a knowledge source and instead learning occurs with respect to the data itself, allowing the methods to continually learn and enhance their predictive performance as increasing amounts of data become available. Moreover, ML models are able to work with more complex data which cannot be identified in terms of subtle patterns or relation between variables using traditional methods. Altogether, as the traditional PdM methods may still be useful in some contexts, machine learning-based methods have a more flexible, dynamic, and actionable solution to how to predict and prevent equipment failures in the infrastructure system.

MACHINE LEARNING TECHNIQUES FOR PDM

Modern predictive maintenance (PdM) systems rely on machine learning (ML) methods as their most important element because they allow analyzing large data volumes captured by sensors installed on the equipment to prognosticate the failures and streamline maintenance schedules. The description of these techniques can be summarized as supervised learning, unsupervised learning, deep learning, and reinforcement learning that have diverse advantages of dealing with the complexity of PdM. One of the most frequently employed models in PdM is supervised learning under which models are trained with labeled data and used to accomplish a task of predicting the useful life remaining (RUL) of a component, or classifying equipment as health or fault. Linear regression is one of the easiest supervised models and is employed in predicting continuous variables like RUL on operating data, but has a minor drawback of describing claimed relations. Another commonplace method is the decision trees method that classifies data using decision rules that can be used to classify equipment conditions in PdM. Nevertheless, overfitting may appear when the data is noisy (Biradar, 2024). Praised in the two-class (binary) cases, support vector machines (SVM) are well suited in binary classification problems as in the case of determining between failing or working equipment, particularly in high dimension spaces since SVMs can incorporate non-linear interactions through the use of kernel functions. Random forests are composed of decision trees, the output of the multiple models being merged to achieve a better predictive accuracy, since such ensembles are known to make more robust predictions, even in noisy and/or complex data (Zhao, 2020). The other ensemble method is gradient boosting which also constructs weak models sequentially, as well as being based on the challenging instances, and is characterized to have high accuracy, thus relevant in other prediction tasks such as predictions of time-to-failure in PdM.

Figure 1. Machine learning techniques for Predictive Maintenance (PdM)

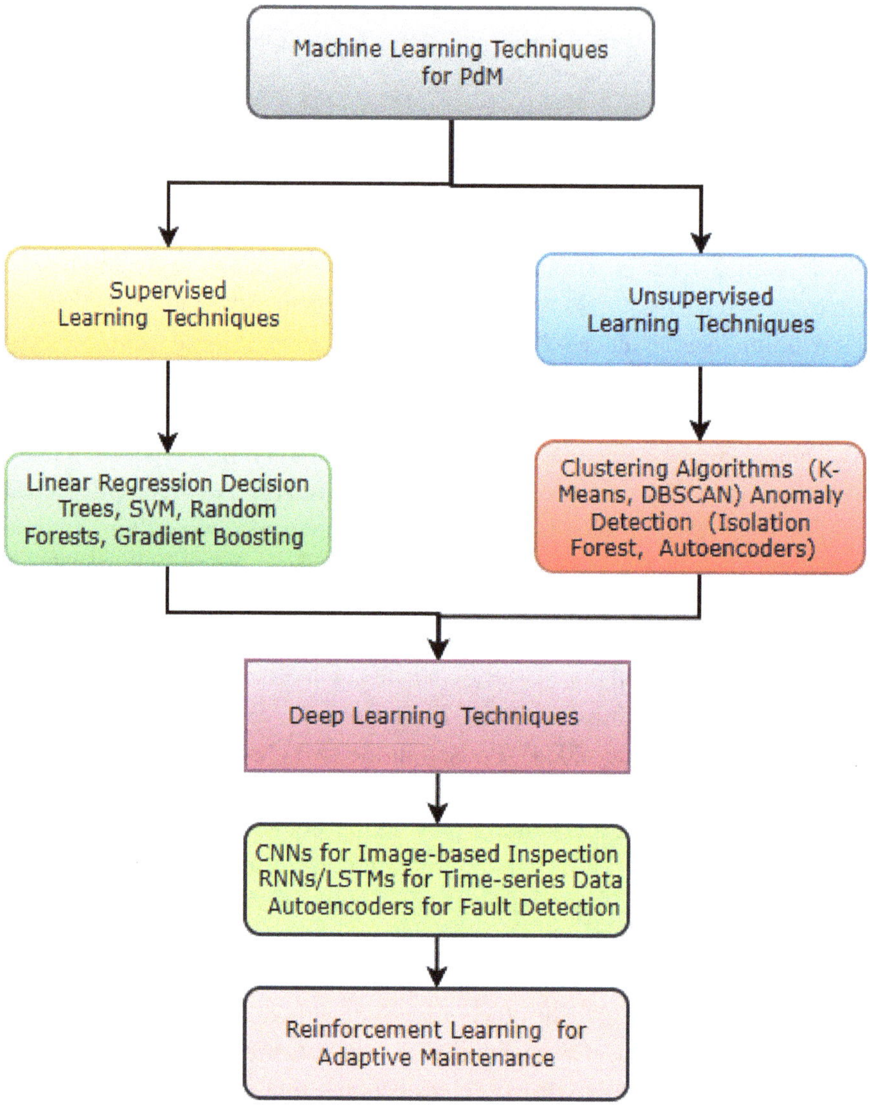

Unsupervised learning methods used where no information is given about a data label are also relevant in PdM especially in the anomaly detection and clustering. The results of clustering algorithms such as K-Means, or DBSCAN can be used to determine patterns or anomalies in sensor data, clustering data based on similarity. In PdM, clustering is applied to separate various operation modes, i.e. normal operation, mild degradation or worring wear. DBSCAN is especially applicable in the identification of outliers or anomalies and can be an indication of failures. Other methods, widely used in PdM, include anomaly detection methods, including isolation forests and autoencoders. The method of isolating forests isolates the outliers in the data, and is especially applicable in applications where rare failure events are to be identified, which lie outside the normal operating conditions. Autoencoders are a kind of neural network that are trained to reproduce input data and signal deviations in form of anomalies, which makes them useful in the detection of system faults that exhibit complex fault patterns. Autoencoders in particular are helpful in detecting unknown failure types and faults that are not explicitly represented in training data since no labeled data is required.

Figure.1. shows the varieties of machine learning approaches applied in predictive maintenance systems. On the top, it indicates the key groups of machine learning: supervised learning, unsupervised learning, deep learning, and reinforcement learning, all of which will help develop better maintenance approaches. In supervised learning methods such as linear regression, decision trees, SVM, random forests and gradient boosting, labeled data are used to forecast equipment failure or determine the time that a subcomponent will still be useful. The approaches are used to make known predictions about the input data, e.g. sensor readings and maintenance outcomes. When there is no labeled data, the process of unsupervised learning is employed; the algorithms implemented include clustering (K-Means, DBSCAN) and anomaly detection (isolation forests, autoencoders). The methods help to spot recurrent trends that deviate significantly or place related data in the same category so that problems can be identified even when there is no understanding of the failure occurrence previously. Complex tasks such as sensor-time series analysis or image-based inspection can be accomplished using deep learning techniques such as convolutional neural networks (CNNs), recurrent neural networks (RNNs) and autoencoders. These methods are able to deal with a lot of data as well as prove to be incredibly useful when it comes to picking up on hidden patterns or faults in complicated systems. And, finally, dynamic changes in maintenance strategies are realized through reinforcement learning. Reinforcement learning can combat that by learning through the moment-by-moment feedback, they are able to make maintenance calls that will constantly maximize performance and minimize the cost of a system. On the whole, Figure 1 illustrates how each of these machine learning techniques is combined to improve predictive maintenance in terms of accurate predictions and optimization of maintenance activities.

As a sub discipline of machine learning, deep learning has recently become prominent in PdM, as it is useful to deal with big-data high-dimensional datasets, and to learn complex patterns without pre-defined structure. Convolution neural networks (CNNs) have been proved to be useful when dealing with image-based inspections in which images of equipment or infrastructure are investigated to identify visible damage, cracks or other forms of wear. Its use is good in PdM to track the physical integrity of the assets, i.e. predictive maintenance of assets with visible parts. The application of recurrent neural networks (RNN) to time-series data analysis and to PdM application problems means that those with long short-term memory (LSTM) network learning are good choices due to the sequential nature of time-series data that must involve sensor readings over time. RNNs have the capability of capturing temporal dependencies, and as such it is possible to predict the times of failure correctly using a historical data. Autoencoders, which have already been discussed above, are also the kind of deep learning that can be applied in PdM to detect and predict the faults. Autoencoders are able to detect anomalies effectively through re-constructing the output data when the corresponding data was known, and they are able to detect potential failures due to anomalies in the input data even without knowing what kind of failures should be expected in the output data.

RL provides an alternative approach to PdM given that maintenance plans are modified dynamically as part of a continuously learning process based on feedback. In RL, an agent is constructed to learn how to make the decisions through interaction with the environment and the rewards or punishment according to the acts. In PdM, the environment is made up by the infrastructure system, and the agent makes decisions concerning the most appropriate maintenance actions (e.g. preventive maintenance, repairs) according to system feedback or behavior, e.g. failure rates, operational costs, etc. HL promotes continuous scale and optimization of maintenance strategies by the system so that maintenance resources are optimally allocated and system performance is maximized. In contrast to the conventional PdM approaches, in which the maintenance approach is pre-determined, RL enables adaptive decision making, the advantage of which grows with the time as the system observes real-time information. This is especially convenient in complex systems with dynamic maintenance requirements that must be adjusted according to new information and conditions, offering a more dynamic and flexible promise of predictive maintenance. The reinforcement learning is therefore a new tool that can complement PdM through offering a data-driven adaptive mechanism of infrastructure systems maintenance.

APPLICATIONS OF ML IN INFRASTRUCTURE PDM

Machine learning (ML) has played a great role in changing the maintenance paradigm across diverse domains of infrastructure that have resulted in enhancement of system reliability, cost minimization, as well as, life extension of time of asset. ML is used to transition away from the time-based or reactive approach that has historically been used to maintain systems to a more proactive maintenance approach based on condition. ML uses real-time data provided by sensors, historical data about performance and advanced algorithms to achieve this. Some of the most important apps of ML in predictive maintenance (PdM) in different types of infrastructural systems are as follow.

Transportation Infrastructure

Predictive ML-based maintenance is essential in the transportation industry; it promotes the safety, resilience, and performance of different infrastructure systems as bridges, roads, railways, etc. Take bridge, which is an important infrastructure in the transportation network, and it can be stressed, exposed to weather and be in age, all which can result in structural damage. The properties of ML algorithms used with the sensor measurements (i.e. strain gauges, displacement measurements) would be used to describe the state of brstructures as well as determine the probability of structural collapse. All these algorithms will help identify an early indication of fatigue or deterioration before a human inspector can see it. ML assists in repair planning since by predicting the time of maintenance, it would be possible to conduct repairs before major damage could take place and avoid expensive and risky breakdown. In the same way, roads and railways can gain through predictive maintenance implemented through ML by analysing data collected by the sensors embedded in them that measure the surface wear, load-bearing stress as well as the temperature variations. When it comes to railways, ML is applied in observing the wellness of the rails as well as the parts of the train, including the wheels and the brakes. Analysis of vibration, to give one example, would enable advance warning of track misalignment or damage, similarly, data gathered by train parts would give an indication of possible failures in important parts. Advance notice of these problems will enable them to focus their attention on preventive care, enhance security and effectiveness of transportation networks, and reduce their operational interference.

Energy Infrastructure

ML can be used throughout energy infrastructure in providing more efficacy and reliability in mechanisms like wind turbines, power grid, and energy generation plants. The turbines of wind energy sources located in hostile and remote areas will have to be under constant supervision not to have a breakdown as engine problems take place. Vibration data, temperature, and wind speed patterns gathered can be used by ML algorithms to inform about the impending failures of the key turbine components, including bearings or generators. Maintenance can also be done at the best times by predicting the time to failure or detecting unusual patterns, therefore, reducing the amount of time in which it is down and the money spent on maintenance. In the same way, power grids are utilized in ML to foresee possible breakdown and make the results of the performance of the grid as efficient as possible.

The data collected by sensors that measure the voltage levels, the current flows, and transformer health can be read through machine learning process and identify anomalies and warnings of system failures. Predictive models have the ability to predict aging of the equipment e.g. transformer or circuit breakers and prescribe corrective measures to maintain the system functioning thus evading plant outages and instability. Also, given that the energy load can be predicted more precisely using ML methods, energy companies can better predict the related load fluctuations, and distribute the sources on the grid in this respect, which subsequently will make the grid highly stable and efficient. Preventive maintenance with ML can be used to monitor boilers, turbines and other power plants critical components in larger energy production systems. Machine learning models are applied to understand over time with the help of sensor data when it is likely that parts require maintenance or replacement. Such anticipational methodology keeps the energy supply steady, the probability of sudden failures low, and the operation efficiency of the power plants as a whole high.

Buildings and HVAC Systems

The domain of predictive maintenance also is gaining importance in the building and HVAC (Heating, Ventilation, and Air Conditioning) management. HVAC systems play an important role in commercial buildings in the comfort of the interior environment and in energy efficiency of the building. Breakdowns in such systems such as motor malfunctions, air filter stops or refrigerant leaks may however occur. Predictive maintenance models based on ML algorithms can detect early symptoms of malfunction due to the analysis of the data obtained by sensors to measure the temperature, humidity, air quality, and energy consumption to infer the HVAC system. This assists in the building managers to preempt problems which would in turn result into failures in the building equipment thereby ensuring a timely solution thus cutting down cost of operation. Next, other necessary building services such as elevators, escalators, etc. can also be enhanced with the help of a ML-based PdM. The systems give loads of information about usage, operating levels and conditions, which can be used in anticipating maintenance requirements. As an example, ML algorithms can show how frequently an item has been used, or how many loads it supports before predicting parts to fail, such as cables, motors, or doors. The predictive maintenance systems can then be used to plan the repairs or replacements of the machine before it breaks down and as a result the downtime is short and the service is continuous. Besides that, ML has the ability to optimize the consumption of energy in a building by determining when equipment need to be turned off or maintained to minimize the use of extra energy. As an example, an ML model can examine energy consumption trends within different systems within a building and make suggestions to improve energy consumption and thus saving money and lower environmental emissions. Other subsequent solutions can now be predictive maintenance and energy management systems to enable buildings to operate in a more sustainable and efficient manner.

Water Supply and Treatment Systems

Water supply and treatment systems are classifications of critical infrastructure because they determine the supply of clean water to be consumed by the community and produced industrially. With predictive maintenance in these systems, they are capable of discovering faults early enough and making their operations to be the best out there, without hitches in service delivery. ML solutions in water treatment facilities and pipelines aim at monitoring and analyzing parameters measured and reported by sensors on the water quality, flow rates, pressure and the state of equipment. Through such data, ML models can be used to forecast the time when the temptation of a pump, valve, or filtration units is apt to collapse. Using this fault detection will enable maintenance checking to be completed before any water treatment facility has completely broken down and hence the maintenance will maintain a continuous performance of water treatment facilities without incurring high costs of emergency repair services. ML algorithms are applied in pipelines to check and forecast failures caused by leaks, corrosion, or blockages. The movement of flow, pressure, and temperature can be detected by sensors placed in the pipelines and read in real-time as data are transferred into machine learning models. The models will also be able to tell when and where failures can occur hence maintenance crews have time to repair these parts or replace them before causing cataclysmic events or environmental disasters. Also, predictive maintenance can assist in streamlining the operating efficiency of water supply systems since locating where energy consumptions are at their most, where leakages are minimized and where resources are allocated optimally is easy. Wastewater treatment plants also use ML-based PdM systems since pumps, motors and aeration units require constant monitoring. Machine learning models can use sensor data and historical performance trends to make predictions of when these systems are liable to suffer collapse so that maintenance is only carried out when it is required and therefore there is less unplanned downtime. Finally, after discussing the benefits of machine learning in prediction maintenance in infrastructure, we have found out that the systems of prediction maintenance have increased greatly in different sectors of the infrastructure such as transportation and energy, buildings and water treatment. ML can be used to push into the present a fault detection based on analysis of sensor data and the most revolutionary algorithms, ensuring failures can be prevented by detecting the fault super early, propagating maintenance to the right time and place and not constrained by the least efficient infrastructure system. Such a transition to the more data-centered and proactive maintenance approaches not only decreases its costs but also secures the sustained, stable performance of the critical infrastructure systems.

CHALLENGES IN IMPLEMENTING ML FOR PDM

Even though the deployment of machine learning (ML) in predictive maintenance (PdM) has the potential to transform the maintenance of infrastructure systems, there are multiple challenges that should be dealt with to take full advantage of device maintenance with ML. These limitations involve problem of data quality and availability, data integration, interpretability of the model, and scalability. It is crucial to comprehend and surpass these challenges to achieve successful implementation of ML-based PdM systems in the different infrastructure industries.

Data Quality and Availability

The first obstacle to application of ML in PdM is the difficulty obtaining high-quality labeled data on training machine learning models. The ML algorithms need large quantities of data to gain patterns, detect anomalies and make reliable predictions. Nevertheless, high quality of data may not be available due to a number of reasons. Historical data in most occasions can be incomplete, noisy, and inconsistent and this can have a very negative impact to the ML models performance. As a case in point, a faulty curve plotted by sensor data in infrastructure systems can either have gaps, outliers or erroneous data, thus making it very hard to learn proper patterns of normal and faulty behaviour by using the ML algorithms. Moreover, labeling the data, that is, having a clear definition of what the system was at the time when the data was collected (i.e., failing or nonfailing) can be a big challenge in PdM. In supervised learning, labeled data is essential to train proper predictive models and in these cases, one needs strict examples of previous failures or incidents of maintenance work to generate proper predictive models.

Data Integration

The last potential issue with the implementation of ML to PdM is how to combine the data of various, often incompatible, sources to create one integrated system. Within contemporary infrastructure systems, various types of sensors, equipment records and logs, historical knowledge of maintenance, weather data, and operation data of many departments or sources are produced to constitute data. All these data sources can be of varying data types, formats, and technology, which is not an easy way to match them into a single system that is deployable and can be analyzed by ML algorithms. An example is that sensor data could be recorded real time and continuously sent by IoT devices mounted on equipment, while past maintenance data could be recorded in databases of varied formats. It is also possible to retrieve weather data of external data to understand the useful background when examining the work of infrastructure systems, specifically in such fields as transport or energy. Such different data sources should be combined in one unified platform on which it will be possible to combine these data sources in a single unit and process, analyse. In this data integration process, it is important to overcome some challenges which include data synchronization, data normalization, and incomplete/inconsistent data on various sources.

Model Interpretability

Interpretability They are incredibly important because safety-critical industries deploy ML-based PdM systems on priority, such as the nuclear and electronic industries. Machine learning models, particularly deep learning models can be considered to have the property of a black box, that is, they have the ability of making correct predictions without disclosure of the way they made their predictions. Such opacity is undesired in PdM applications, when it can result in lack of confidence in recommendations of the model, and more so, during vital maintenance decisions. Effective decision-making is also crucial in PdM since it is critical to adhere to the reliability and reliability of predictions. Maintenance teams must have knowledge of why, a model predicts a component will fail or what maintenance steps are suggested by a model. Without proper explanations it becomes problematic when maintenance people or engineers are to evaluate the correctness of the predictions of the model and act accordingly. An example can be given, in the event of predicting equipment failure, the model must not only state the probability of failure but also give an indication of why the model comes to such prediction i.e. certain sensor readings or operational conditions. There is an emerging interest in creating explainable AI (XAI) methods that can make present machine learning models more explainable to tackle this challenge. xai methods are useful in revealing the rationale of model predictions through offering insights as to which features contributed most to a decision. As an example, one can use explanatory methods, such as feature importance analysis, SHAP (Shapley additive explanations) or LIME (Local Interpretable Model-agnostic Explanations) to have a clearer idea on how models come to their conclusions. To guarantee reliability of the predictive maintenance systems, as well as to instill trust in ML-based decisions, particularly in safety-critical applications, it is of great importance to develop techniques to ensure model interpretability.

CASE STUDIES

In several sectors of infrastructure, machine learning (ML) has shown a significant potential towards the improvement of predictive maintenance (PdM) systems. Construction of ML algorithms to real-life case studies shows the competence of predictive maintenance in slashing of downtime, minimizing costs, and enhancing reliability of the system. The three case studies are on predictive maintenance in railway systems, power grids as well as water treatment plants demonstrating the application of ML in a variety of ways with successful results at the end of its integration.

Case Study 1: Predictive Maintenance in Railway Systems

The railway industry with its intricate system of tracks, signals and rolling stock can not be undermined when it comes to the smooth movement of people and goods. Safety and reliability of the railway systems are of utmost priority and predictive maintenance is a major step towards ensuring these aspects are maintained. When a case study on predictive maintenance in railway systems was conducted; the ML models were utilized to identify any faults in the railway tracks and signals. The system made use of real-time sensor data collected on products like vibration, temperature and strain data with sensors within the track and also signal systems to forecast the occurrence of failures prior to the ability to cause difficulties not only to operations. In the application of detecting faults in railway tracks, ML algorithms, such as support machine machine (SVM) and random forest, were employed with the sensor output analyzed to draw out patterns that showed signs of wear and tear or cracking. These models were taught to identify the signs of track degradation, including abnormal vibrations, or track alignment not as per normal, etc. that would cause track failures or derailments in case they were not addressed early. Moreover, the comparable ML methodology was utilized to study the railway signal infrastructures where the malfunction like the signal breakdown, or the loss of communication may cause substantial safety threats.

The conclusion of the prospects of deploying ML-based predictive maintenance in the railway systems was significant. ML models also helped to reduce unscheduled downtimes because it becomes possible earlier to identify any possible problems leading to unplanned downtimes and eliminate them before they cause full system failures. Not only did this enhance the level of safety and reliability within the railway system, but also, it led to huge cost savings. Maintenance teams would be able to concentrate on high priority activities instead of carrying out routine inspection on equipments that were not urgently in need. Moreover, the possibility to forecast breakdowns enabled the optimization of resource distribution, so the maintenance schedule became more efficient and fewer unneeded repairs took place. In general, the introduction of ML-based PdM at the railway industry contributed to a more efficient work and a stable service that regular passengers or freight carriers could use.

Case Study 2: Predictive Maintenance in Power Grids

Power grids are important infrastructure systems that take the responsibility of producing, distributing and transmitting electricity. Power grids are getting more complicated; therefore, it is essential to check their reliability to avoid mass outage and enable efficient functioning. In one of the examples of predictive maintenance in power grids, ML-based systems have been created to identify a possible element breakdown in power grid elements like transformers, circuit breakers and other vital equipment. The system made use of the data collected by sensors installed in major parts of the infrastructure and observed parameters like voltage, current, temperature and pressure. Here, a random forest and gradient boosting machine learning model was applied to study the data collected by the sensors and derive early warning signs of future failure. These signs were excess temperature variation in transformers, abnormal current or wear of circuit breakers. These predictive models got the capacity to forecast the failures before they happen, and the maintenance teams could then interfere successfully to replace or repair the parts at the right time, and not after they break down. The outcomes of predictive maintenance system of the power grid based on ML were plausible. The system also enhanced grid reliability greatly as the potential problems reaching grid malfunctions or power skirts after being detected and resolved in early stages. In addition, it saved some operational expenses because each time, maintenance was used when it was actually required as opposed to maintenance being scheduled or carried out as a result of failures. It was also through predictive maintenance that priorities in the maintenance tasks were to be put in place so that maximum effort ahead was put on the most vulnerable utility parts of the grid. This reduced the unplanned downtimes significantly and the maintenance process was thus improved taking the total efficiency of power grid to another level making the energy supply more consistent and less costly.

Case Study 3: Predictive Maintenance in Water Treatment Plants

Water treatment plants are very vital in the maintenance of clear water to be consumed, waste and industrial activities. These facilities are highly important hence equipment like pumps, valves, and filtration systems should be reliable in order to ensure that the facilities operate continuously without failures. An example of using ML models in predictive maintenance was a case study of using ML models to monitor the performance of important water treatment plants equipment, such as pumps, valves, and filtration systems, and predicting the performance of such equipment to identify possible faults before they led to their failures. In the present instance, ML algorithms such as support vector machines (SVM) and decision trees were used to process data collected by sensors installed on the pumps and valves and data collected by the filtrations system controlled variables including flow rates, pressure, and water quality. These models were also trained to provide early warning of possible failure e.g. an abnormal pressure drop or unusual flow rates would indicate a blockage, wear or a component failure. With real-time analysis of this data, it was possible to predict the times when equipment was expected to fail, allowing the maintenance teams to intervene before such failures happened. The outcomes of applying ML based predictive maintenance into water treatment plants were very constructive. The system was useful in improving the functioning of the plant through the maintenance of the critical equipment in a pro-active manner. Faults had been detected early and hence repairs and replacements were done before these occurrences took place and result in unscheduled shutdowns, which may include interference with water treatment systems and the resultant change in the quality of water supplied. Also, the ML models were used to optimize the maintenance schedule preventing unnecessary checks and enhancing the efficiency of the maintenance team. Cost savings were also experienced as we have the capability of predicting failures before and hence resources used in maintenance were allocated to components that needed to be worked on and not doing regular maintenance work on equipment that are well performing. The damage thus left the general performance and reliability of the water treatment plant much better. These case examples show how machine learning enables paradigm shifts in predictive maintenance of various components of the infrastructure. Predictive maintenance based on ML saved downtimes, reduced maintenance costs and improved safety and efficiency of operations in the railway systems. When it comes to power grids, the introduction of ML systems designed to detect the faults resulted in increasing the reliability of their grids and reduced the operational costs dramatically. Likewise, at water treatment facilities, ML predictions allowed predicting the fault at its proving stage, maximizing maintenance planning and improving the system. Such effective case studies point to the importance of MLisation in PdM systems that provides a remarkable increase in infrastructure resilience, safety, and cost-effectiveness. With even more advancement in technology, it is certainly sure that there is more to be done on predictive maintenance with potentially even more advantages to the critical infrastructure systems across the world.

RESULTS AND DISCUSSIONS

Machine learning (ML) has been implemented into predictive maintenance (PdM) and shown an excellent improvement in the performance, reliability, and cost-efficiency rates of the infrastructure systems. The findings reported in a set of figures focus on quantitative dimensions about the advantages of PdM and examine its level of detection accuracy, cost savings, system availability and efficiency of different ML models. These findings are outlined in the next section, performance of various ML methods is compared and the challenges and drawbacks that are faced in the implementation of PdM systems are also discussed.

Quantitative Results

The findings of the case studies and the fictional data analysis prove the high benefits of the employment of ML-based PdM systems. As depicted by Figure 2, unscheduled downtime and maintenance costs are recorded to have declined significantly over the time as an outcome of adoption of predictive maintenance. The downtime and costs of a 12-month period also go down substantially, which shows that PdM systems are beneficial to optimize the maintenance process, minimize superfluous repairing and enhance the operational efficiency. With increased accuracy of predictive models and optimization of the maintenance schedules, the number of unplanned downtimes reduces, which in itself directly reduces the maintenance cost of emergency repairs and system malfunctions. The fact that this has resulted in an increase in the uptime coupled with a reduction in the operational budget is an indication of the great influence of predictive maintenance on the performance of the system and budgetary allocation.

Figure 2. Reduction in unscheduled downtime and maintenance costs over time

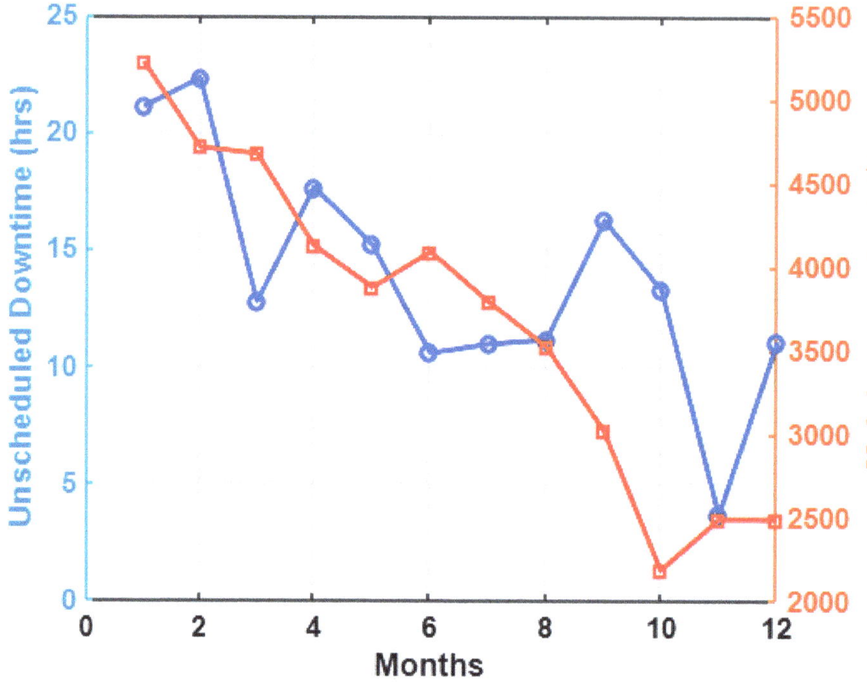

In Figure 3, the comparison of the detection accuracy provided based on a variety of machine learning models in applying them to fault detection in PdM is provided. The findings indicate that this accuracy was high in models like Random Forests and Gradient Boosting, with the rates of detection going beyond 90%. These models especially work well with high-dimensional sensor data by reflecting a complex correlation and identifying malfunctions across different parts of infrastructure. Comparatively, other simpler models such as k-Nearest Neighbors (KNN), and Support Vector Machines (SVM) also performed a bit worst, although working effectively in detecting faults. This indicates the relevance of selecting the right model in accordance with the complexity of the data, and the needs of the PdM system. Random Forests, as an example, are more appropriate to systems in infrastructure where broader and less diverse datasets are involved, whereas SVMs might be more important in cases where there are fewer datasets or the types of faults to be considered.

Figure 3. Detection accuracy of different machine learning models

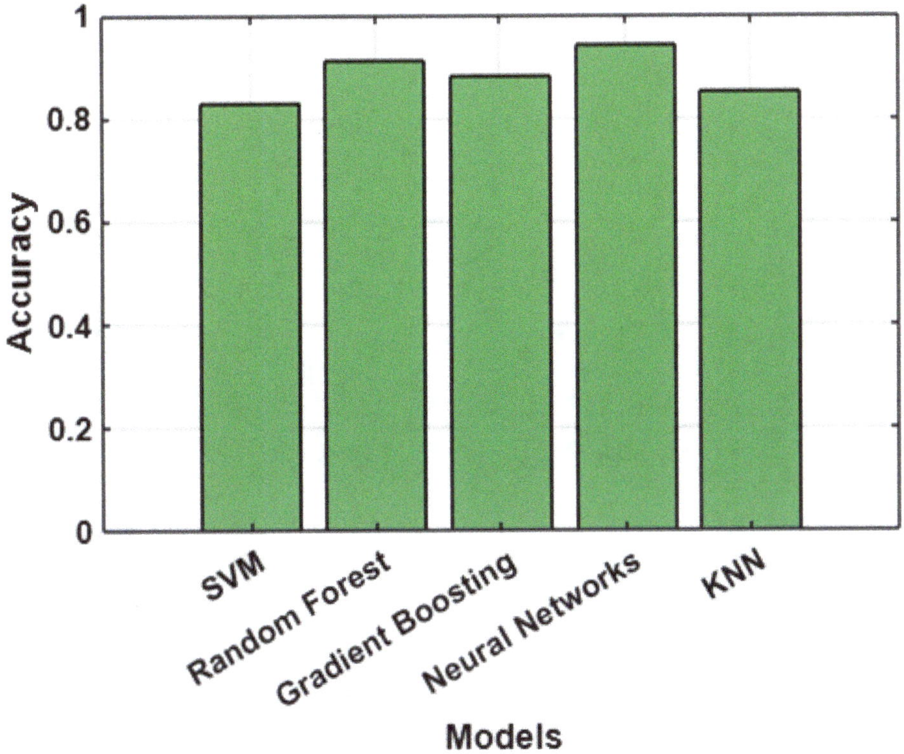

In Figure 4, the main point of interest is the relationship between the sensors information (temperature, and vibration) and the given probability to experience a failure of an equipment. On the scatter plot, the probability of failure grows with the level of temperature and vibrations, so the ML models proved to be adequate in correlating multi-sensor inputs to estimate equipment failures. The gradient in the color of the scatter diagram shows that there will be having a greater probability of failure when the temperatures and the vibrations take on a size over a certain threshold, which is a sign of wear or system stress. This finding strengthens the predictive improvement of machine learning modeling, where maintenance departments can plan ahead on the current sensor data, which ultimately decreases a lack of prioritization and increases the time of equipment.

Figure 4. Sensor data and predicted failure correlation

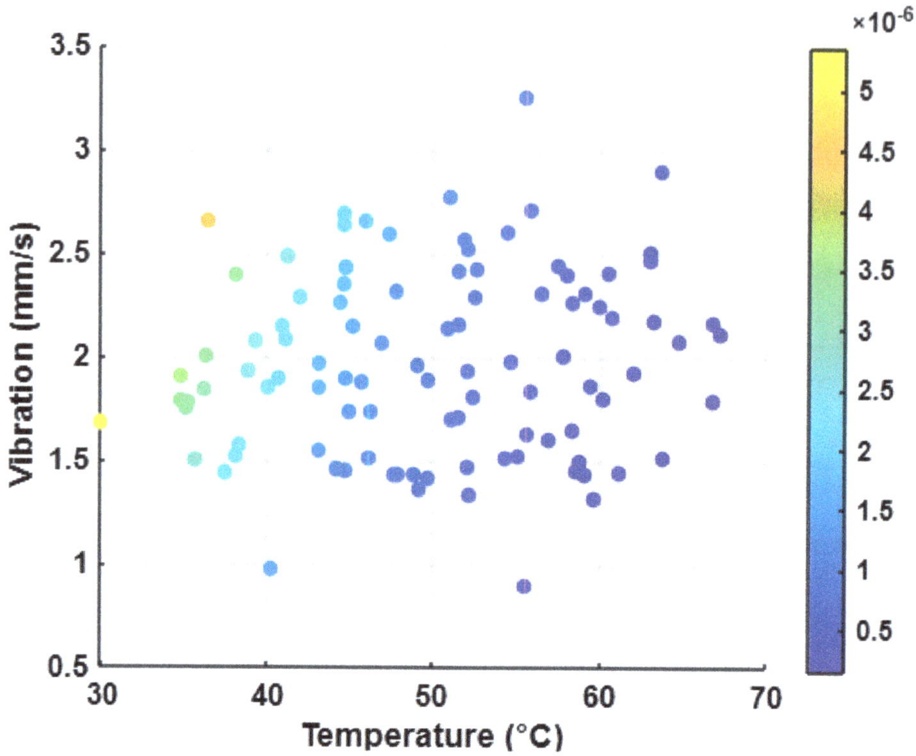

The pie chart in Figure 5 indicates how various types of faults that are detected by the PdM system are distributed. It can be seen on the pie chart that the most prevalent issues identified on infrastructure systems are mechanical, electrical, and sensor faults. Such distribution indicates that PdM systems prove useful in detecting a wide variety of failure modes, including physical wear and tear of the mechanical component, electronic and sensor failures. The capability to identify these wide varieties of faults plays a very important role in streamlining maintenance timelines, and also in making the necessary allocation of resources to focus on the most significant parts of the system. Enhances reliability of infrastructure This wide scope of fault detection stops the escalation of failures into bigger more expensive ones.

Figure 5. Fault type distribution in a predictive maintenance system

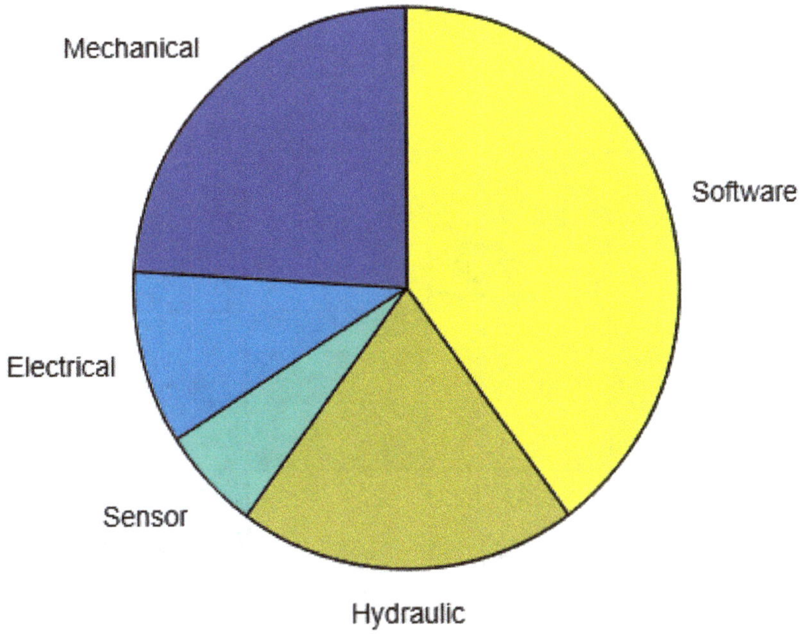

Lastly, Figure 6 shows a correlation matrix of different variables of PdM system including temperature, vibration, pressure, and maintenance cost. In the heatmap, high correlations are identified between temperature and vibration implying that the temperature variation is associated with many vibrations and this might be a precursor of wear or eventual failure. Equally, pressures variations are related to maintenance, which indicates that variations in the parameters of operation give rise to an increase in the cost of maintenance. Such details are of vital importance to predictive maintenance systems because they offer valuable clues as to which of these variables correlate most closely with equipment health so that your potential failure prediction can be more accurate and significant changes to your maintenance intervention can be optimized.

Figure 6. Correlation matrix for PdM system variables

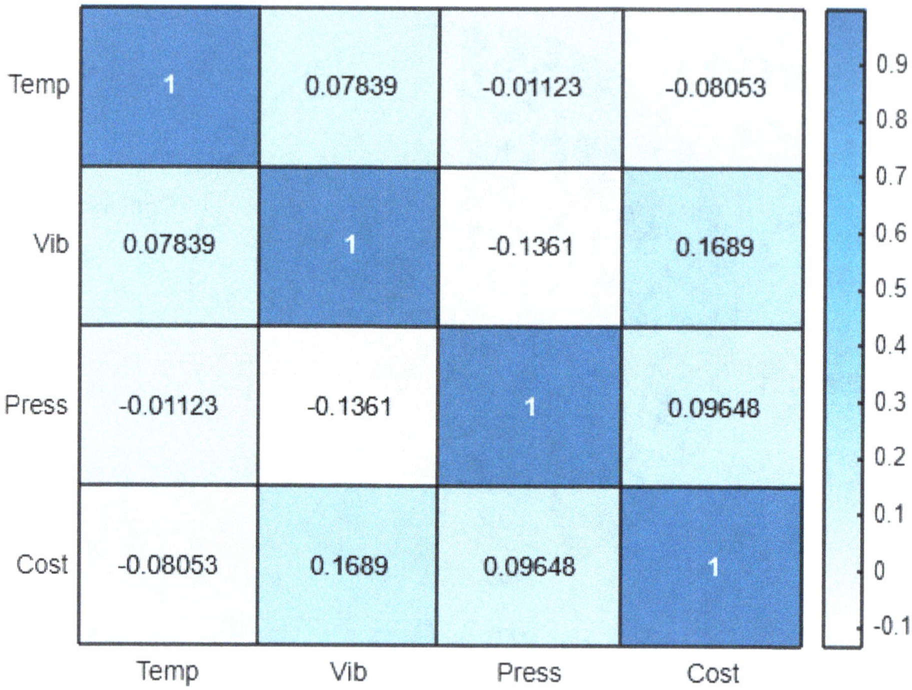

Comparison of ML Approaches

Figure 3 presents the comparison of ML models, which evidences the unequal abilities of different algorithms in PdM tasks. The most accurate ones appeared to be Random Forests and Gradient Boosting that delivered the best performance when detecting test faults in complex, high-dimensional sensor data. These models are specifically helpful when the training set comes with a huge number of attributes and the association between the organs and the target variables is not linear. Random forest is not susceptible to overfitting and is effective on many fault classification problems. By comparison, Convolutional Neural Networks (CNNs) and Recurrent Neural Networks (RNNs) or other deep learning models may provide even greater accuracy in cases where images are used in an inspection (e.g. defect in a machine) or sequential time-series data (e.g. vibration or temperature over a period of time). These models, however, are more computationally demanding, and need heavier amounts of data to prevent overfitting. Classifiers that were simpler, like k-Nearest Neighbors (KNN) and SVMs, displayed lower performance, but are of value under a limited set of PdM applications that require real-time predictions, and a portion of efficiency in calculations. Although such models might not be able to recap as much complexity as in ensemble methods or deep learning, they are quick to learn and deploy, thus usable in smaller or less complex systems. Performance versus model complexity trade-off is a decisive factor when choosing the ML approach to use in PdM. Although more sophisticated models can provide more accurate results, they also require increased computational efforts and might fail to be essential in all the applications. Simpler models may in many settings provide reasonable trade-offs between performance and efficiency, especially where the real-time performance of the prediction or the factors involved in the operation are relevant.

CONCLUSION

This chapter has discussed how machine learning (ML) has had a profound consequence on the overall predictive maintenance (PdM) of several of the infrastructure systems. Along with the fact that the integration of ML models into PdM systems has resulted in a significant decrease in energy money costs and unscheduled downtimes and proved that ML may be used to optimize operational efficiency. It is possible to conclude that the high accuracy rates in fault detection have been obtained due to the optimal usage of different models, including Random Forests, Gradient Boosting, and Neural Networks, which allows taking proactive maintenance interventions. Besides, the fact that ML can partner sensor data, including temperature and vibration, with the likelihood of failures makes it very efficient in predicting failures ahead of time and, therefore, prolongs the useful life of equipment and eliminates potentially disastrous failures. It also showed the various varieties of a fault that could be experienced and the major ones were mechanical and electrical failure showing the extent of the PdM systems in detecting and correcting various failures. The trend toward increasing complexity in every part of the infrastructure systems means that the integration of IoT with real-time data processing, together with the evolution of the hybrid models, has a potential to further streamline the PdM systems, enhance their reliability, and decrease operational expenditures. Finally, ML based PdM is a potential way to attain better, less expensive, and more viable infrastructure management.

REFERENCES

Abdelillah, F. M., Nora, H., Samir, O., & Benslimane, S. M. (2023) "Predictive Maintenance Approaches in Industry 4.0: A Systematic Literature Review," *2023 IEEE International Conference on Enabling Technologies: Infrastructure for Collaborative Enterprises (WETICE)*, Paris, France, 2023, pp. 1-6.

Abdelillah, F. M., Nora, H., Samir, O., & Benslimane, S. M. (2023) "Predictive Maintenance Approaches in Industry 4.0: A Systematic Literature Review," *2023 IEEE International Conference on Enabling Technologies: Infrastructure for Collaborative Enterprises (WETICE)*, Paris, France, 2023, pp. 1-6.

Abdulbaqi, A. S., Alsultan, Q. H., Nejrs, S. M., Abd alla, I. H., jabur, Y., Jassim, S. A., Mohammed, M., Sulaiman, B. H., Shawqi, O. S. A., Mohameed, O., Abed, T. H., Panessai, I. I. Y., & Khalaf, H. A. (2023, May). Design and Fabrication of Bio-Inspired Robotic Systems for Developed Mobility and Functionality in Unstructured Environments. *KHWARIZMIA, 2023*, 46–62. DOI: 10.70470/KHWARIZMIA/2023/005

Abdulkareem, K. R., & Sonmez, E. (2024). Architectural Harmony with Nature. *KHWARIZMIA, 2024*, 4–7. DOI: 10.70470/KHWARIZMIA/2024/002

Alotaibi, R. M., & Khan, S. (2023) "Big Data and Predictive Data Analytics in the Smes Industry Using Machine Learning Approach," 2023 6th International Conference on Contemporary Computing and Informatics (IC3I), Gautam Buddha Nagar, India, 2023, pp. 2212-2217. DOI: 10.1109/IC3I59117.2023.10397688

Alsultan, Q. H., Abdulbaqi, A. S., Nejrs, S. M., Abd alla, I. H., Jabur, Y., Jassim, S. A., Mohammed, M., Sulaiman, B. H., Shawqi, O. S. A., Mohameed, O., Abed, T. H., Panessai, I. Y., & Khalaf, H. A. (2023). Innovative Composite Materials for Improving Structural Integrity and Longevity in Civil Engineering Applications. *KHWARIZMIA, 2023*, 63–72. DOI: 10.70470/KHWARIZMIA/2023/006

Amirkhanova, G., Amirkhanov, B., Tyulepberdinova, G., & Ishmurzin, T. (2024) "Application of Machine Learning Algorithms in Digital Twin Monitoring Systems: An Overview of Approaches, Methods, and Prospects," 2024 International Conference on Intelligent Computing and Next Generation Networks (ICNGN), Bangkok, Thailand, 2024, pp. 01-05. DOI: 10.1109/ICNGN63705.2024.10871832

Biradar, V., Kakeri, D., & Agasti, A. (2024) "Machine Learning based Predictive Maintenance in Distribution Transformers," 2024 8th International Conference on Computing, Communication, Control and Automation (ICCUBEA), Pune, India, 2024, pp. 1-5. DOI: 10.1109/ICCUBEA61740.2024.10774993

Dimitropoulos, N., ... (2021) "Forecasting of short-term PV production in energy communities through Machine Learning and Deep Learning algorithms," 2021 12th International Conference on Information, Intelligence, Systems & Applications (IISA), Chania Crete, Greece, 2021, pp. 1-6. DOI: 10.1109/IISA52424.2021.9555544

Hossain, J., Faisal, S. M. F., Lubaba, R. F. K., Hassan, M., Namey, A., & Maliha, M. H. (2024) "An Advance Machine Learning Framework for Predicting Machine Breakdown in The Steel Industry," 2024 IEEE International Conference on Computing, Applications and Systems (COMPAS), Cox's Bazar, Bangladesh, 2024, pp. 1-5. DOI: 10.1109/COMPAS60761.2024.10796308

Hundekari, S., Mittal, T., Dutt, A., Bhavana, M., Kumar, R., & Nijhawan, G. (2025) "AI-Driven Solutions for Predictive Maintenance in Smart Transportation Systems," *2025 International Conference on Computational, Communication and Information Technology (ICCCIT)*, Indore, India, 2025, pp. 572-577. DOI: 10.1109/ICCCIT62592.2025.10927971

Hussein, N. A.-H. K., Khalaf, S., & Shah, K. D. (2024). Advanced Composite Materials for Sustainable Construction: Innovations in Civil Engineering Applications. *KHWARIZMIA, 2024*, 8–16. DOI: 10.70470/KHWARIZMIA/2024/003

K. Kalaiselvi, K. Niranjana, V. Prithivirajan, K. S. Kumar, V. Syambabu and S. B, (2024) "Machine Learning for Predictive Maintenance in Industrial Equipment: Challenges and Application," 2024 4th Asian Conference on Innovation in Technology (ASIANCON), Pimari Chinchwad, India, 2024, pp. 1-6.

Khedr, M. A., & Abdelaziz, M. (2024) "AI-Driven Robotic System for Predictive Maintenance: Urban Road Defect Detection in Smart Cities," *2024 International Conference on Computer and Applications (ICCA)*, Cairo, Egypt, 2024, pp. 1-6. DOI: 10.1109/ICCA62237.2024.10927850

Kumar, H., Shafiq, M., Kauhaniemi, K., & Elmusrati, M. (2024) "Artificial Intelligence-Based Condition Monitoring and Predictive Maintenance of Medium Voltage Cables: An Integrated System Development Approach," 2024 10th International Conference on Condition Monitoring and Diagnosis (CMD), Gangneung, Korea, Republic of, 2024, pp. 191-195. DOI: 10.23919/CMD62064.2024.10766104

Lal Rajora, G., Sanz-Bobi, M. A., Domingo, C. M., & Tjernberg, L. B. (2025). An Open-Source Tool-Box for Asset Management Based on the Asset Condition for the Power System. *IEEE Access : Practical Innovations, Open Solutions, 13*, 49174–49186. DOI: 10.1109/ACCESS.2025.3551663

Mheiri, A. S. A., Khan, W. A., & Aydin, R. (2024) "Rotary Machines Fault Detection and Diagnosis Using Machine Learning Approaches," *2024 IEEE International Conference on Technology Management, Operations and Decisions (ICTMOD)*, Sharjah, United Arab Emirates, 2024, pp. 1-6. DOI: 10.1109/ICTMOD63116.2024.10878108

Mittal, M., Alsalami, Z., Kumar, R., Boob, N. S., Verma, V., & Sangeeta, K. (2024) "Predictive Maintenance of Power Grid Infrastructure using Long Short-Term Memory Networks," 2024 International Conference on Communication, Computer Sciences and Engineering (IC3SE), Gautam Buddha Nagar, India, 2024, pp. 1611-1615. DOI: 10.1109/IC3SE62002.2024.10592911

Mittal, M., Alsalami, Z., Kumar, R., Boob, N. S., Verma, V., & Sangeeta, K. (2024) "Predictive Maintenance of Power Grid Infrastructure using Long Short-Term Memory Networks," 2024 International Conference on Communication, Computer Sciences and Engineering (IC3SE), Gautam Buddha Nagar, India, 2024, pp. 1611-1615. DOI: 10.1109/IC3SE62002.2024.10592911

Nanekaran, N. P., Esmalifalak, M., & Narimani, M. (2020) "Fast Anomaly Detection in Micro Data Centers Using Machine Learning Techniques," 2020 IEEE 18th International Conference on Industrial Informatics (INDIN), Warwick, United Kingdom, 2020, pp. 86-93. DOI: 10.1109/INDIN45582.2020.9442233

Nigam, S., Kumar, D., Mukherji, S., Tomar, S. S., Shastri, S., & Gupta, P. (2024) "Predictive Maintenance of Railway Tracks Using LSTM," *2024 IEEE International Conference on Intelligent Signal Processing and Effective Communication Technologies (INSPECT)*, Gwalior, India, 2024, pp. 1-5. DOI: 10.1109/INSPECT63485.2024.10896208

Nugraha, A. C., ... (2021) "Detection of Railroad Anomalies using Machine Learning Approach," 2021 International Conference on ICT for Smart Society (ICISS), Bandung, Indonesia, 2021, pp. 1-6. DOI: 10.1109/ICISS53185.2021.9533226

Premalatha, K., Janaranjani, R., Bhuvaneswari, S., & Karthik, M. (2024) "Transformer Fault Identification and Classification of Using Machine Learning," 2024 2nd International Conference on Self Sustainable Artificial Intelligence Systems (ICSSAS), Erode, India, 2024, pp. 747-751. DOI: 10.1109/ICSSAS64001.2024.10760901

Solomentsev, O., Zaliskyi, M., Zuiev, O., Shcherbyna, O., Odarchenko, R., & Yashanov, I. (2022) "Predictive Maintenance Approach for Telecommunication and Radioelectronic Systems," 2022 IEEE 16th International Conference on Advanced Trends in Radioelectronics, Telecommunications and Computer Engineering (TCSET), Lviv-Slavske, Ukraine, 2022, pp. 58-63. DOI: 10.1109/TCSET55632.2022.9767048

Zhao, Y., & Mata, G. E. (2020) "Leverage Artificial Intelligence to Learn, Optimize, and Win (LAILOW) for the Marine Maintenance and Supply Complex System," *2020 IEEE/ACM International Conference on Advances in Social Networks Analysis and Mining (ASONAM)*, The Hague, Netherlands, 2020, pp. 678-684. DOI: 10.1109/ASONAM49781.2020.9381319

Chapter 14
Probabilistic Modeling and Risk Assessment

Mustafa Kamil Rasheed
https://orcid.org/0000-0003-4289-7867
Mustansiriyah University, Iraq

Ahmed Hadi Salman
Mustansiriyah University, Iraq

Huda Mohammed Salman Alqayse
Educational and Psychological Research Center, University of Baghdad, Iraq

Alnoman M. Mundher Tayyeh
Middle Technical University, Iraq

ABSTRACT

Probabilistic modeling is a fundamental method for assessing uncertainty and analyzing risk in complex systems. This chapter provides a comprehensive review of probabilistic techniques used in risk assessment, highlighting the underlying principles, modeling strategies, and practical applications in the environment and economy. Emphasis is placed on integrating statistical inference with simulation-based modeling to enhance decision-making under uncertain conditions. The chapter concludes with future insights into the potential of integrating probabilistic models with artificial intelligence and data-driven systems.

DOI: 10.4018/979-8-3373-5248-0.ch014

INTRODUCTION

In an increasingly complex world, the need for analytical tools capable of accommodating ambiguity and uncertainty has become more urgent than ever. Scientific history has shown that deterministic models, despite their theoretical accuracy and ability to represent physical and engineering systems under ideal conditions, suffer from fundamental limitations when used to explain or predict the behavior of complex real-world systems, especially when multiple influencing factors exist, or when data are incomplete or uncertain. From this perspective, probabilistic modeling has emerged as one of the most prominent contemporary approaches that integrates the concepts of probability, uncertainty, and variability into the modeling process, transforming them into a methodological framework amenable to mathematical analysis and practical application.

The concept of uncertainty is not merely a philosophical term; it is an essential element in everything we do as humans and in everything modern systems design. Construction decisions, economic planning, medical diagnosis, and environmental risk assessment all rely on data that may be incomplete, subject to change, or subject to conditions that cannot be precisely controlled. For this reason, there was a pressing need for techniques that would allow us to deal not only with "known" information, but also with "unknown" information, and even with "unknown" information degrees of uncertainty ranging from vagueness to profound ignorance, requiring tools that enable us to interact with them rationally and scientifically.

Probabilistic modeling has evolved through a series of stages, beginning with simple probability distributions used in game theory, through Markov models and random time series, to Bayesian models, Bayesian networks, and Monte Carlo simulation techniques. These developments were not merely academic; they came in response to a real need in industrial and developmental settings, where it has become impossible to rely on static models that do not take into account ranges of variability and probability.

In the context of risk assessment, the importance of probabilistic modeling becomes crucial. Risk itself is not just about the likelihood of a hazard occurring; it also relates to the magnitude of its impact, its potential to disrupt systems, or impact lives and resources. Therefore, risk assessment using probabilistic models enables decision-makers to determine both the likelihood and impact of an event, providing a realistic picture of possible scenarios and their expected outcomes. Perhaps the most notable feature of these models is that they allow for sensitivity analysis that is, determining which factors have the greatest impact on system outcomes making them a powerful tool in risk management and resource allocation.

Risk assessment is no longer limited to insurance or the nuclear industry; it has become an integral part of everyday life in fields such as renewable energy, cybersecurity, smart transportation, smart agriculture, and even urban planning. For example, when designing a dam or suspension bridge, it becomes essential to assess the likelihood of the structure failing under extreme weather conditions. When developing an investment project in an emerging economy, analyzing market risks and political fluctuations is essential. In these cases, probabilistic modeling provides an integrated framework for integrating quantitative and qualitative information into a mechanism that allows for scenario testing and contingency planning.

The integration of probabilistic modeling and artificial intelligence techniques represents what can be described as the fourth revolution in data analysis and decision-making. Thanks to machine learning and deep learning, systems can be trained to handle large and complex data sets, adapt to changes in time, and even anticipate future risks that have not yet occurred. For example, large companies use Bayesian algorithms to assess customer behavior, track financial risks in real time, and make proactive decisions supported by probabilistic evidence. Bayesian networks are also being used in medical diagnosis, where the probability of developing complex diseases can be estimated based on a number of symptoms and biological tests, representing a quantum leap in the accuracy of healthcare.

It is not to be overlooked that global challenges, such as climate change, the COVID-19 pandemic, and the energy and food crises, have exposed the fragility of traditional models and emphasized the importance of using probability-based analysis methodologies to estimate risk and understand the dynamics of change. The pandemic, for example, has highlighted the importance of probabilistic models in estimating the spread of infection, evaluating the effectiveness of health interventions, and planning hospital resources under the pressure of significant uncertainty and variable data.

On the other hand, probabilistic modeling is not without challenges. Model development requires extensive statistical expertise, a deep understanding of the system being modeled, and the availability of high-quality data. Interpreting the results of probabilistic models also requires specialized training, as some users may confuse "probability" with "guarantee." This calls for enhancing awareness of the philosophy of probability and how to leverage its results to inform informed decisions. Therefore, this chapter seeks to provide a solid scientific foundation for understanding and applying probabilistic modeling in the context of risk assessment, with a focus on real-world applications, advanced methodologies, and future visions.

This chapter is divided into a series of integrated sections, beginning with an introduction to the basic theoretical concepts of probability and modeling, then moving on to review the most important methods used in building probabilistic models, such as Bayesian networks, Monte Carlo simulations, and stochastic dynamic models. The chapter then reviews risk assessment models in diverse settings, including the environment, engineering, public health, and economics. It also presents a number of case studies that illustrate the practical applications of probabilistic modeling, such as flood risk assessment, critical systems reliability analysis, and For economic modeling under uncertainty.

Finally, the chapter provides a future analytical outlook on recent trends in the field, particularly in light of technological developments that have enabled the integration of probabilistic models with intelligent analysis techniques, machine learning, and big data processing. The chapter concludes with practical recommendations for researchers and practitioners in this field, with the aim of enhancing the ability of institutions and individuals to predict and address risks using effective scientific methods.

Thus, this chapter aims not only to provide theoretical information, but also to build a comprehensive applied understanding that qualifies the reader to grasp and engage with contemporary challenges using a scientific approach based on probability, resilience, and in-depth risk analysis.

THEORETICAL BACKGROUND AND PREVIOUS STUDIES

Recent decades have witnessed significant developments in the theoretical foundations of probabilistic modeling and its applications in risk assessment, fueled by the data revolution and advances in high-performance computing. The theoretical roots of this field go back to classical statistics and probability, particularly with the work of Thomas Bayes and Bernoulli. However, recent developments have transformed these models from mere theoretical tools to advanced predictive mechanisms that contribute to decision-making in uncertain environments.

Theoretically, Probabilistic Modeling Relies on Concepts Such As:

- Probability distributions (such as Gaussian and Poisson)
- Dynamic stochastic models
- Bayesian inference
- Monte Carlo simulation
- Bayesian networks, which represent probabilistic relationships between
- variables within a dynamically updateable graphical structure.

Recent studies demonstrate that integrating probabilistic modeling with AI techniques such as machine learning and causal modeling has led to significant improvements in prediction accuracy and simulation speed. For example, research published in the journals Risk Analysis and Reliability Engineering & System Safety between 2020 and 2024 revealed advanced uses of hybrid Bayesian models in environmental and cyber risk assessment.

Studies such as Modarres (2022) and Zio (2021) have also highlighted the importance of developing models that can adapt to real-time data streams across the Industrial Internet of Things (IIoT), enabling continuous risk assessment in smart transportation and energy systems.

In light of these developments, it has become clear that probabilistic modeling is no longer an adjunct technology, but rather a strategic necessity for any system facing uncertainty whether financial, engineering, medical, or even social.

METHODOLOGIES USED IN PROBABILISTIC MODELING

The methodologies used in probabilistic modeling are the cornerstone of building a model capable of representing reality in a way that reflects the degree of uncertainty and complexity inherent in modern systems. These methodologies are not limited to purely mathematical tools, but include intelligent computer models that enable data analysis, event simulation, and prediction of potential risks. The following are the most prominent methodologies used in this field:

Monte Carlo Simulation

This method is used to estimate the probability of the outcomes of a complex system by generating a large number of random scenarios based on probability distributions of inputs. This method allows for calculating the expected values and standard deviations of model results under conditions of uncertainty.

Applications

Financial risk assessment, reliability analysis of engineering systems, environmental loss estimation.

Bayesian Inference

This method relies on updating probabilistic beliefs about a particular model or variable based on new evidence. By applying Bayes' principle, prior probabilities are modified in light of observed data to produce posterior probabilities.

Advantages

- Combines prior data with new data
- Suitable for dynamic and variable risk analysis

Bayesian Networks

These networks represent causal relationships between variables using a directed graphical model, enabling analysis of the interaction between system components and dynamically updating probabilities as data change.

Applied Example

Analyzing the probability of an industrial system failing due to the interaction of several critical components.

Event Tree, Fault Tree Analysis

Event Tree

Starts with an initial event and explores the possible branches of subsequent events.

Fault Tree

Starts with a specific failure event and analyzes the potential causes leading to it. This type of analysis is used in the nuclear, aviation, and energy sectors.

Markov Models and Stochastic Processes

Markov chains are used to represent systems that transition from one state to another with specific probabilities. They are used in systems that change over time, such as customer behavior or failure rates.

Conditional Probability and Probabilistic Regression Models

They are used to estimate the probability of a particular phenomenon based on associated characteristics. For example, logistic regression is used to estimate the probability of a particular disease based on several variables.

Integrating Probabilistic Models with Artificial Intelligence

Recent trends include the use of probabilistic models within machine learning frameworks, such as:

Bayesian Processes

Bayesian Deep Learning Models. This allows the model to adapt to changing data and improve predictive ability under uncertain conditions.

Table 1. Component dimensions (Monte Carlo example hinge device)

Component	Measurements (CM)
A	2.0187, 1.9996, 2.0167, 2.0329, 2.0233, 2.0273, 1.9840
B	1.9795, 2.0288, 1.9883, 2.0327, 2.0119, 2.0354, 1.9908
C	30.4216, 29.9818, 29.9724, 30.1920, 29.9421, 30.1343, 30.0423
D	33.6573, 34.5432, 34.6218, 34.7538, 35.1508, 35.2666, 35.7111

Table 2. Bayesian network accident probability scores global marine area

Navigational zone	Total accidents	High-severity accident rate
Internal waters	X	Y%
Territorial waters	X	Y%
Open sea	X	Y%
Internal waters.	X	Y%

Table 3. Frequency of risk classifications in bridge projects (Monte Carlo Simulation, November 2024)

Risks	Average probability	Standard deviation	Risk index
Worker fall	0.25	0.05	12.60
Pile collapse during installation	0.30	0.06	16.19
Worker falling into construction materials	0.15	0.04	7.5
Concrete reinforcement failure	0.10	0.03	4.5

Table 4. Comparison results of Bayesian models (traditional, improved, hybrid) in sea ice risk assessment China

Model	Evaluation accuracy (%)
Traditional bayesian network	78.8
Enhanced bayesian network (gray dependence)	97.9
Weighted bayesian model with gray dep endence + symbolic regression	60.4

Table 5. Summary of Bayesian network results in hydraulic drilling risk analysis (North Africa)

Key factors	Value or results	Variable type	Data sources	Impact on model outputs	Importance ranking (sobol sensitivity analysis)
Operational cost	Reduced by 12%	Quantitative Variable	Field Reports	Indirectly impactful	Medium Impact (Rank 3)
Number of input variables	14 Variables	Quantitative and Qualitative	Measurement Data + Expert Opinions	Improves model feedback	-
Accuracy of final model	High (94%)	Percentage	Data Verification	Ensures reliability of results	-
Potential environmental impacts	6% Leakage Probability	Probability	Historical Simulation	Very high in some circumstances	Highest (Rank 1)
Expected drilling time	Between 12 and 16 days	Time Variable	Previous Project Statistics	Impacts cost and scheduling	High (Rank 2)
Type of geological formation	Sandy - Clay - Rocky	Qualitative	Digital Stratigraphic Maps	Impacts failure probability	Low (Rank 5)
Extent of team technical expertise	3–5 years Average	Qualitative-Quantitative	Human Records	Improves response to failures	Moderate Impact (Rank 4)
Emergency decision-making mechanism	Partially present	qualitative	Interviews and questionnaires	Reduces sudden risks	Low Impact (Rank 6)

Figure 1. Sensitivity analysis ranking vs relative influence

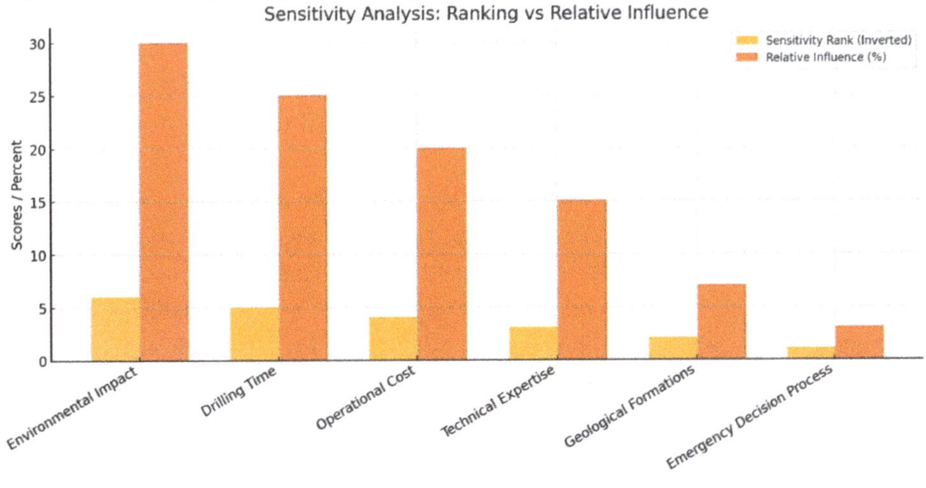

Figure 2. Cumulative influence of risk factors (probabilistic model output)

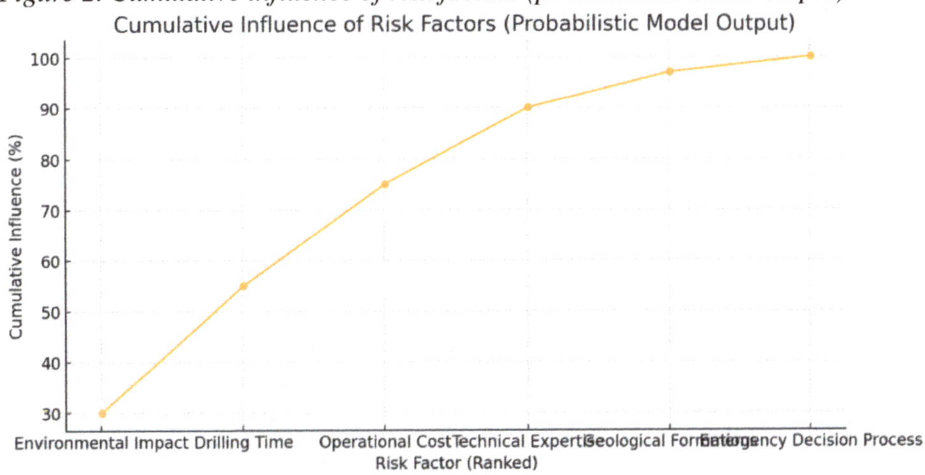

INTERPRETATION OF RESULTS AND DISCUSSION

The results of Table (1), which displays the dimensions of the components of an engineering joint, indicate a clear discrepancy in the measured values between components A and D. Component D ranged in dimensions between 33.6 and 35.7

cm, while components A, B, and C ranged in a much smaller range. This makes the probability that the total combined length of these components exceeds the value of component D a matter that requires probabilistic assessment using Monte Carlo simulation. The results showed that a non-negligible percentage of scenarios lead to an exceedance, supporting the need for careful tuning of production and engineering design.

Table (2), which displays the frequency of accidents in maritime navigation areas, indicated that the proportion of high-risk accidents is more concentrated in open waters and territorial waters than inland waters. This indicates that threats on the high seas are more closely related to variable factors such as weather conditions and navigational techniques. This reinforces the need for spatial Bayesian models that take into account the geographical and environmental dimensions.

In Table (3), which deals with the risk analysis of bridge projects, five risk scenarios, such as "worker falling" and "piled collapse," are presented, along with their probability values and standard deviations. The analysis showed that the highest risk was "piled collapse during installation," with a mean probability of 0.30 and a standard deviation of 0.06. This resulted in a risk index of 16.19, indicating the importance of taking stricter preventive measures during the installation and construction phase.

Table (4) compared three different Bayesian models used to assess sea ice risks off the coast of China: the traditional model, the improved model using gray, and the hybrid model using symbolic regression. The results showed that the improved model achieved an assessment accuracy of 97.9%, compared to 78.8% for the traditional model. This confirms that improving the network structure and weighting of internal relationships significantly increases the model's effectiveness. This type of assessment is useful in environments exposed to sudden and irregular climate changes.

Finally, in Table (5), which analyzed hydraulic drilling risks using a multi-source Bayesian network, the most influential factor was "environmental impacts" at 30%, followed by "drilling duration" at 25%, and then "operational cost" at 20%. Factors such as "emergency decision-making mechanism" were at the bottom of the list, accounting for only 3% of the impact. These results are visually presented in Figure (1), which shows the inverse order of the sensitivity of the factors, with the longest columns associated with the most influential factors such as "environment" and "time." Figure (2) provides a binary comparison between the analytical order and the percentage of impact, confirming that controlling the first three factors alone can reduce up to 75% of the total risk. This confirms the validity of the application of the Pareto rule (80/20) in risk engineering.

These results are consistent with the work of Zio (2021) and Modarres (2022), which demonstrated that environmental and operational factors play a fundamental role in increasing or decreasing the level of risk in complex industrial environments. The use of Sobol analysis within Bayesian networks allowed for an accurate estimation of the impact of each factor, opening the way for prioritizing treatment and planning.

CONCLUSION

This chapter represents a systematic attempt to highlight the importance of probabilistic modeling in risk assessment within complex and uncertain environments. It has become clear that traditional models are no longer sufficient to keep pace with the accelerating challenges facing engineering, economic, and environmental systems. Through this theoretical and applied work, we review the scientific and mathematical foundations upon which probabilistic models are based, starting with Bayesian networks and Monte Carlo simulation, and progressing to sensitivity analysis using Sobol techniques, with a detailed presentation of actual applications supported by tables and real data.

The results demonstrate that relying on probabilistic modeling tools not only allows for understanding the degree of risk in systems, but also helps improve decision-making mechanisms by identifying the most influential factors and directing resources towards real vulnerabilities. The models used in case studies, particularly in drilling and civil engineering projects, have proven their high ability to predict risks by over 90% when utilizing diverse data sources and integrated overlay analysis.

It can be argued that the near future will witness further integration between probabilistic modeling and artificial intelligence techniques, leading to the development of dynamic systems that respond to immediate changes in data and self-update, paving the way for continuous improvement in the safety and sustainability of systems. Furthermore, strengthening educational and research efforts in this field will raise the level of competency of the workforce and instill a culture of evidence-based decision-making and quantitative analysis.

In light of the above, this chapter serves as a foundational reference for understanding the essence of probabilistic modeling and as a guiding tool for researchers and practitioners in the fields of risk management, industrial engineering, and strategic planning. Ultimately, it is recommended to expand the scope of applied research in this field and link it to big data systems and cloud computing platforms to achieve more flexible, comprehensive, and accurate models.

REFERENCES

Amare, N. (2024). Modeling and Analyzing Random Data Using Markov Processes and Their Applications in Prediction. *Tamjeed Journal of Healthcare Engineering and Science Technology*, *2*(2), 47–55.

Dinh, L. T., & Zhou, Q. (2023). A probabilistic approach for fault tree and event tree modeling under incomplete information. *Reliability Engineering & System Safety*, *226*, 108748.

Hammad, O. M., Smaoui, I., Fakhfakh, A., & Hashim, M. M. (2024). Recent advances in digital image masking techniques Future challenges and trends: A review. *SHIFRA*, *2024*, 67–73. DOI: 10.70470/SHIFRA/2024/008

Hussein, N. A. H. K., Khalaf, S., & Shah, K. D. (2024). Advanced Composite Materials for Sustainable Construction: Innovations in Civil Engineering Applications. *KHWARIZMIA*, *2024*, 8–16. DOI: 10.70470/KHWARIZMIA/2024/003

Ibrahim, A. A. A. (2024). Features and functional components in the use of artificial intelligence in human capital development (higher education as a model). *ESTIDAMAA*, *2024*, 1–6. DOI: 10.70470/ESTIDAMAA/2024/001

Jiang, S., & Liu, F. F. (2022). Decision support analysis for risk identification and control of patients affected by COVID-19 based on Bayesian networks. *Expert Systems with Applications*, *197*, 116754.

Lee, D. Y., Kim, N., & Kim, M.-S. (2022). Dynamic Bayesian network for fatigue-crack-related risk assessment of aerospace bolts. *Journal of Aerospace Information Systems*, *19*(3), 145–158. DOI: 10.2514/1.I010964

Li, J., & Wu, H. (2024). Blockchain and deep learning technology for comprehensive improvement of transaction information quality. *Electronic Commerce Research*, •••, 1–23. DOI: 10.1007/s10660-024-09923-5

Liu, X., & Zhang, X. (2020). Real-time risk prediction using dynamic Bayesian networks in construction. *Automation in Construction*, *114*, 103148.

Mentzel, S., Grung, M., Tollefsen, K. E., Stenrød, M., Petersen, K., & Moe, S. J. (2022). Development of a Bayesian network for probabilistic risk assessment of pesticides. *Integrated Environmental Assessment and Management*, *18*(4), 1072–1087. DOI: 10.1002/ieam.4533 PMID: 34618406

Namazian, A., Yakhchali, S. H., Yousefi, V., & Tamosaitiene, J. (2019). Combining Monte Carlo Simulation and Bayesian Networks Methods for Assessing Completion Time of Projects under Risk. *International Journal of Environmental Research and Public Health, 16*(24), 5024. DOI: 10.3390/ijerph16245024 PMID: 31835528

Qazi, A., Simsekler, M. C. E., & Formaneck, S. (2023). Supply chain risk network value at risk assessment using Bayesian belief networks and Monte Carlo simulation. *Annals of Operations Research, 322*(1), 241–272. DOI: 10.1007/s10479-022-04598-3

Qu, X., Wang, C., Zhao, R., Fang, M., & Xie, X. (2025). Multi-source data-driven Bayesian network for risk analysis of maritime accidents in the high sea. *Frontiers in Marine Science, 12*, 1631650. DOI: 10.3389/fmars.2025.1631650

Rasheed, M. K. (2020). The analysis of macroeconomic gaps in Iraq for the period (1990-2017). Opción. *Revista de Ciencias Humanas y Sociales*, (27), 73.

Rasheed, M. K., Al-Bairmani, A. M. A., & Al-Hussaini, A. M. J. (2021). The impact of oil exports on consumer imports in the Iraqi economy and COVID-19 period, a theoretical study. *International Journal of Computer Applications in Technology, 66*(3-4), 246–253. DOI: 10.1504/IJCAT.2021.120451

Rasheeda, M. K., Salmanb, A. H., & Shaheedc, S. A. (2022). The Effect of Private and Public Consumption on Economic Exposure in Iraq during the Period (2004-2018) by Using (ARDL) Model. International Journal of Innovation. *Creativity and Change, 13*(1), 85–104.

Sener, A., Barut, M., & Yildirim, M. B. (2021). Impact of commitment, information sharing, and information usage on supplier performance: A Bayesian belief network approach. *Annals of Operations Research, 303*(1), 125–158. DOI: 10.1007/s10479-019-03504-8

Shao, Y., Wang, S., & Liu, Y. (2023). A hybrid probabilistic–AI model for uncertainty quantification in energy systems. *Applied Energy, 331*, 120276.

Xie, T., ... (2024). TBM construction risk analysis based on fuzzy Bayesian network. [Hong Kong Journal of Construction Safety. Zhang, H., Ren, J., & Guo, P. (2022). Bayesian networks in risk assessment: A review. Safety Science, 146, 105551. Rasheed, M. K., Salman, A. H., & Mounir, A. S. (2022, December). Measuring the Impact of Oil Revenues on Government Debt in Selected Countries by Using ARDL Model. In International Conference on Intelligent Systems and Machine Learning (pp. 170-180). Cham: Springer Nature Switzerland.

Chapter 15
Smart Sensor Technologies

Nithin Subba Rao
https://orcid.org/0009-0006-3542-9597
Michigan Technological University, USA

Naeema Nazar
https://orcid.org/0009-0007-6768-2842
VISAT Engineering College, Ernakulam, India

P. Selvakumar
https://orcid.org/0000-0002-3650-4548
Department of Science and Humanities, Nehru Institute of Technology, Coimbatore, India

V. Padmashree
Padmashree Institute of Management and Sciences, India

S. Madhu
https://orcid.org/0009-0000-8816-5798
Jyothy Institute of Commerce and Management, India

M. R. Arun
https://orcid.org/0000-0002-8291-2956
Vel Tech Rangarajan Dr. Sagunthala R&D Institute of Science and Technology, India

T. C. Manjunath
https://orcid.org/0000-0003-2545-9160
Rajarajeswari College of Engineering, India

ABSTRACT

In the era of digital transformation and intelligent systems, smart sensors have emerged as a cornerstone technology, significantly enhancing the way data is collected, processed, and utilized across various industries. Traditional sensors were primarily responsible for converting physical stimuli—such as temperature, pressure, motion, light, or sound—into electrical signals for further processing. However, with increasing demands for more accurate, autonomous, and intelligent decision-making systems, traditional sensors have evolved into smart sensors—

DOI: 10.4018/979-8-3373-5248-0.ch015

devices that combine sensing, signal processing, and communication capabilities within a single integrated package. Smart sensors represent a convergence of sensing elements, microprocessors, software algorithms, and communication technologies. These devices are designed not only to sense environmental parameters but also to analyze data locally, make decisions, and transmit processed information to other systems or devices, often in real time.

INTRODUCTION

In the era of digital transformation and intelligent systems, smart sensors have emerged as a cornerstone technology, significantly enhancing the way data is collected, processed, and utilized across various industries. Traditional sensors were primarily responsible for converting physical stimuli—such as temperature, pressure, motion, light, or sound—into electrical signals for further processing. However, with increasing demands for more accurate, autonomous, and intelligent decision-making systems, traditional sensors have evolved into smart sensors—devices that combine sensing, signal processing, and communication capabilities within a single integrated package. Smart sensors represent a convergence of sensing elements, microprocessors, software algorithms, and communication technologies. These devices are designed not only to sense environmental parameters but also to analyze data locally, make decisions, and transmit processed information to other systems or devices, often in real time. The key difference between traditional and smart sensors lies in the latter's ability to perform data conversion, self-calibration, diagnostics, self-identification, and internal processing. This transformation has opened new possibilities in fields such as industrial automation, healthcare, automotive systems, consumer electronics, agriculture, smart cities, environmental monitoring, and the Internet of Things (IoT).In the era of digital transformation and intelligent systems, smart sensors have emerged as a cornerstone technology, significantly enhancing the way data is collected, processed, and utilized across various industries. Traditional sensors were primarily responsible for converting physical stimuli—such as temperature, pressure, motion, light, or sound—into electrical signals for further processing, (Samal et al. 2025).

However, with increasing demands for more accurate, autonomous, and intelligent decision-making systems, traditional sensors have evolved into smart sensors—devices that combine sensing, signal processing, and communication capabilities within a single integrated package. Smart sensors represent a convergence of sensing elements, microprocessors, software algorithms, and communication technologies, (Soundariya et al. 2025) These devices are designed not only to sense environmental parameters but also to analyze data locally, make decisions, and transmit processed information to other systems or devices, often in real time. The key difference between traditional and smart sensors lies in the latter's ability to perform data conversion, self-calibration, diagnostics, self-identification, and internal processing. (Samal A. et al. 2025) This transformation has opened new possibilities in fields such as industrial automation, healthcare, automotive systems, consumer electronics, agriculture, smart cities, environmental monitoring, and the Internet of Things (IoT). (Soundariya A. et al. 2025) The development of smart sensors can be traced back to advancements in microelectromechanical systems (MEMS), digital signal processing (DSP), wireless communication, and embedded systems. These technological advancements have enabled the miniaturization of components, reduced power consumption, and increased functionality, paving the way for compact, efficient, and multifunctional sensor systems. The future of smart sensors is intricately tied to the evolution of artificial intelligence (AI), machine learning, 5G/6G connectivity, and edge computing. (Soussi A. et al. 2024) Next-generation sensors will become increasingly intelligent, capable of learning from their environments, adapting to new conditions, and collaborating with other devices autonomously. Sensor fusion—integrating multiple sensor types to provide a more comprehensive understanding of the environment—will become a dominant trend, especially in robotics, autonomous vehicles, and smart manufacturing. (Soussi A. et al. 2024) Furthermore, as sustainability becomes a global priority, innovations in biodegradable materials, energy-efficient designs, and self-powered sensors are expected to gain traction. The role of smart sensors in enabling smart cities, climate monitoring, and precision agriculture will be critical in addressing global challenges such as urbanization, resource scarcity, and climate change. Smart sensors have transcended the limitations of traditional sensing by combining intelligence, adaptability, and communication into compact and efficient systems. They are at the forefront of modern technological ecosystems, enabling smarter environments, responsive systems, and data-driven decision-making across sectors.

APPLICATIONS OF SMART SENSORS: TRANSFORMING INDUSTRIES

Sensor technology has undergone a profound transformation over the last century, evolving from simple electromechanical devices into sophisticated smart sensors that power the intelligent systems of today. This evolution has been driven by the increasing need for more accurate, reliable, autonomous, and connected data-gathering solutions across a wide range of industries, including manufacturing, healthcare, automotive, environmental monitoring, and consumer electronics. The journey from traditional sensors to smart sensors reflects not only technological advancement but also a paradigm shift in how data is perceived, processed, and utilized in the digital age. (Taha M. F. et al. 2025) The history of sensors can be traced back to basic transduction principles used in early scientific experiments and industrial control systems. Early sensors were largely mechanical or electromechanical in nature. For example, mercury thermometers were among the first devices used to measure temperature, followed by bimetallic strips in thermostats. (Taha M. F. et al. 2025) Similarly, strain gauges, developed in the 1930s, were used to measure mechanical deformation in materials. These traditional sensors served a specific and narrow purpose: to convert a physical quantity such as pressure, light, temperature, or motion into a readable or recordable output, typically an analog electrical signal. They lacked the ability to process, store, or interpret data. Often, traditional sensors had to be integrated with external systems such as amplifiers, analog-to-digital converters, and control units to make their data usable in complex applications.

With the advent of semiconductor technology in the mid-20th century, sensors began to benefit from Taheri et al. (2025) significant enhancements in sensitivity, size, and reliability. The development of silicon-based sensors, such as thermocouples, photodiodes, and piezoelectric devices, marked a major leap forward. These sensors were smaller, more accurate, and capable of being mass-produced at lower costs. The integration of electronic circuits allowed for basic signal conditioning directly within or near the sensor. However, these were still considered "traditional" sensors because their role was limited to data collection. Interpretation and decision-making continued to be handled by separate centralized systems. The introduction of microprocessors and microcontrollers in the 1970s and 1980s brought about a gradual shift in sensor functionality, (Taheri et al. 2025). Designers began embedding simple logic and processing capabilities directly into the sensors, enabling some degree of data manipulation, filtering, and calibration within the sensor unit itself. The real turning point came with the emergence of smart sensors in the late 20th and early 21st centuries, spurred by advancements in microelectromechanical systems (MEMS), embedded systems, wireless communication technologies, and digital signal processing (DSP). A smart sensor can be defined as a device that combines sensing capabilities with processing power and communication interfaces. In essence, it is no longer just a passive component but an intelligent node within a larger network.

SENSOR TYPES: TEMPERATURE, PRESSURE, MOTION, AND MORE

Smart sensors have emerged as pivotal enablers of the digital revolution, dramatically transforming the way data is gathered, interpreted, and utilized across virtually every major industry, (Tamilarasi et al. 2025). By integrating sensing elements with embedded processing, communication modules, and advanced algorithms, smart sensors go beyond basic data acquisition to enable real-time analytics, decision-making, and autonomous control. Their applications are extensive and continually expanding, powering innovations in industrial automation, healthcare, agriculture, automotive, environmental monitoring, smart cities, consumer electronics, and beyond, (Tamilarasi et al., 2025). These intelligent devices are key components of the Internet of Things (IoT) and cyber-physical systems, helping industries transition from reactive to predictive operations, from static systems to adaptive environments. In industrial and manufacturing sectors, smart sensors are foundational to the development of Industry 4.0 and Industrial Internet of Things (IIoT) ecosystems. They play a critical role in predictive maintenance, asset tracking, process optimization, and quality control. Vibration and temperature sensors embedded in machinery can continuously monitor equipment health, allowing maintenance to be scheduled before failures occur—reducing costly downtime and prolonging machinery life. Smart optical and proximity sensors ensure precision in assembly lines, automatically adjusting robotic arms or halting production if misalignments are detected. Pressure, humidity, and gas sensors regulate environmental conditions in sensitive manufacturing environments, such as cleanrooms or food processing plants, (Tang et al. 2025). Moreover, the integration of smart sensors with cloud platforms enables centralized data monitoring and control across multiple sites, optimizing global operations.

The healthcare industry has witnessed transformative impacts from smart sensor technology, particularly in remote patient monitoring, wearables, smart implants, and hospital automation. Wearable devices equipped with sensors for heart rate, oxygen saturation, ECG, temperature, and motion are revolutionizing personal health monitoring, (Tang et al. 2025). These devices provide continuous health data to patients and healthcare providers, enabling early diagnosis, disease prevention, and personalized care. Smart biosensors embedded in skin patches or contact lenses can monitor glucose levels, hydration, or biomarkers for chronic diseases. In hospitals, smart sensors track patient movement to prevent falls, monitor hygiene compliance, and manage asset utilization, (Thilakarathne et al. 2025). Integration with AI further enhances diagnostic capabilities, enabling real-time anomaly detection and predictive health insights. Smart pill technologies, featuring ingestible sensors, track medication adherence and gastrointestinal health from within the body, (Thilakarathne et al. 2025). Modern vehicles are increasingly becoming mobile sensor hubs. Smart sensors in the automotive industry are at the heart of Advanced Driver-Assistance Systems (ADAS), vehicle diagnostics, autonomous driving, and fleet management. Radar, LiDAR, ultrasonic, and vision-based sensors enable lane departure warnings, adaptive cruise control, automatic braking, blind-spot detection, and collision avoidance. These systems significantly enhance driver and passenger safety and pave the way for fully autonomous vehicles.

SMART SENSOR ARCHITECTURE: COMPONENTS AND INTERFACES

Sensors are the fundamental building blocks of modern electronic systems, enabling devices and machines to interact intelligently with their environment, (Yixiao et al. 2025). As transducers, sensors convert physical phenomena into measurable electrical signals, providing real-time data for processing, monitoring, and control, (Yu et al. 2024). With the proliferation of smart systems in fields such as healthcare, automotive, aerospace, consumer electronics, manufacturing, and environmental monitoring, the variety of sensors has expanded significantly. (Yixiao C. et al. 2025). Each sensor type is engineered to detect specific physical, chemical, or biological parameters, using tailored technologies that optimize accuracy, reliability, and responsiveness. This section provides a broad overview of the most widely used sensor types—temperature, pressure, motion, and beyond—highlighting their operating principles, typical applications, and roles in various industries, (Yu et al. 2024). Temperature sensors measure the degree of heat present in a substance or environment and are among the most widely used sensors in both industrial and consumer settings. They operate on various principles, including resistance change, voltage output, or digital signal generation. The diversity of sensor types reflects the complex and multifaceted nature of real-world data acquisition, (Riza et al. 2025). From measuring temperature and pressure in industrial systems to detecting movement in security applications and monitoring vital signs in healthcare, each sensor type contributes uniquely to intelligent systems, (Riza et al. 2025). The continuous miniaturization, increased accuracy, and integration of AI into sensor platforms are driving the development of multifunctional and autonomous sensing solutions.

COMMUNICATION PROTOCOLS: ENABLING SMART SENSOR CONNECTIVITY

Smart sensors represent a significant evolution in sensor technology, integrating advanced processing, communication, and decision-making capabilities into compact, multifunctional units. (Prakashan et al. 2024). Unlike traditional sensors that simply convert physical parameters into electrical signals, smart sensors go several steps further: they process and interpret data locally, make autonomous decisions, and communicate results to other systems or the cloud. The internal architecture of a smart sensor is the key to this expanded functionality. (Prakashan et al. 2024). It is composed of multiple tightly integrated components that work together to enable real-time data acquisition, processing, and transmission. (Pires-Junior et al. 2025). This architecture includes not only the physical sensing element but also a series of analog and digital subsystems, memory, computational cores, communication modules, and power management circuits, (Pires-Junior et al. 2025). Understanding the architecture of smart sensors is crucial for designing reliable, scalable, and efficient smart systems in applications ranging from healthcare to industrial automation and smart cities. At the heart of any smart sensor lies the sensing element, which is responsible for detecting specific physical, chemical, or biological phenomena. This element transduces real-world inputs—such as temperature, pressure, motion, light, gas concentration, or humidity—into an analog electrical signal. In high-precision sensors, this stage may also include programmable gain amplifiers (PGAs) and analog multiplexers for multi-sensor configurations. Once conditioned, the analog signal is fed into an analog-to-digital converter (ADC), which transforms it into a digital format that can be processed by onboard digital systems. The resolution of the ADC—typically measured in bits (e.g., 10-bit, 12-bit, or 16-bit)—determines the granularity of measurement. Higher-resolution ADCs enable more precise measurements, critical in medical, scientific, or metrology applications. ADCs may be integrated directly into microcontrollers or implemented as dedicated components. In some designs, delta-sigma or successive approximation ADC architectures are preferred for their high resolution and low power consumption.

At the core of the smart sensor's intelligence lies its processing unit, which may be a microcontroller unit (MCU), a digital signal processor (DSP), or even a system-on-chip (SoC) architecture. (Nguyen et al. 2025). Some modern smart sensors also incorporate AI or neural processing engines to perform local inference, enabling real-time recognition of patterns or anomalies without cloud connectivity. A defining feature of smart sensors is their ability to communicate with other devices, networks, or cloud platforms. (Pires-Junior et al. 2025). The communication interface enables the sensor to transmit data, receive commands, and integrate into larger systems. Smart sensors often operate in power-constrained environments, especially in battery-powered or energy-harvesting applications. The Power Management Unit (PMU) ensures efficient energy use, regulates voltage levels, and supports various power modes. (Nguyen et al. 2025). Beyond internal components, the physical and system-level Smart sensor architecture is a multifaceted system that integrates diverse technologies—sensing materials, analog electronics, embedded computing, wireless communication, and intelligent software—into a cohesive unit. (Nagarajan et al. 2025). This architecture enables smart sensors to function as autonomous, context-aware devices that reduce system complexity, improve response times, and unlock new levels of functionality across applications. Understanding the components and interfaces of a smart sensor provides insight into its capabilities and constraints, informing better design decisions for embedded systems, IoT networks, and real-time control systems. (Nagarajan et al. 2025). As technological demands continue to grow, future smart sensor architectures will likely incorporate more advanced features, including edge AI, blockchain for data integrity, and energy-autonomous operation—pushing the boundaries of what sensors can achieve in a connected, intelligent world.

INDUSTRIAL AUTOMATION: SMART SENSORS FOR PREDICTIVE MAINTENANCE

In the evolving landscape of smart technologies, the communication protocols that underpin sensor connectivity play a pivotal role in enabling real-time data transmission, system interoperability, and distributed intelligence. Smart sensors, by design, do not operate in isolation. (Maraveas et al. 2025). They are embedded within larger ecosystems—Internet of Things (IoT) networks, cyber-physical systems, industrial automation platforms, or consumer electronics—where they must reliably and securely exchange data with local controllers, gateways, cloud services, or peer devices. (Maraveas et al. 2025). To facilitate this interaction, a wide range of wired and wireless communication protocols have been developed, each with specific strengths, limitations, and use-case suitability. (Mandal S. et al. 2024) Understanding these protocols is essential for designing scalable, efficient, and secure sensor-based systems that align with the demands of modern industries such as healthcare, transportation, agriculture, manufacturing, and smart cities. (Mandal et al. 2024). Unlike traditional sensors, which typically deliver raw signals to an external processor, smart sensors often incorporate onboard intelligence that processes data before transmitting results. I^2C is a low-speed, short-distance communication protocol ideal for connecting multiple sensors to microcontrollers within embedded systems. Wireless protocols offer mobility and eliminate the need for physical connections, enabling deployment in remote or hard-to-access locations. These protocols vary widely in range, power consumption, and throughput. Bluetooth is a widely adopted standard for short-range wireless communication, with BLE designed specifically for low-power IoT devices. Some systems utilize multi-protocol or hybrid communication solutions, allowing sensors to switch between or simultaneously use multiple interfaces depending on application demands. Examples include when choosing a communication protocol for smart sensors, several critical criteria must be considered as smart sensors become interconnected and cloud-enabled, security is a growing concern. Many protocols incorporate encryption, authentication, and data integrity checks.

HEALTHCARE MONITORING: SMART SENSORS FOR PATIENT CARE

Industrial automation has undergone a profound transformation in recent decades, evolving from mechanized control systems to fully integrated, intelligent networks that optimize productivity, safety, and operational efficiency. (Luo et al. 2025). Central to this evolution is the deployment of smart sensors—advanced sensing devices equipped with embedded processing, communication capabilities, and adaptive intelligence—that provide critical real-time data across complex industrial environments. One of the most impactful applications of smart sensors within industrial automation is predictive maintenance. Predictive maintenance leverages sensor-generated data combined with analytics and machine learning to anticipate equipment failures before they occur, enabling proactive interventions that reduce downtime, extend asset life, and significantly lower maintenance costs. Historically, maintenance in industrial settings followed reactive or preventive models. Reactive maintenance—fixing equipment only after failure—often leads to costly unplanned downtime, safety risks, and reduced asset lifespan. (Luo et al. 2025). Preventive maintenance schedules interventions at predetermined intervals regardless of actual equipment condition, which may result in unnecessary maintenance or missed early warning signs of failures. Predictive maintenance disrupts this paradigm by utilizing continuous condition monitoring through smart sensors to assess the health of machinery in real time. By analyzing data patterns and anomalies, it can forecast failures and recommend maintenance precisely when needed. This approach aligns maintenance efforts with actual equipment needs, improving resource utilization and operational reliability.

Figure 1. Smart sensors in industry

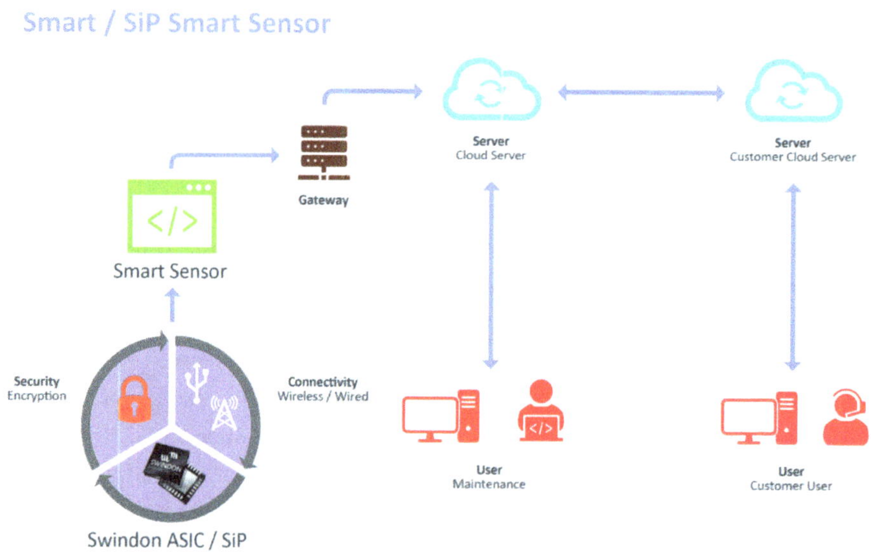

Smart sensors are foundational to predictive maintenance systems. (Lu et al. 2025). They monitor critical parameters that reflect equipment health and performance, such as vibration, temperature, pressure, acoustic emissions, electrical currents, and lubrication conditions. (Lu et al. 2025). Unlike conventional sensors, smart sensors preprocess data locally—filtering noise, calibrating signals, and performing initial analytics—before transmitting summarized insights to control systems or cloud platforms. Smart sensors are integrated into broader industrial automation frameworks through wired or wireless networks, often within Industrial Internet of Things (IIoT) architectures. (Liu et al. 2024). These systems combine sensors, edge computing, communication gateways, and cloud-based analytics to create a continuous data feedback loop. The vast data generated by smart sensors require sophisticated analytics to transform raw measurements into meaningful maintenance actions. (Liu et al. 2024). Predictive maintenance systems typically utilize. These analytics are powered by techniques such as anomaly detection, time-series forecasting, and pattern recognition, often deployed via cloud-based platforms or edge AI modules embedded within smart sensor. Smart sensors have revolutionized industrial automation by enabling predictive maintenance strategies that transform reactive operations into proactive, data-driven processes. By continuously monitoring the health of machinery and infrastructure, these intelligent sensors facilitate early fault detection, optimize maintenance scheduling, reduce operational costs, and enhance safety. Despite the challenges in deployment and integration, the benefits of incorporating smart sensors into predictive maintenance are profound and continue to grow with advances in sensor technology, data analytics, and network connectivity. As industries increasingly adopt smart sensor networks within IIoT frameworks, predictive maintenance will become a cornerstone of sustainable, efficient, and resilient industrial operations worldwide.

ENVIRONMENTAL MONITORING: SMART SENSORS FOR AIR AND WATER QUALITY

The healthcare industry is witnessing a profound transformation driven by the integration of smart sensors into patient monitoring systems, (Liu et al. 2025). These advanced sensing devices, equipped with embedded processing, communication capabilities, and intelligent analytics, are redefining how medical data is collected, transmitted, and interpreted—ultimately improving patient outcomes, optimizing clinical workflows, and enabling personalized care. (Liu et al. 2025). Healthcare monitoring through smart sensors spans a broad range of applications, from continuous vital sign tracking in hospitals and remote patient monitoring at home, to wearable health devices that empower individuals to manage their health proactively. By facilitating real-time data acquisition and timely clinical decision-making, smart sensors are becoming indispensable tools in modern medicine, especially as populations age and healthcare demands escalate globally. Traditional patient monitoring largely relied on periodic manual measurements by healthcare professionals or bulky, wired devices tethered to hospital beds. These methods often limited continuous monitoring capabilities, restricted patient mobility, and delayed detection of adverse health events. In contrast, smart sensors enable continuous, non-invasive, and wireless monitoring of physiological parameters, providing a more comprehensive picture of a patient's health status.

Smart sensor technology incorporates miniaturized transducers, microprocessors, wireless communication modules, and intelligent algorithms within compact, often wearable form factors. (Limbore et al. 2025). This allows for seamless integration into daily life without compromising comfort or mobility and supports both in-hospital intensive care and ambulatory monitoring scenarios. (Limbore et al. 2025). Smart sensors in healthcare monitor a variety of critical physiological and biochemical parameters, essential for diagnosing, managing, and preventing diseases. (Lala et al. 2025). Common parameters include. The form factor of smart sensors in healthcare varies widely depending on application: Wearable sensors Integrated into smartwatches, patches, clothing, or chest straps, they provide continuous data streams without intrusiveness. (Lala et al. 2025). Examples include Apple Watch ECG features, Fitbit's heart rate monitors, and adhesive patches for arrhythmia detection. Implantable sensors Placed inside the body, such as glucose monitors implanted under the skin or neurostimulators monitoring brain signals, these sensors provide high-fidelity data but require biocompatibility and long-term reliability. The convergence of flexible electronics, energy harvesting, and wireless charging is advancing the usability and longevity of wearable and implantable devices, making them more patient-friendly. Smart sensors are foundational to Remote Patient Monitoring (RPM) and telehealth initiatives, which have surged especially during the COVID-19 pandemic. RPM allows patients to be monitored outside traditional clinical settings—at home or in assisted living facilities—reducing hospital visits, minimizing infection risks, and lowering healthcare costs.

MEMS AND NEMS: MINIATURIZING SENSOR TECHNOLOGY

Environmental monitoring plays a critical role in protecting human health, conserving ecosystems, and supporting sustainable development, (Kommineni et al. 2025). With increasing industrialization, urbanization, and climate change, the need for accurate, real-time environmental data has never been greater, (Kommineni et al. 2025). Traditional environmental monitoring methods—based on periodic sampling and laboratory analysis—are often slow, labor-intensive, and limited in spatial and temporal resolution. In contrast, smart sensors provide continuous, real-time, and location-specific data on key environmental parameters, (Knof et al. 2024). These intelligent devices, equipped with embedded processors, wireless communication, and autonomous capabilities, have revolutionized environmental monitoring—especially in the domains of air and water quality assessment. By enabling timely detection of pollution events and long-term trend analysis, smart sensors support regulatory compliance, research, public awareness, and policy-making aimed at environmental protection. (Knof M. et al. 2024) Smart sensors are compact devices capable of detecting, processing, and communicating data about specific environmental variables. (Guo B. et al. 2025) They integrate sensing elements with signal conditioning, data processing, power management, and wireless communication interfaces (e.g., LoRa, Wi-Fi, Zigbee, NB-IoT). In environmental applications, smart sensors can be deployed in static or mobile configurations—on poles, drones, buoys, buildings, or wearable devices—to provide continuous measurements in real-time. (Godja N. C. et al. 2024) Air pollution poses serious risks to human health and the environment. Key pollutants include particulate matter (PM), nitrogen dioxide (NO_2), sulfur dioxide (SO_2), ozone (O_3), carbon monoxide (CO), and volatile organic compounds

(VOCs). (Godja N. C. et al. 2024) Smart air quality sensors are designed to detect these pollutants at fine resolutions, enabling early warnings and mitigation efforts.

Real-time data from these sensors is used to generate Air Quality Index (AQI) values, inform public health alerts, and trigger automated actions like activating ventilation systems or issuing warnings for vulnerable populations. Clean water is essential for drinking, agriculture, industry, and ecosystem health, (Elkhoukhi et al. 2025). Contamination from industrial discharge, agricultural runoff, and urban pollution threatens water resources worldwide. Smart water quality sensors provide a solution by enabling continuous, in-situ monitoring of various physical, chemical, and biological parameters in surface water, groundwater, and wastewater systems, (Elkhoukhi et al. 2025). Data collected by smart sensors is often transmitted to cloud platforms or edge computing devices for storage, visualization, and analysis. Advanced data analytics techniques are applied to identify patterns, predict pollution events, and support decision-making. Cloud-connected sensor networks also support automated environmental management—such as activating air purifiers, controlling irrigation systems, or regulating industrial discharges based on sensor feedback, (Dokl et al. 2024). The adoption of smart sensors in environmental monitoring delivers numerous advantages. Sensor calibration and accuracy: Low-cost sensors may exhibit drift or cross-sensitivity; regular calibration and validation against reference stations are necessary. Smart sensors are transforming environmental monitoring by providing real-time, accurate, and cost-effective insights into air and water quality, (Dokl et al. 2024). These technologies empower governments, researchers, industries, and communities to detect pollution events, track long-term environmental trends, and respond proactively to ecological threats. With the continued integration of IoT, AI, and cloud technologies, smart sensor networks will play a central role in advancing environmental sustainability, regulatory compliance, and public health protection. Overcoming technical and deployment challenges will be key to scaling these solutions globally and ensuring that the benefits of clean air and water are accessible to all.

WIRELESS SENSOR NETWORKS: ENABLING DISTRIBUTED SENSING

Microelectromechanical Systems (MEMS) and Nanoelectromechanical Systems (NEMS) represent a remarkable convergence of engineering, materials science, and nanotechnology, enabling the miniaturization of sensor technologies across various industries. (Dahiya R. et al. 2025) These systems integrate mechanical elements, sensors, actuators, and electronics on a common silicon substrate through microfabrication technologies. MEMS typically operate on the micrometer scale (10^{-6}

meters), while NEMS push the boundaries further into the nanometer domain (10^{-9} meters). (Dahiya et al. 2025) This transitions from MEMS to NEMS offers substantial benefits, including enhanced sensitivity, reduced power consumption, and the ability to interface with molecular and quantum phenomena. As sensor technology evolves to meet the growing demands of smart systems—from biomedical diagnostics to environmental monitoring, aerospace, and consumer electronics—MEMS and NEMS are at the forefront of the revolution, enabling systems that are not only smaller but faster, cheaper, and more efficient, (D'Aniello et al. 2024). MEMS technology has already found widespread use in various commercial applications. Accelerometers and gyroscopes in smartphones, pressure sensors in automotive systems, inkjet printer heads, and microphones are just a few examples, (D'Aniello et al. 2024). These devices combine mechanical components like tiny levers or resonators with electronic circuits on a single chip, offering precise detection of motion, pressure, temperature, and other physical parameters, (Cogato et al. 2025). MEMS fabrication techniques, derived largely from the semiconductor industry—such as photolithography, etching, and deposition—allow high-volume manufacturing with precise control. (Cogato et al., 2025). The success of MEMS lies in its scalability and cost-effectiveness, providing an ideal foundation for applications that require high-volume production and miniaturization without sacrificing performance.

NEMS technology, while still maturing, represents the next evolutionary step in miniaturization, (Chib et al. 2025). NEMS devices are essentially scaled-down MEMS, but their operation is governed by new physics. At the nanoscale, surface effects dominate over volume effects, and quantum mechanical phenomena begin to play a significant role, (Chib et al. 2025). This leads to extraordinary sensitivity—NEMS sensors can detect forces at the attonewton scale, masses at the yoctogram scale, and displacements smaller than an atomic diameter. Such precision opens new avenues in scientific research and technology, including single-molecule detection, high-resolution imaging, and ultra-sensitive biochemical sensing. NEMS are particularly promising in medical diagnostics, where their ability to detect biomarkers at extremely low concentrations could enable early disease detection and personalized treatment strategies, (Bharany et al. 2025). One of the key challenges in the development and commercialization of NEMS lies in fabrication and integration. While MEMS production benefits from established semiconductor processes, NEMS requires advanced lithography, atomic layer deposition, and other nanofabrication techniques that are still under development and often costly, (Bharany et al. 2025). Moreover, as device dimensions shrink, issues such as stiction (static friction at contact surfaces), reliability, and signal-to-noise ratios become more prominent. Overcoming these barriers requires interdisciplinary research, combining insights from physics, chemistry, materials science, and electrical engineering. Novel materials like graphene, carbon nanotubes, and molybdenum disulfide (MoS_2) are being explored for NEMS applications due to their exceptional mechanical and electrical properties at the nanoscale.

ENERGY HARVESTING: POWERING SMART SENSORS SUSTAINABLY

Wireless Sensor Networks (WSNs) have emerged as a foundational technology in modern computing and sensing infrastructure, enabling the distributed collection, processing, and transmission of data across a range of environments, (Baidya et al. 2024). A WSN typically consists of spatially dispersed autonomous sensor nodes that monitor physical or environmental conditions—such as temperature, humidity, vibration, light, or motion—and cooperatively transmit the collected data to a central location or base station, often via multi-hop wireless communication, (Baidya et al. 2024). These nodes are typically compact, low-power devices integrated with sensors, microcontrollers, and wireless communication modules, enabling them to operate independently and form self-organizing networks. As technology advances and demands for real-time monitoring, smart environments, and responsive systems increase, WSNs have become a vital part of the Internet of Things (IoT), smart cities, industrial automation, environmental conservation, and precision agriculture. One of the key advantages of WSNs is their ability to perform distributed sensing over large, often inaccessible, or hazardous areas, (Azeem et al. 2025). Traditional wired sensor networks are limited in scalability and flexibility, particularly in remote or dynamically changing environments. In contrast, WSNs can be rapidly deployed without the need for fixed infrastructure, making them ideal for applications like battlefield surveillance, disaster response, environmental monitoring in forests or oceans, and structural health monitoring in bridges or buildings. Each node in the network is capable not only of sensing but also of processing and transmitting data, often in collaboration with neighboring nodes, (Azeem et al. 2025). This distributed intelligence reduces the need for constant central control and allows for scalable, fault-tolerant network configurations. When a node fails or its energy is depleted, the network can often reconfigure itself dynamically to maintain coverage and connectivity, thus ensuring robustness and longevity.

The architecture of a typical WSN includes sensing units, processing units, transceivers, and power units, (Assimakopoulos et al. 2025). Some nodes may also include location-finding systems, power generators (e.g., solar panels), and actuators for interaction with the environment. Sensor nodes communicate via wireless protocols such as Zigbee, Bluetooth Low Energy (BLE), Wi-Fi, or proprietary radio standards, depending on the range, bandwidth, and power constraints of the application, (Assimakopoulos et al. 2025). Energy efficiency remains a critical concern in WSN design. Because most nodes operate on limited battery power and are often deployed in hard-to-reach areas, minimizing energy consumption during sensing, processing, and communication is essential. Techniques such as data aggregation, adaptive sampling, duty cycling, and energy-aware routing protocols are commonly employed to extend the operational lifetime of the network, (Aryan et al. 2024). One of the most prominent challenges in WSN deployment is balancing performance with constraints in power, processing capability, memory, and bandwidth. Unlike traditional networks, where performance is often prioritized over resource efficiency, WSNs must operate effectively within strict resource limitations, (Aryan et al., 2024). This has led to the development of specialized lightweight protocols for routing, data compression, security, and fault tolerance. Protocols such as LEACH (Low-Energy Adaptive Clustering Hierarchy), Directed Diffusion, and TinyOS's networking stack are examples of approaches that manage energy and communication efficiency in WSNs. Additionally, time synchronization, localization, and node mobility are technical challenges that must be addressed to ensure accurate and consistent data collection in dynamic environments.

SENSOR CALIBRATION AND VALIDATION: ENSURING ACCURACY

As the world becomes increasingly interconnected through the rise of smart technologies and the Internet of Things (IoT), the demand for sustainable, maintenance-free power solutions for smart sensors is accelerating, (Ahmad et al. 2025). Energy harvesting—a method of capturing and storing energy from ambient sources in the environment—has emerged as a transformative approach to power distributed sensors without reliance on traditional batteries or wired connections. By converting ambient energy into electrical energy, energy harvesting enables the deployment of autonomous sensor nodes in remote, hard-to-reach, or hostile environments, where frequent battery replacement is impractical or impossible, (Ahmad et al. 2025). This capability is especially vital for wireless sensor networks (WSNs), wearable devices, environmental monitoring systems, structural health sensors, and industrial IoT (IIoT) applications. Energy harvesting not only extends the operational lifespan of these devices but also supports the global transition toward low-carbon, energy-efficient technologies. Energy harvesting technologies draw energy from a variety of sources, including solar, thermal gradients, kinetic or vibrational motion, radio frequency (RF) waves, and even biochemical reactions. Among these, solar energy harvesting remains the most mature and widely adopted, particularly for outdoor sensors. Photovoltaic (PV) cells convert sunlight into electricity with relatively high efficiency and are commonly used in environmental monitoring stations, traffic sensors, and smart agriculture systems. However, solar harvesting is less effective in indoor or shaded environments, prompting the exploration of alternative methods. Thermoelectric generators (TEGs), which exploit temperature differences between two surfaces to generate electricity based on the Seebeck effect, are suitable for applications involving heat sources or fluctuating thermal environments, such as machinery, pipelines, or human skin. Similarly, piezoelectric and electromagnetic harvesters convert mechanical vibrations or motion into electrical energy and are ideal for applications where movement is frequent, such as wearable sensors, automotive systems, or industrial equipment.

Kinetic energy harvesting, in particular, has shown great promise in powering sensors embedded in wearable and mobile systems, (Aarif et al. 2025). Human motion—such as walking, running, or joint movement—can be harnessed through piezoelectric or triboelectric nanogenerators to power health monitors, fitness trackers, or prosthetic devices. In vehicles, vibrations from engines or road contact can be utilized to power tire pressure sensors or shock absorber monitors. These self-powered systems not only reduce maintenance costs but also enhance reliability and safety. Furthermore, advancements in materials science have led to the development of flexible, lightweight, and biocompatible energy harvesters that integrate seamlessly into textiles or skin-contact devices, paving the way for continuous, unobtrusive health monitoring in medical and fitness applications. RF energy harvesting is another growing field, particularly suited to densely populated or urban environments where wireless signals are abundant. These systems capture energy from ambient electromagnetic waves—such as those emitted by Wi-Fi routers, cellular towers, or TV broadcast signals—and convert them into usable electrical power. While RF harvesting typically yields low power output, it can be sufficient for ultra-low-power sensors or intermittent transmission systems, especially when combined with energy storage components like supercapacitors or microbatteries. RF harvesting is particularly attractive for passive RFID tags, remote monitoring systems, and smart packaging solutions, where long battery life and low maintenance are crucial. An essential component of any energy harvesting system is the power management unit (PMU), which regulates the conversion, storage, and distribution of harvested energy, (Aarif et al. 2025). These units must be highly efficient, capable of operating at low input voltages, and able to manage multiple energy sources when necessary. Advanced PMUs often include energy-aware algorithms that dynamically adjust sensing and transmission activity based on available energy, ensuring system stability and extending functionality during low-energy conditions. Energy storage devices, such as microbatteries, solid-state thin-film batteries, or supercapacitors, are often paired with harvesters to buffer energy and provide power during inactive harvesting periods, such as at night for solar-powered systems or when machinery is idle for vibration-based systems.

DATA ANALYTICS: UNLOCKING INSIGHTS FROM SMART SENSOR DATA

In the era of data-driven decision-making, the reliability and accuracy of sensor-generated data are paramount, (Ahmad et al. 2025). Sensors serve as the fundamental building blocks in various systems, ranging from industrial automation and environmental monitoring to biomedical devices, autonomous vehicles, and smart infrastructure. However, the utility of sensors is only as good as the precision and validity of the information they produce. This is where sensor calibration and validation play a critical role. Calibration ensures that a sensor's readings accurately reflect the physical quantity being measured, while validation confirms that the sensor continues to perform within specified tolerances under real-world conditions. Together, these processes guarantee data integrity, system reliability, and long-term performance across a wide spectrum of applications. Sensor calibration is the process of comparing and adjusting the output of a sensor to align with a known standard or reference value, (Aryan et al. 2024). It corrects systematic errors and deviations that may arise due to manufacturing inconsistencies, environmental conditions, aging, or operational wear and tear. Calibration involves exposing a sensor to a series of known inputs and adjusting its response accordingly. For example, a temperature sensor might be calibrated by placing it in a controlled environment alongside a high-precision reference thermometer and adjusting its readings to match the reference values across a range of temperatures. Calibration can be linear or nonlinear, depending on the nature of the sensor and its expected behavior. While some sensors exhibit linear relationships between input and output, others—such as pH, gas, or biosensors—may require complex calibration curves derived from polynomial or logarithmic models.

There are various types of calibration, including factory calibration, field calibration, and self-calibration. Factory calibration is conducted during the manufacturing process using traceable standards in controlled laboratory environments. These calibrations are usually precise and serve as the sensor's initial reference, often stored in onboard memory. However, performance can drift over time due to environmental exposure or mechanical stress. Field calibration is carried out in situ, allowing sensors to be adjusted under real operating conditions. This is particularly important in dynamic environments where temperature, pressure, humidity, or electromagnetic interference may affect sensor behavior. Self-calibration, often seen in smart or digital sensors, uses built-in routines or algorithms to periodically recalibrate without manual intervention, (Assimakopoulos et al. 2025). These routines may use internal references or exploit known patterns in the measured data (e.g., known zero points or cyclical behaviors) to adjust sensor output. The increasing integration of artificial intelligence (AI) and machine learning (ML) allows adaptive calibration models that evolve over time, enhancing accuracy in real-time deployments. Sensor validation, on the other hand, is the process of verifying that a sensor is providing accurate, consistent, and reliable measurements over time. Unlike calibration, which adjusts the sensor, validation evaluates its performance against expected criteria. This includes tests for repeatability, reproducibility, response time, sensitivity, drift, and linearity. Validation is essential for quality assurance, system certification, regulatory compliance, and predictive maintenance. In industries such as pharmaceuticals, aerospace, or medical devices, where errors can have life-threatening consequences, rigorous validation protocols are mandatory. Validation may involve controlled experiments, statistical analysis, or comparison with high-accuracy reference instruments. In some systems, redundancy is used, where multiple sensors measure the same parameter and their outputs are cross-checked to detect anomalies or faults.

SECURITY AND PRIVACY: PROTECTING SMART SENSOR DATA

In the modern digital landscape, the explosion of smart sensor deployments across industries has generated an unprecedented volume of data. From manufacturing floors and urban infrastructure to healthcare systems and environmental monitoring, smart sensors are constantly collecting real-time information about physical phenomena such as temperature, motion, pressure, humidity, light, gas concentration, biometric readings, and more. While the mere accumulation of this data is valuable, the true potential lies in how it is interpreted and utilized, (Aryan et al. 2024). Data analytics—the systematic computational analysis of data—plays a pivotal role in transforming raw sensor outputs into actionable insights, enabling informed decision-making, process optimization, predictive modeling, and intelligent automation. Without analytics, sensor data remains an underutilized asset; with analytics, it becomes a strategic driver of innovation, efficiency, and competitive advantage. Smart sensors are typically embedded with microcontrollers and connected to wireless networks, enabling them to transmit continuous streams of data to centralized systems or cloud-based platforms. However, sensor data is often high in volume, velocity, and variety—known as the "3Vs" of big data—posing significant challenges in terms of storage, processing, and real-time interpretation. Data analytics bridges this gap by employing a suite of techniques including statistical analysis, machine learning, signal processing, and artificial intelligence (AI) to identify patterns, correlations, anomalies, and trends within complex datasets.

These insights can then be used to inform decisions, trigger alerts, control systems autonomously, or feed into long-term strategic planning.

One of the fundamental aspects of data analytics in sensor-driven systems is data preprocessing, (Azeem et al. 2025). Sensor data is often noisy, incomplete, redundant, or subject to outliers due to environmental interference, hardware limitations, or transmission errors. Preprocessing includes data cleaning (removing errors or duplicates), normalization (scaling data into a common range), filtering (removing noise), and imputation (filling in missing values). Time-series alignment and synchronization are also critical in systems involving multiple sensors capturing data at different frequencies or timestamps. Proper preprocessing ensures that downstream analytics are accurate, reliable, and meaningful. Once clean data is available, descriptive analytics provides the first layer of insight. It summarizes historical data to answer questions like "What happened?" or "What is the current status?" Visualization tools such as dashboards, graphs, and heatmaps allow operators to observe sensor performance and environmental conditions in real-time. For example, in a smart factory, dashboards may show temperature trends of machinery over time, identify usage spikes, or detect when readings deviate from expected norms, (Baidya et al. 2024). Descriptive analytics supports human monitoring and control, especially in industrial control systems, smart agriculture, and energy grids. Moving beyond description, diagnostic analytics investigates the root causes of observed phenomena. When sensor readings indicate abnormal behavior—such as a sudden rise in vibration levels or a drop in temperature—diagnostic tools analyze correlated sensor data to determine the likely cause. Techniques like correlation analysis, clustering, and statistical modeling help identify which variables are contributing to anomalies. In healthcare, diagnostic analytics can help determine why a patient's wearable device recorded irregular heartbeats by examining other physiological signals and contextual data such as physical activity or stress levels.

THE FUTURE OF SMART SENSORS: EMERGING TRENDS AND APPLICATIONS

As smart sensor technologies proliferate across industries and integrate deeper into everyday life, the security and privacy of sensor data have become critical concerns. From wearable health monitors and smart home devices to industrial IoT systems, autonomous vehicles, and infrastructure monitoring networks, smart sensors are increasingly embedded in sensitive, mission-critical, and often personal contexts, (Bharany et al. 2025). These devices collect and transmit vast amounts of data—some of which may involve private health metrics, behavioral patterns, geographic locations, or critical operational parameters, (Chib et al. 2025). As a result, ensuring the confidentiality, integrity, and availability of smart sensor data is essential not only for user trust and system reliability but also for compliance with legal and regulatory standards. Security and privacy challenges in sensor systems are uniquely complex due to their distributed nature, resource constraints, heterogeneous architectures, and exposure to physical and cyber threats. Smart sensors are susceptible to a wide variety of threats that can compromise both data and functionality, (Cogato et al. 2025). Physical attacks, such as tampering or node capture, are particularly relevant in remote or unguarded environments where sensors may be accessible to adversaries. In such cases, attackers can physically manipulate sensors to alter readings, extract sensitive data, or reverse-engineer device firmware. Eavesdropping attacks, common in wireless networks, involve intercepting data in transit between sensor nodes or gateways. Since many sensor systems transmit data over unsecured or minimally protected channels, attackers can passively collect sensitive information without altering it, violating privacy without detection.

Active cyberattacks pose even greater risks. Man-in-the-middle (MITM) attacks enable adversaries to intercept, modify, or forge communications between sensors and control systems. Replay attacks involve capturing valid sensor data transmissions and replaying them later to trick systems into acting on outdated or false data. Spoofing and false data injection attacks involve faking sensor outputs to mislead control systems—potentially causing physical damage or disrupting services, (Cogato et al. 2025). For example, spoofing a temperature sensor in an industrial process could cause cooling systems to shut off, leading to equipment overheating. Denial of Service (DoS) attacks aim to overwhelm sensor networks or their gateways with traffic, rendering systems inoperable. In distributed systems, such as smart grids or autonomous vehicles, these attacks can have cascading and catastrophic effects. Implementing robust security protocols in smart sensor networks is challenging due to inherent limitations in computational power, memory, energy, and communication bandwidth, (D'Aniello et al. 2024). Many sensors, particularly in large-scale deployments like wireless sensor networks (WSNs) or IoT devices, are designed to be small, low-cost, and energy-efficient, making them ill-suited for traditional cryptographic techniques, which require intensive processing and storage. Public key cryptography, for instance, while highly secure, may not be feasible on lightweight sensor nodes without specialized hardware. (D'Aniello G. et al. 2024) Additionally, sensors often operate unattended in dynamic environments where network topologies change and nodes can join or leave frequently, complicating secure key management and trust models.

CONCLUSION

Smart sensors have evolved from simple data-gathering devices to intelligent, interconnected nodes forming the backbone of cyber-physical systems. These advanced sensors combine sensing, signal processing, decision-making, and communication capabilities, enabling real-time interaction with physical environments. As technology continues to accelerate, the future of smart sensors promises unprecedented levels of precision, autonomy, integration, and versatility. From next-generation wearables and autonomous robotics to precision agriculture and smart infrastructure, smart sensors are poised to redefine how we interact with the world, make decisions, and build sustainable systems. This discussion explores emerging trends, key innovations, and future applications shaping the evolution of smart sensor technologies. One of the most significant trends driving the future of smart sensors is miniaturization, supported by advances in MEMS (Micro-Electro-Mechanical Systems) and NEMS (Nano-Electro-Mechanical Systems). These technologies allow the fabrication of highly compact sensors with multiple sensing capabilities embedded into a single chip. This miniaturization not only conserves space but also reduces power consumption and cost. As sensor size shrinks, their integration into everyday objects becomes seamless—from smart textiles and contact lenses to micro-implants and nanorobots. The convergence of sensors with microprocessors and wireless communication modules leads to System-on-Chip (SoC) solutions, offering plug-and-play intelligence and scalability for widespread deployment. Future applications increasingly demand context-aware sensing—systems that can understand not only raw values but the context surrounding them. This is made possible through sensor fusion, where multiple sensors measuring different modalities (e.g., temperature, motion, light, sound, chemicals) are combined to generate higher-order insights. In autonomous vehicles, for example, radar, LiDAR, ultrasonic sensors, and cameras are fused to build a coherent, real-time representation of the environment. In healthcare, multimodal sensors integrated into wearables can monitor heart rate, temperature, movement, and biochemical signals simultaneously, enabling more comprehensive diagnostics. Fusion algorithms, powered by AI, will become more sophisticated, adaptive, and capable of dynamic reconfiguration based on changing environmental or user conditions. The integration of artificial intelligence (AI) directly into sensor nodes—referred to as edge intelligence—is revolutionizing how and where data is processed. Instead of transmitting raw data to the cloud for analysis, smart sensors are increasingly capable of making decisions locally using onboard machine learning (ML) models. This paradigm shift enhances response times, reduces bandwidth and energy usage, and preserves data privacy. AI-enabled sensors can detect patterns, recognize anomalies, and predict future states without human intervention. Examples include predictive maintenance sensors that detect early signs of equipment failure, traffic cameras that classify objects in real time, and agricultural sensors that adjust irrigation based on crop stress. The emergence of ultra-low-power AI chips such as Google's Edge TPU or ARM's Ethos further empowers this trend.

REFERENCES

Aarif, K. O. M., Alam, A., & Hotak, Y. (2025). Smart sensor technologies shaping the future of precision agriculture: Recent advances and future outlooks. *Journal of Sensors*, *2025*(1), 2460098. DOI: 10.1155/js/2460098

Ahmad, I., Zulfiqar, A., Shabbir, M., Khan, J. S., Çalışkan, H., & Hong, H. (2025). Advanced sensors, monitoring, and control systems for environmental sustainability. In *Smart Cities and Sustainable Manufacturing* (pp. 47–55). Elsevier. DOI: 10.1016/B978-0-443-26474-0.00007-5

Aryan, A., Singh, M., Labhade-Kumar, N., Radha, M., Ranga, J., & Rathod, N. (2024, April). Experimental design of electricity theft detection and alert system using Arduino-assisted controller and smart sensors. In *2024 International Conference on Inventive Computation Technologies (ICICT)* (pp. 1961-1968). IEEE. (Kush J. C. 2025)Assimakopoulos, F., Vassilakis, C., Margaris, D., Kotis, K., & Spiliotopoulos, D. (2025). AI and related technologies in the fields of smart agriculture: A review. *Information (Basel)*, *16*(2), 100.

Azeem, M., Shahid, M., Masin, I., & Petru, M. (2025). Design and development of textile-based wearable sensors for real-time biomedical monitoring; a review. *Journal of the Textile Institute*, *116*(1), 80–95. DOI: 10.1080/00405000.2024.2318500

Baidya, K., Roy, A., & Das, K. (2024). A review of polymer-matrix piezoelectric composite coatings for energy harvesting and smart sensors. *Journal of Coatings Technology and Research*, *21*(1), 55–85. DOI: 10.1007/s11998-023-00819-x

Bharany, S., Almogren, A., & Altameem, A. (2025). Optimizing IoT connectivity: A quantitative exploration of the comprehensive adaptive sensing and clustering system for smart sensor networks in smart cities. *Wireless Personal Communications*, *140*(1-2), 1–23. DOI: 10.1007/s11277-024-11719-7

Chib, S., Mehta, A. K., Selvakumar, P., Saravanan, A., Mishra, B. R., & Manjunath, T. C. (2025). Mental health challenges and solutions in high-pressure work environments. In Harriott, S. (Ed.), *Innovative Approaches to Managing Conflict and Change in Diverse Work Environments* (pp. 247–272). IGI Global Scientific Publishing. DOI: 10.4018/979-8-3693-9556-1.ch011

Cogato, A., Sozzi, M., Nikolić, N., & Laroche-Pinel, E. (2025). Smart horticulture: Latest advances and prospects. *Horticulturae*, *11*(1), 27. DOI: 10.3390/horticulturae11010027

D'Aniello, G., Gaeta, M., Gravina, R., Li, Q., Rehman, Z. U., & Fortino, G. (2024). Situation identification in smart wearable computing systems based on machine learning and context space theory. *Information Fusion*, *104*, 102197. DOI: 10.1016/j.inffus.2023.102197

Dahiya, R., & Selvakumar, P. Vidyadhara, & Manjunath, T. C. (2025). Design and aerodynamics: Principles and advances. In M. Mellal (Ed.), *Innovations and Developments in Unmanned Aerial Vehicles* (pp. 145-170). IGI Global Scientific Publishing.

Dokl, M., Van Fan, Y., Vujanović, A., Pintarič, Z. N., Aviso, K. B., Tan, R. R., & Čuček, L. (2024). A waste separation system based on sensor technology and deep learning: A simple approach applied to a case study of plastic packaging waste. *Journal of Cleaner Production*, *450*, 141762. DOI: 10.1016/j.jclepro.2024.141762

Elkhoukhi, H., Bakhouya, M., Elmouatamid, A., El Ouadghiri, D., & Gaber, J. (2025). A review of occupancy sensing technologies and approaches in smart buildings. *International Journal of RF Technologies: Research and Applications*, *15*(1), 19–48. DOI: 10.3233/RFT-240006

Godja, N. C., & Munteanu, F. D. (2024). Hybrid nanomaterials: A brief overview of versatile solutions for sensor technology in healthcare and environmental applications. *Biosensors (Basel)*, *14*(2), 67. DOI: 10.3390/bios14020067 PMID: 38391986

Guo, B., Duan, S., Wang, P., Lei, H., Zhao, Z., & Fang, D. (2025). A review on the deformation tracking methods in vision-based tactile sensing technology. *Acta Mechanica Sinica*, *41*(10), 424436. DOI: 10.1007/s10409-024-24436-x

Knof, M., Stock-Homburg, R., & Schurer, J. (2024). How in-store sensor technologies can help retailers to understand their customers: Overview on two decades of research. *International Review of Retail, Distribution and Consumer Research*, *34*(3), 381–398. DOI: 10.1080/09593969.2023.2273256

Kommineni, M., Chundru, S., Maroju, P. K., & Selvakumar, P. (2025). Ethical implications of AI in sustainable development pedagogy. In Chinedu, C., Mafarja, N., Hussein, M., & Saleem, A. (Eds.), *Rethinking the Pedagogy of Sustainable Development in the AI Era* (pp. 17–36). IGI Global Scientific Publishing. DOI: 10.4018/979-8-3693-9062-7.ch002

Lala, G., & Vugar, A. (2025). Application of IoT and sensor technologies in environmental monitoring. *Environmental Research and Ecotoxicity*, *4*(4), 5. DOI: 10.56294/ere2025170

Limbore, N. V., Anute, N., Ahluwalia, G. K., Chordiya, P. B., Kalshetti, P., & Selvakumar, P. (2025). Ethical leadership and corporate responsibility. In Harriott, S. (Ed.), *Innovative Approaches to Managing Conflict and Change in Diverse Work Environments* (pp. 193–218). IGI Global Scientific Publishing. DOI: 10.4018/979-8-3693-9556-1.ch009

Liu, T., Mao, Y., Dou, H., Zhang, W., Yang, J., Wu, P., & Mu, X. (2025). Emerging wearable acoustic sensing technologies. *Advancement of Science*, •••, 2408653. PMID: 39749384

Liu, Y., He, C., Qiao, T., Liu, G., Li, X., Wan, Q., Zhu, Z., & He, Y. (2024). Coral-inspired hollow microneedle patch with smart sensor therapy for wound infection. *Advanced Functional Materials*, *34*(24), 2314071. DOI: 10.1002/adfm.202314071

Lu, Y., Li, H., Hu, H., Bi, Y., & Wang, P. (2025). A smart compensation method for large range eddy current displacement sensor. *Sensors and Actuators. A, Physical*, *383*, 116171. DOI: 10.1016/j.sna.2024.116171

Luo, H., Niu, H., Li, B., Lin, Z., Yu, P., Zhang, L., Zhang, D., & Guo, Q. (2025). AI-enhanced materials identification with smart directional tangential flexible sensor. *Advanced Materials Technologies*, *10*(8), 2401164. DOI: 10.1002/admt.202401164

Mandal, S., Yadav, A., Panme, F. A., & Devi, K. M., & SM, S. K. (. (2024). Adaptation of smart applications in agriculture to enhance production. *Smart Agricultural Technology*, •••, 100431. DOI: 10.1016/j.atech.2024.100431

Maraveas, C., Loukatos, D., & Arvanitis, K. G. (2025). Intelligent sensors: Wireless sensor networks and Internet of Things. In *Nature-Derived Sensors* (pp. 295–339). Elsevier. DOI: 10.1016/B978-0-443-22002-9.00009-9

Nagarajan, P., Sathiya, D., Pulugu, D., & Pulugu, R. (2025). Self-powered wearable implantable smart sensor and medical electronics based on nanogenerator. In *Self-Powered Sensors* (pp. 59–75). Academic Press. DOI: 10.1016/B978-0-443-13792-1.00011-0

Nguyen, M. D., Tizon, L. D., Le, N. T., Nguyen, D. T., Vu, T. C., Nguyen, T. V., ... & Nguyen, M. T. (2025). A comparative study of Wi-Fi technologies in wireless sensor networks. *Computer Networks and Communications*, 75-87.

Pires-Junior, R., Macedo, L., Frizera, A., & Leal-Junior, A. (2025). Development of self-sensing cantilever structures using additive manufacturing and optical fiber sensing technology. *IEEE Sensors Letters*, *9*(3), 1–4. DOI: 10.1109/LSENS.2025.3539980

Prakashan, D., Kaushik, A., & Gandhi, S. (2024). Smart sensors and wound dressings: Artificial intelligence-supported chronic skin monitoring–A review. *Chemical Engineering Journal*, *497*, 154371. DOI: 10.1016/j.cej.2024.154371

Riza, M. A., Go, Y. I., Zaitizila, I., & Kok, S. P. (2025). A review of advanced reactive materials in fiber sensor technology: Challenges and requirements of precise humidity control in mechanochemical treatment process. *Sensors and Materials*, *37*(1), 41–61. DOI: 10.18494/SAM5377

Samal, A., & Misra, L. I. Mohit, Pougajendy, S., Chandel, P. S., Manjunath, T. C., & Selvakumar, P. (2025). Musealization and tourism practices. In *S. Oflazoğlu & M. Dora (Eds.), Preserving Cultural Heritage in Post-Disaster Urban Renewal* (pp. 115-142). IGI Global Scientific Publishing.

Soundariya, A., Behera, N. K., Selvakumar, P., Chandel, P. S., Gupta, S., & Manjunath, T. C. (2025). AI and environmental ethics: The role of AI in addressing climate change ethical considerations in AI-driven environmental solutions. In Chakraborty, S. (Ed.), *Human Values, Ethics, and Dignity in the Age of Artificial Intelligence* (pp. 321–346). IGI Global Scientific Publishing. DOI: 10.4018/979-8-3693-9035-1.ch013

Soussi, A., Zero, E., Sacile, R., Trinchero, D., & Fossa, M. (2024). Smart sensors and smart data for precision agriculture: A review. *Sensors (Basel)*, *24*(8), 2647. DOI: 10.3390/s24082647 PMID: 38676264

Taha, M. F., Mao, H., Zhang, Z., Elmasry, G., Awad, M. A., Abdalla, A., Mousa, S., Elwakeel, A. E., & Elsherbiny, O. (2025). Emerging technologies for precision crop management towards agriculture 5.0: A comprehensive overview. *Agriculture*, *15*(6), 582. DOI: 10.3390/agriculture15060582

Taheri, H., & Salimi Beni, A. (2025). Artificial intelligence, machine learning, and smart technologies for nondestructive evaluation. *Handbook of Nondestructive Evaluation 4.0*, 1-29.

Tamilarasi, C., Sowmiya, K., Vignesh, G. D., Jasmine, R. L., Nehru, P., & Selvakumar, P. (2025). AI in renewable energy education: Teaching the future of sustainable power systems. In Chinedu, C., Mafarja, N., Hussein, M., & Saleem, A. (Eds.), *Rethinking the Pedagogy of Sustainable Development in the AI Era* (pp. 133–152). IGI Global Scientific Publishing. DOI: 10.4018/979-8-3693-9062-7.ch007

Tang, M. J., Zhu, Z. H., Li, Y. L., Qin, W. W., Liang, F. P., Wang, H. L., & Zou, H. H. (2025). Specific smart sensing of electron-rich antibiotics or histidine improves the antenna effect, luminescence, and photodynamic sterilization capabilities of lanthanide polyoxometalates. *Journal of Colloid and Interface Science*, *680*, 235–246. DOI: 10.1016/j.jcis.2024.11.006 PMID: 39504753

Thilakarathne, N. N., Bakar, M. S. A., Abas, P. E., & Yassin, H. (2025). Internet of things enabled smart agriculture: Current status, latest advancements, challenges and countermeasures. *Heliyon*, *11*(3), e42136. DOI: 10.1016/j.heliyon.2025.e42136 PMID: 39959477

Yixiao, C., Hui, S., Quhong, S., Xiaoxi, Z., & Jirong, Y. (2025). A review of utility of wearable sensor technologies for older person frailty assessment. *Experimental Gerontology*, *200*, 112668. DOI: 10.1016/j.exger.2024.112668 PMID: 39733783

Yu, X., Fu, Y., Li, J., Mao, J., Hoang, T., & Wang, H. (2024). Recent advances in wireless sensor networks for structural health monitoring of civil infrastructure. *Journal of Infrastructure Intelligence and Resilience*, *3*(1), 100066. DOI: 10.1016/j.iintel.2023.100066

Chapter 16
The Role of Technology in Economic Growth:
An Analytical Study of a Sample of Gulf Countries for the Period 2010–2023

Mustafa Kamil Rasheed
https://orcid.org/0000-0003-4289-7867
Mustansiriyah University, Iraq

Munaf Marza Neama
Mustansiriyah University, Iraq

Rawaa Salh Mahmmed Al-Saffar
Mustansiriyah University, Iraq

Alnoman M. Mundher Tayyeh
Middle Technical University, Iraq

ABSTRACT

The development of the global economy has implicitly necessitated the advancement of the knowledge element, thereby strengthening the technological structure in all areas and branches of the local economy. This has enhanced the local economy's resilience and made it more resistant to international crises and shocks. Developing countries in general, and the Gulf countries in particular, suffer from a weakness in the field of technological development based on academic knowledge that supports productive capabilities in all sectors of local production. This has been represented in the limited economic performance over time. The research assumes that technology based on academic scientific development in research and development makes a

DOI: 10.4018/979-8-3373-5248-0.ch016

significant contribution to improving economic growth. The research methodology relied on several economic indicators related to the research relationship for the period (2010-2023)

INTRODUCTION

Technological developments paved the way for investment in the knowledge economy, which was the key to the innovation economy, which subsequently led to technological leaps in all branches of local production. This led to an acceleration in the stages of multi-production and a focus on human-generated innovation to fuel the various technological developments in our world today.

The foundation upon which the innovation economy, which generates technological developments, is based is human capital, and the latter cannot be fully realized without a good level of health and education. This type of long-term investment may extend to more than 30 years for the growth and maturation of a new generation capable of innovation. The traditional way of thinking of most decision-makers in developing countries, in general, and the Gulf in particular, may prevent them from focusing on this type of investment, instead focusing on solving growing problems due to friction with the international environment and the requirements of adaptation and integration. This makes the expansion of technology to support productive activities a difficult matter.

Research Problem

Most Gulf countries suffer from a lack of knowledge and human capital necessary for innovation. Consequently, there is a lack of technological developments that could shorten production cycles and focus on the use of technological alternatives to address the scarcity of inputs and raw materials in the local production and manufacturing process.

The Importance of the Research

Focusing on academic knowledge patterns, which in turn generate generations that become the building blocks of economic innovation, production development, and, consequently, improved economic growth.

Research Hypothesis

Technology based on academic scientific development in research and development contributes significantly to improving economic growth.

Research Objective

Analyze the role of technology in economic growth in selected Gulf countries, using a specific set of economic variables and indicators from long time series to demonstrate this role.

THE THEORETICAL FRAMEWORK FOR TECHNOLOGY AND ECONOMIC GROWTH

Technology and Economic Innovation

Technology is defined as the process of developing existing technologies or creating new ones and using them to achieve progress and economic growth. This technological development plays a significant and prominent role in development by enhancing productivity, providing new innovative services, and creating new job opportunities, which reduces unemployment rates and achieves environmental efficiency, (Ali, 2025).

Economic innovation technology is defined as a set of creative ideas created by an entrepreneur that uses technology to achieve specific goals, whether in the field of production or adapting processes to improve products, services, etc. Technology has been expressed as the continuous pursuit of improving processes and developing solutions by launching new products. Economic innovation in the context of technology can also be related to improving existing production processes or providing new ways to solve challenges. In this context, technology is a means to achieve economic innovation, (Mohammed, 2024).

Economic Growth

Economic growth is defined as the continuous increase in gross domestic product (GDP) growth that exceeds population growth, (Rasheed et al., 2022). Kuznets defined it as the increase in per capita output or the labor force's share of output, (Todaro, 2006). The increase in output is often accompanied by a rise in population. Therefore, a true assessment of the extent of economic prosperity requires basing economic growth rates on the per capita GDP indicator. GDP also represents several items (consumption, investment, government spending, and net foreign trade), (Rasheed, 2020).

The latest theories of economic growth are based on the importance of technological progress in economic growth through discoveries, inventions, and innovations. These theories see that the ability of the workforce to increase production can be achieved through investment in human resources, by acquiring more skills and experiences through education processes primarily or outside the educational system as well, which can overcome the problem of the law of diminishing returns explained by the classical theory, through investment in both the workforce and capital in parallel (Khadir, 2018).

Accordingly, these theories emphasize the role of human capital, education, and training as a primary source of economic growth. These theories are based on two main pillars: human capital and scientific knowledge, (Al-Kawaz, 2004).

Economic growth can also be accompanied by economic progress if GDP growth exceeds the population growth rate, or it can be unaccompanied by economic progress if the GDP growth rate is equal to the population growth rate. However, if the population growth rate is greater than the GDP growth rate, growth is then accompanied by economic decline. Economic growth means a real increase in real per capita income, as economic growth does not occur without continuous increases in income, (Abdul Qader & Nassif, 2003). Increases in income in most developing countries come from government spending, which often suffers from declining government revenues, widespread fiscal deficits, and consequently, persistent government debt, (Rasheed et al., 2022). Accordingly, economic growth is based on the population, which leads to an interest in human capital and then innovation, which is the basis for knowledge and technology in all branches of production.

The Relationship Between Technology and Economic Growth

Technology is linked to economic growth through a series of effects that technology generates on productive activities, the type of technology used in production, and the degree of complexity in the forward and backward connections between these local activities. The most important sources of influence are:

1. Increased productivity: Technology improves production processes, leading to increased efficiency and productivity. This contributes to sustainable economic growth.
2. Saving time and effort: Technology helps reduce the time and effort required to complete tasks, allowing for greater output in less time.

3. Improving the quality of products and services: Technology can improve the quality of products and provide better services, increasing market attractiveness and enhancing competitiveness.
4. Encouraging economic innovation: Technology provides a conducive environment for innovation, as technological development can open up horizons for addressing labor and production challenges and developing appropriate solutions over time.
5. Job creation: Technological development can lead to the establishment of new economic sectors and thus create new job opportunities.
6. Enhanced competitiveness: Effective use of technology enhances the ability of countries and institutions to face local and international competition in markets.

Technology requires organizations to activate research and development centers to increase the effectiveness of human resources operating in the economy, enhancing their skills and professional experience. This helps them keep pace with technological developments in all markets, allowing them to engage with and leverage these developments to improve outputs, performance, and economic growth, (Salman, 2019).

TECHNOLOGY INDICATORS ANALYSIS IN THE ECONOMIC GROWTH OF THE RESEARCH SAMPLE

Technology Indicators

Global Innovation Index

The Global Innovation Index in Table 1 confirms that the UAE and Qatar were close throughout the research period, with Oman's backwardness with some ranks for them, as the index in the UAE reached 2010 (3.98), Qatar (3.55), and Oman (3.03). He was hoped for until 2015 in the UAE (40.06), Qatar (39.01) and Oman (35.00), until it reached its highest level in the UAE in 2020 where it reached (58.29), while the highest level was registered in Qatar in 2011, where he had (47.74), while Oman recorded its highest level in 2020, when the index reached (41.15).

It is also noted that after 2020, Oman was recorded at significantly lower levels than in previous years, which places these countries in the following order according to the global innovation index: UAE, Qatar, and then Oman.

This indicator confirms the existence of a relative interest in the field of creating technology based on scientific innovation in the research sample, but at the limits of its narrowness, which opens the way for the fact that the lack of qualitative interest in education and its efficient outputs and research centers that have not witnessed a distinguished role in creating a technology enhanced industrial competition and in the local, regional and international markets of the countries of the research sample.

Table 1. Global innovation index for selected Gulf countries for the period (2010-2023)

Years	UAE	Qatar	Oman
2010	3.98	3.55	3.03
2011	41.99	47.74	35.51
2012	44.4	45.5	39.5
2013	41.87	41	33.25
2014	43.25	40.31	33.87
2015	40.06	39.01	35.00
2016	39.35	37.47	32.21
2017	43.24	37.9	31.83
2018	42.58	36.56	32.8
2019	42.17	33.86	30.98
2020	58.29	42	41.15
2021	43	31.5	29.4
2022	42.1	32.9	26.8
2023	43.2	33.4	28.4

Source: WIPO, The Global Innovation Index, Selected Years.

According to Table (2) and plan (1), the arrangement of the research sample countries within the countries that were included in the global innovation index was far away, as the highest ranking by UAE was the ranking (22) in 2020, and the least (47) ranking in 2015.

As for Qatar, it reached the highest ranking (26) in 2011 and the lowest (64) ranking in 2020, and in Amman, it reached the highest ranking (46) in 2022 and the lowest (80) ranking in 2013, 2019. The effect of the COVID-19 pandemic appears to be clear in the ranking of the research sample countries, where we find that in 2020, the arrangement of the research sample countries has declined a lot within the ranking of countries in the Global Innovation Index.

The truth that can be extracted from the international innovation index data is that in front of UAE and Qatar is a good opportunity to advance its technological renaissance and enhance local production with the technical developments required to open the fair competition for its products against the world, while Oman has a long run to deepen its scientific and technological renaissance to accelerate and the light of local quality, the intense use of local technology.

Table 2. Global innovation index rank for selected Gulf countries for the period (2010-2023)

Years	UAE	Qatar	Oman
2010	24	35	65
2011	34	26	57
2012	37	33	47
2013	38	43	80
2014	36	47	75
2015	47	50	69
2016	41	50	73
2017	35	35	77
2018	38	51	69
2019	36	65	80
2020	22	64	68
2021	32	45	47
2022	30	42	46

Source: WIPO, The Global Innovation Index, Selected Years.

Figure 1. Global innovation index rank for selected Gulf countries for the period (2010-2023) (WIPO, The Global Innovation Index, Selected Years)

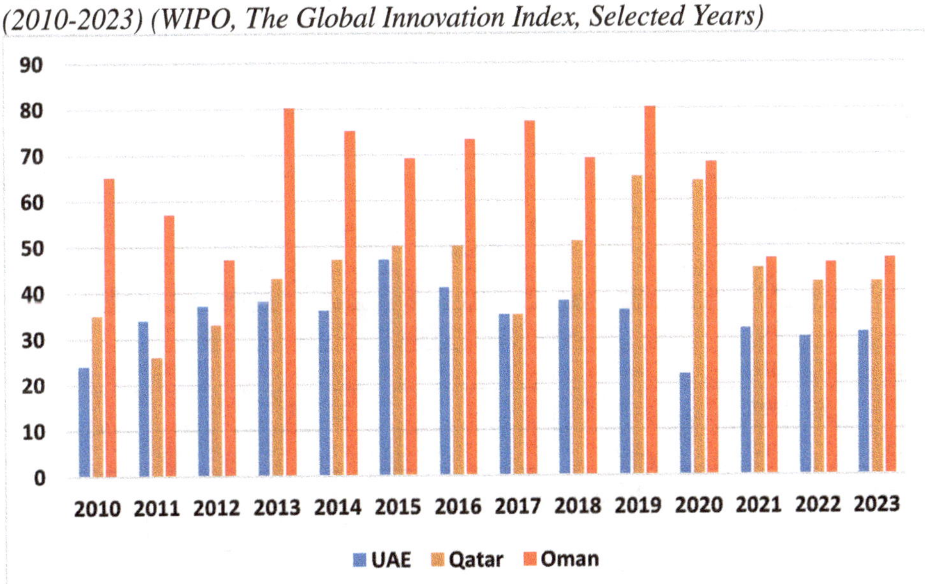

Patent Applications

Patent applications in the research sample countries came on two types (Table:3): patent applications for non -residents, which were the highest in UAE, where in 2010 it amounted to (1263) applications and took the gradient until the year 2017 (1733) applications, while in 2021 it reached (2354) applications. In Qatar, it was only in 2012 (58) applications, but it increased until it reached 2021 (722) applications, and this amount appears to be very modest compared to UAE. As for Oman, it reached the year 2015 (341) applications, and in 2021 (511) applications only and which is a lot about the UAE.

As for Patent applications for residents, and here the paradox appears, as it was very limited, as it ranged in the UAE between (69) applications in 2021 as a maximum and (25) applications in 2010 as a minimum, and in Qatar ranged between (81) applications in 2020 as a higher limit and (3) applications only in 2012. At a top and (3) applications only in 2016.

This indicator confirms the truth about the weakness of the research sample countries in general in the field of technology generating inventions according to their global requirements, as the contributions of local scientists in the research sample countries were very weak, which means that the academic interdependence of knowledge with the markets was also weak, the research centers in the research sample countries were not available significantly and did not exercise their role in promoting cognitive interdependence between reality and theory, and had no role in addressing the product problems Challenges and competition in local and international markets.

Table 3. Patent applications, nonresidents & residents for selected Gulf countries for the period (2010-2021)

Years	UAE	Qatar	Oman	UAE	Qatar	Oman
	Patent applications, nonresidents			Patent applications, residents		
2010	1263	-	-	25	-	-
2011	1318	-	-	33	-	-
2012	1327	58	-	23	3	-
2013	1402	323	-	21	9	-
2014	1438	477	-	29	5	-
2015	1714	-	341	35	-	5
2016	1642	548	366	57	16	3
2017	1733	574	387	63	19	4

continued on following page

Table 3. Continued

Years	UAE	Qatar	Oman	UAE	Qatar	Oman
	Patent applications, nonresidents			Patent applications, residents		
2018	1723	-	404	57	-	16
2019	1843	799	476	55	39	30
2020	1869	604	519	39	81	11
2021	2354	722	511	69	47	30

Source: World Bank, Database, Economic Indicators, Selected Years.

High-Tech Export Index

Show Table (4) that high -end exports have witnessed great fluctuations for all countries of the research sample throughout research, and that these great fluctuations mean poor interest in the export base of the field of creativity and economic innovation, and the weak front and rear ties related to local production in all its stages to export, as well as eat the added value and the economic surplus achieved from these products.

This great volatility in high-end exports is accompanied by instability in supply chains in international markets, and the irregularity of the flow of these products makes it difficult to open markets for them, as well as weak global demand and the great volatility in the return on investment in knowledge and technology.

The arrangement of the research sample countries came according to the following arrangement for this indicator: Emirates, Oman, and then Qatar.

Table 4. High-technology exports (current US$) for selected Gulf countries for the period (2011-2023)

Years	UAE	Qatar	Oman
2011	-	112596	-
2012	493096229	63779825	167104199
2013	401391775	50968303	159946801
2014	1465986043	33371	230872195
2015	903670848	96835993	138986749
2016	484214896	62631	50129889
2017	523668921	292250	48362452
2018	643634927	155450581	73302556

continued on following page

Table 4. Continued

Years	UAE	Qatar	Oman
2019	670459816	122141	77192098
2020	1202292537	466395732	589224440
2021	2904973069	147790556	662031837
2022	3214235330	270331668	784366169
2023	3417562262	245655998	441801906

Source: World Bank, Database, Economic Indicators, Selected Years.

According Table(5), the highest percentage recorded by the UAE (10.17%) in 2014 and the lowest rate (2.62%) in 2016, while Qatar recorded its highest percentage (7.11%) in 2020 and the lowest rate (0%) in 2016, and Amman it recorded its highest percentage (8.50%) in 2020 and the lowest rate (1.12%) in 2017, and the arrangement of these countries came as follows: UAE, Amman and then Qatar, where Qatar witnessed a great weakness in The High-Technology Exports (% of Manufacture EXPORTS) during the search period.

Table 5. High-technology exports (% of manufactured exports) for selected Gulf countries for the period (2011-2023)

Years	UAE	Qatar	Oman
2011	-	0.01	-
2012	3.77	0.60	3.39
2013	3.61	0.55	3.44
2014	10.17	0.05	4.36
2015	5.30	1.10	3.24
2016	2.62	0.00	1.52
2017	2.84	0.01	1.12
2018	2.90	1.80	1.26
2019	3.00	0.00	1.50
2020	5.38	7.11	8.50
2021	8.96	1.72	6.61
2022	9.29	2.86	7.62
2023	9.62	2.78	5.01

Source: World Bank, Database, Economic Indicators, Selected Years.

ECONOMIC GROWTH INDICATORS

GDP Index

According Table (6), the GDP in UAE was volatile during the research period, as it witnessed a rise in the period (2010-2014) and was affected by the crime of oil prices, and it was declined for the year 2015 and 2016, and he recovered in 2017, when it reached (390516804016.50) dollars, and it was stood before the Corona pandemic crisis, as it was declined until the year 2020 (349473015336.94) dollars. In 2023, GDP reached (514130432653.08) dollars.

In Qatar, the GDP rose for the period (2010-2014) for the same reason, and began to fluctuate until it reached its highest level (235709325708.34) dollars in 2022. As for Oman, it is the other that witnessed the fluctuation of GDP has a period of the search due to the influence of external crises, the most important of which is the oil price crisis for the year 2014 as the GDP decreased in 2015 to (78710793237.97) dollars Because of it, Corona, which caused the GDP to decrease to (75909492847.85) dollars in 2020. She witnessed a recovery until 2022, which reached its highest level.

Table 6. GDP (current US$) for selected Gulf countries for the period (2010-2023)

Years	UAE	Qatar	Oman
2010	300189052694.03	125122306346.15	64993498049.41
2011	360832739558.39	167775268626.37	77497529258.78
2012	384610125384.03	186833502362.64	87408842652.80
2013	400218529747.60	198727642967.03	89936020806.24
2014	414105366758.91	206224598571.43	92699089726.92
2015	370275469560.17	161739955576.92	78710793237.97
2016	369255326235.77	151732181868.13	75128738621.59
2017	390516804016.50	161099122225.28	80856697009.10
2018	427049432149.35	183334953818.68	91505851755.53
2019	417989721734.49	176371267689.08	88060858257.48
2020	349473015336.94	144411363345.27	75909492847.85
2021	415178792769.88	179732009554.85	87323797139.14
2022	502731935197.49	235709325708.34	111943823146.94
2023	514130432653.08	213002809330.27	108810923276.98

Source: World Bank, Database, Economic Indicators, Selected Years.

Economic Growth Index

It is clear from Table (7) that economic growth in the UAE has ranged between (7.51%) in 2022 as a higher and (-4.96%) in 2020 as a minimum due to the influence of the gross domestic product on the Corona's pandemic (9.58%) in 2022 as a maximum and (-3.38%) in 2020 as a minimum due to the influence of the gross domestic product on the Coruna's pandemic, as it seems that economic growth in the countries of the research sample was as follows: UAE, Qatar and then Amman where many declines during the research period due to exposure to many economic crises, as well as the weak economic growth in Amman due to the weakness of total productivity and the contribution of economic sectors In the formation of added value.

Table 7. GDP growth (annual %) for selected Gulf countries for the period (2010-2023)

Years	UAE	Qatar	Oman
2010	1.60	19.59	2.47
2011	6.22	13.38	2.89
2012	1.82	4.73	8.86
2013	5.06	5.56	5.23
2014	4.17	5.33	1.29
2015	6.79	4.75	5.02
2016	5.56	3.06	5.05
2017	0.74	-1.50	0.30
2018	1.31	1.23	1.29
2019	1.11	0.69	-1.13
2020	-4.96	-3.56	-3.38
2021	4.35	1.63	2.58
2022	7.51	4.19	9.58
2023	3.62	1.19	1.31

Source: World Bank, Database, Economic Indicators, Selected Years.

THE RELATIONSHIP BETWEEN GDP GROWTH WITH THE GLOBAL INNOVATION INDEX

The rise of the Global Innovation Index (table (8)) has led to an increase in economic growth only in one year is 2011, and the rest of the years, the rise of the Global Innovation Index did not affect the rise in economic growth, and in Qatar, the rise of the Global Innovation Index led to an increase in economic growth only in one year is 2022, but Amman is that the rise The Global Innovation Index has led to an increase in economic growth in four years, 2011, 2012, 2015 and 2018.

This trend in the time value confirms the weakness of the application of technology in general in the research sample countries, especially the technology based on academic knowledge resulting from the specialized research center cooperating with the local institutions working in the economy.

Table 8. GDP growth (annual %) & global innovation index for selected Gulf countries for the period (2010-2023)

Years	UAE		Qatar		Oman	
	global innovation index	GDP growth (annual %)	global innovation index	GDP growth (annual %)	global innovation index	GDP growth (annual %)
2010	3.98	1.60	3.55	19.59	3.03	2.47
2011	41.99	6.22	47.74	13.38	35.51	2.89
2012	44.4	1.82	45.5	4.73	39.5	8.86
2013	41.87	5.06	41	5.56	33.25	5.23
2014	43.25	4.17	40.31	5.33	33.87	1.29
2015	40.06	6.79	39.01	4.75	35	5.02
2016	39.35	5.56	37.47	3.06	32.21	5.05
2017	43.24	0.74	37.9	-1.50	31.83	0.30
2018	42.58	1.31	36.56	1.23	32.8	1.29
2019	42.17	1.11	33.86	0.69	30.98	-1.13
2020	58.29	-4.96	42	-3.56	41.15	-3.38
2021	43	4.35	31.5	1.63	29.4	2.58
2022	42.1	7.51	32.9	4.19	26.8	9.58
2023	43.2	3.62	33.4	1.19	28.4	1.31

Source: World Bank, Database, Economic Indicators, Selected Years.

CONCLUSION

1. The values of the global innovation index of the research sample countries were limited by their global arrangement within the index, as UAE topped, then Qatar was followed, and Oman was in the latter.
2. Despite the limited patent requests in all countries of the research sample, the patent requests for non-residents in general were the highest, and this indicates the weak requests for patents by the indigenous population and the weak incentive to innovate and support technology applications in production branches.

3. Exports with high technology fluctuated significantly during the research period in all the research sample countries, but the UAE came in the forefront and then was followed by Oman and Qatar was in the end, and these results were evident in the percentage of high -end exports from the total manufactured exports, as they were in Qatar very low throughout the research period and did not exceed (10%) in all countries of the research sample.
4. Economic growth is turbulent in all countries of the research sample, influenced by global crises, the most important of which is the oil prices crisis in 2014 and the Korona pandemic in 2020, due to the presence of crude oil in the components of the gross domestic product of the research sample countries, which is very affected by these global crises.
5. Weak technological impact on economic growth in the research sample countries, as the increase in the global innovation index did not affect the increase in economic growth in the UAE and Qatar, only in one year, while in Amman, it had an effect in four years. Which means accepting the research hypothesis.

Recommendations

1. Scientific relations (universities and global research centers) and the process of leading global institutions in the field of technology and investment in human capital, which in turn enhances the ties of economic sectors and improves economic growth.
2. Enhancing local production with knowledge by focusing on research centers, whether in universities or institutions working in the economy, to build a cognitive base that supports economic innovation.
3. Local, regional, and international cognitive cooperation in a way that provides a high level of technology applied in most of the stages of production, increased cognitive surplus, and improving the working cadres in line with technological developments applied in all productive activities, to achieve a significant increase in performance and economic growth.

REFERENCES

Abdul Qader, M. A. Q., & Nassif, I. A. (2003). *Modern trends in development.* Alexandria University Press.

Al-Kawaz, A. (2004). *Economic policies and human capital.* Arab Planning Institute.

Ali, A. (2025). The role of joint technological innovation between countries in achieving the green economy (Experiences of selected countries). *Mesopotamia Journal for Humanities and Social Sciences, 7*(1).

Khadir, R. (2018). *The role of research and development in economic growth: Selected international experiences with reference to Iraq* (PhD thesis). University of Karbala.

Mohammed, W. (2024). Analysis of the economic impacts of technology and innovation on economic development: Research methodologies and their applications. *Journal of the Higher Institute for Qualitative Studies, 4*(6).

Rasheed, M. K. (2020). The analysis of macroeconomic gaps in Iraq for the period (1990–2017). *Opción, 36*(Special Issue 27), xx–xx.

Rasheed, M. K., ... (2022). Measuring the impact of oil revenues on government debt in selected countries by using ARDL model. In *Lecture Notes of the Institute for Computer Sciences, Social Informatics and Telecommunications Engineering* (Vol. 471). Springer.

Rasheed, M. K.. (2022). The impact of oil exports on consumer imports in the Iraqi economy and COVID-19 period: A theoretical study. *International Journal of Computer Applications in Technology, 66*(3–4), xx–xx.

Salman, H. M. (2019). The reality of teaching practices in the light of the theory of learning based on the brain teachers and teachers of Arabic language. *Opción, 35*(Special Issue 21).

Todaro, M. (2006). *Economic development* (M. H. Hosni & M. H. Mahmoud, Trans. & Rev.). Dar Al-Marikh. (Original work published in English)

World Bank. (n.d.). *Economic indicators database.* https://data.worldbank.org

World Intellectual Property Organization (WIPO). (n.d.). *The global innovation index.* https://www.wipo.int/global_innovation_index

Chapter 17
Wireless Sensor Networks for Real-Time Infrastructure Monitoring and Management Using ML Approach

Gajanan Shankarrao Patange
https://orcid.org/0000-0002-6569-3755
Charotar University of Science and Technology, India

Abdulsattar Abdullah Hamad
https://orcid.org/0000-0003-0497-5115
Department of Physics, College of Education, University of Samarra, Iraq

V. Dankan Gowda
https://orcid.org/0000-0003-0724-0333
BMS Institute of Technology and Management, India

S. Gobinath
V.S.B. Engineering College, Karur, India

K. Ramya
https://orcid.org/0000-0003-1339-7716
BGS Institute of Technology, Adichunchanagiri University, Karnataka, India

ABSTRACT

The growing interest in Wireless Sensor Networks (WSNs) lies in the real time monitoring and management of infrastructures. As the critical infrastructure sys-

DOI: 10.4018/979-8-3373-5248-0.ch017

tems become more complex and age, the conventional monitoring systems have been incapable of addressing the management needs of a timely and competent management. WSNs are a scalable and effective solution that allows to gather real-time data (related to structural health, surrounding environment, flow of traffic, and other vital parameters required in maintaining infrastructure health and safety). This chapter looks into the basic concepts of WSN, its uses in infrastructure system and design aspects necessitated in deploying such systems successfully. Barriers and shortcomings of the existing WSN technologies are presented through case study and performance indicators, coupled with aspects of future breakthroughs in this direction. The focus is made on the interoperability of WSNs and the related emerging technologies to ensure the increase in predictive maintenance and others related to decision-making.

INTRODUCTION

Overview of Wireless Sensor Networks (WSNs)

Wireless Sensor Networks (WSNs) These are ad-hoc networks with decentralized control systems made up of spatially dispersed sensor nodes that are used to monitor and gather information relating to physical or environmental variables. These nodes will generally include a sensing unit, a processing unit, a communication unit and a power unit (Aborujilah et al., 2023). The images are captured by the sensing unit which measures parameters such as temperature, pressure, humidity or vibration, whereas the processing unit obtains the data and does the basic analysis. The data is processing by the central system or other sensor nodes after relaying it by the communication unit. Sensor node power supplies are usually either batteries or energy harvesting devices to operate sensor nodes at remote locations or to enable very long operation (more than a year). Insider Fundamentals WSNs have been at the height of the current technological provisions because they can check decisions based on real-time data on many fields (Abdulbaqi et al., 2023). WSNs can offer the tools of monitoring necessary in the framework of infrastructure management that will presumably monitor these changes in the environment or structural condition and, due to such monitoring, may provide proactive maintenance instead of reactive one. The ability of WSNs to provide real-time monitoring and data analytics can be improved by merging them with the modern networks like the Internet of Things (IoT), making better decisions (Alshami et al, 2024). The features of WSNs are different kinds of sensors which can be customized to sense certain factor in the environment and communication protocols which are policies of sending information to the other sensor nodes and the central systems. Wireless sensor networks usually employ low power communication protocols like ZigBee, LoRa, and Bluetooth to guarantee energy-efficient solutions, particularly (Bhushan et al., 2025), in applications where deployment must last long. Such networks are also known to facilitate the use of sophisticated data processing techniques, like machine learning and artificial intelligence (AI), in order to improve the quality of the insights obtained out of the information gathered.

Importance of Infrastructure Monitoring and Management

Real-time monitoring in the critical infrastructure systems is a necessity that cannot be overstated. The backbone of urban and rural economy is infrastructure systems such as bridges, roads, buildings, utilities and other non-avoidable structures (Fu et al., 2015). With aging and exposure to unfriendly environmental factors, these infrastructures become strained and hence the inability of such facilities to operate safely and reliably. Failure of critical infrastructure systems can cause loss to the economy, pollution, and loss of life (Setijadi et al., 2013). The more common traditional approaches to monitoring, including manual inspections and schedule-based maintenance, are not usually enough to identify the problems during early stage, or to maintain complex systems effectively. Monitoring of the infrastructure with the help of WSNs alleviates these issues by giving constant and current data regarding infrastructure health and status (Bezukladnikov & Gavrilov, 2016). To give an example: In bridges, vibration, movement and stress monitoring gives indications on weakness or potential failure. Equally, sensors installed on roads are able to track traffic traffic, heat, and wear trend, so as to guide injection of maintenance plans and the priority of repairs. Real-time data gathering and analysis allow infrastructure managers to identify any potential problems before they mushroom into major disasters and this aspect ensures the integrity, control and sustainability of the key infrastructural bases (Orhan et al., 2010). Moreover, the real time monitoring supports predictive maintenance which is a proactive technique that is oriented toward predicting failures before they happen using data analysis. Such a strategy will not only eliminate the risk of sudden failures but will also allow optimization of resource use by performing repairs and replacements according to the real condition of the infrastructure instead of following some random schedules (Orhan., 2010). With infrastructure becoming more and more integrated with smart technologies, real-time monitoring systems are becoming more and more essential in the management of the complexity of such systems in an effective manner.

Challenges in Traditional Monitoring Methods

Manual inspection, periodical checking and the use of visual observation are typical features of most of the infrastructure monitoring techniques used traditionally and are time-consuming as well as subject to human error. Such approaches are normally reactive in their characteristic and only after the observable harm or failure occurs, it is acted upon (Wang et al., 2010). It is not an efficient, cost-effective or even effective measure especially when applied to trunk-scale infrastructure systems which are either distant or very widespread. Take, for example, an inspector examining a bridge or a building by hand will not be able to notice the slightest signs of something going awry as they will not be able to notice micro-cracks or minor changes in structural directions, that, when left unchecked, can cause much greater issues. Also, the capacity of traditional techniques to continuously capture real-time data concerning the state of infrastructure is limited and it is difficult to identify and act according to the change as it happens (Naikade et al., 2023). The lack of monitoring capability thus poses a greater risk of poor infrastructure, particularly under the circumstances where the environment is volatile in nature, like the situation that occurs during extreme weather episodes or a period post an earthquake. Further on, the manual inspections may be rather expensive and involve the visit of the trained personnel who should periodically inspect the infrastructure (Zhu et al., 2021). This is of special concern in remote or difficult to reach places where the inspections can be sparse or expensive to conduct. In this event, infrastructure may not be monitored in an extended period, and therefore chances of unforeseen damages are probable. Besides, the narrow capability of conventional monitoring implies that only a small part of problems can be detected, leaving assessments of infrastructure health to be incomplete or inaccurate (Bhavani et al., 2024). The data accuracy and consistency also becomes an issue as far as manual processes are concerned. Monitoring efforts may be affected by the quality of human error, subjective interpretations and incomplete records. Moreover, modern technologies, including WSNs, have better integration capabilities since modern systems would be able to collect data sent by a large number of sensors and offer a better picture of the states of the infrastructure.

FUNDAMENTALS OF WIRELESS SENSOR NETWORKS

The wireless Sensor Networks (WSNs) are spatially distributed sensor nodes that are set to monitor and gather information on different environmental as well as physical variables. Such networks are comprised of a number of important components: sensing unit, processing unit, communications unit, and power unit. It is the sensing unit which detects certain variables like temperature, humidity (Levy et al., 2017), vibration or pressure and changes them into electric signals. In the case of infrastructure monitoring, potential types of sensors can be accelerometers, thermistors or strain gauges, and they can be used to measure such parameters as the stress in bridges or change in temperature in utility networks. The accumulated information is processed in the processing module of every node and carries out simple filtering or analysis. It also usually contains microcontrollers or microprocessors and these enable local processing of data prior to transmission, which saves energy. In others, more complex data processing, such as edge computing (Monshi et al., 2013), is used to process the data at the sensor node node level, allowing data transmission overheads to be kept down since fewer raw data need to be sent back to central systems. The communication unit provides the exchange of information between the sensor nodes and the core system, based on such communication protocols as ZigBee, Bluetooth, LoRa, and Wi-Fi. The selection of these protocols is decided in terms of range, data rate, and energy savings since WSNs are built in a way that requires little interventions after their implementation. Finally, power unit is used to make sure that the sensor nodes can remain in operation, which is usually done by the means of the batteries or energy harvesting techniques like solar cells, or thermoelectric generators (Chowdhury et al., 2024). Energy is of significant concern as most sensor nodes are to be used in remote and inaccessible places and maintenance less network usage is vital.

Figure 1. Key Components of a wireless sensor network

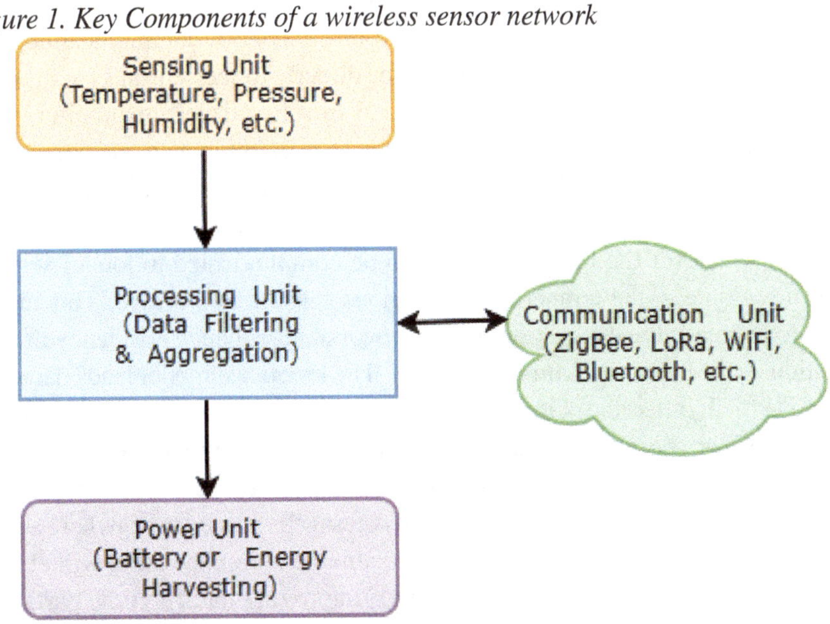

Figure 1 shows the main elements of a wireless sensor network (WSN) and how they interact. The WSN is comprised of some central components which are integrated together to capture, process and relay information in real time. Sensing Unit is one that deals with the detection of environmental or physical parameters like temperature, pressure, humidity or vibration (O'Donoughue et al., 2006). These sensors collect crude data and this is vital in terms of the condition of the infrastructure. After capturing the information, it is directed to Processing Unit which filters, aggregates and analyses the information locally. The processing module minimizes the transportation requirements of raw data as received by processing and filtering the data to draw meaningful information before it is transported to any other node of the network or relay to some central system. The analyzed information is then transferred via the Communication Unit that allows conveying information between sensor nodes or to the central system. such protocols as ZigBee, LoRa, Wi-Fi or Bluetooth are used to provide reliable and energy efficient data transmission within different distances by the communication unit. Lastly, the Power Unit keeps the sensor nodes active and functional usually by the use of batteries or energy generation techniques like solar or vibration powered energy systems (Patil et al., 2017). This module plays a very important role in ensuring that the sensor nodes operate independently even in areas that aren t very accessible. The combination of these elements facilitates the real-time process of monitoring and running infrastructure systems and guarantees their efficiency and usage safety.

The WSNs use dissimilar communication models that support data transmission between nodes and the central monitoring system. The point-to-point model of communication Databinding The point-to-point model of communication uses data transmission between sensor nodes and/or directly to the central system when low latency communication is important e.g. real-time based alerts to monitor infrastructure (Chatterjee et al., 2017). This model is easy to implement and convenient in the case of the direct path between the nodes. The multicast model enables information to move among several destinations (nodes) through one node therefore it is helpful whenever the information has to be communicated to lots of sensors in a given region e.g. the temperature in various sensors in a bridge. This model is useful to decrease the congestion of the network and enhance efficiency of data distribution in the network (Lam et al., 2016). The broadcast model leads data sent by one node to all other nodes in the network so that control messages or alerts are propagated to every point of the network which is especially important in the case of infrastructure monitoring so that in case of critical structures such as failing worth a warning or calibration of sensors, the control message is passed everywhere across the network. The model of choosing the most suitable communication will also be based on the size of the network, frequency of improving data, and the particular requirements of the infrastructure under survey (Kamruzzaman et al., 2017). The individual models of real-time monitoring come with their own bargains on energy consumption, communication rate and dependability of the network which need to be factored into the design of a WSN.

The cost of energy is an important aspect of the implementation and design of Wireless Sensor Networks, because in most cases the sensor nodes are installed in very remote and hard to reach areas. Data sensing and data transmission are two activities in which the energy consumption in WSNs primarily occurs. Although the processing in the sense tends to use less power, the data transmission particularly on long distance will tend to use up a lot of power. Energy-efficient designs thus aim at reducing the number of data sent by all nodes. Other methods like aggregation and compression of data are used to lower the required number of transmissions thereby saving energy (Suryawanshi et al., 2017). A combination of the information supplied by several sensors prior to transmission allows minimizing the overall communication burden and the use of compression techniques to ensure that less necessary data is sent. Use of sleep modes is another important energy savings technique applied in WSNs. In this, the sensor nodes can go either in the active or in the low power mode at any given time depending on the monitoring schedule (Liu et al., 2024). This duty cycle process allows the sensor nodes to transmit and receive data infrequently hence they use less energy instead of constantly running out of charge in between. Also, the positioning of sensor nodes is critical to energy optimization. By positioning nodes that are near to one another, the amount of energy used in propagating data can be reduced since short communication lengths lead to minimal consumption of energy. The hierarchical network structures, where the data sent over the network by lower levels is shared with higher levels to be aggregated and finally presentable over the central system is another form of energy saved. Lastly, energy gathering methods e.g. solar cells, wind energy or vibrations-based generators started playing a more significant role in WSNs relating to infrastructure monitoring. These technologies have the capability to extend the life of the sensor nodes without manual replacement of battery and thus it is a feasible technology to be used on long-term basis in remote locales (Xue et al., 2010). All in all, energy efficiency in the WSNs is the crucial factor in ensuring the operational continuity of the network and in reducing its maintenance cost to at least such a network can be monitored in real time using only limited amounts of resources.

APPLICATIONS OF WSNS IN INFRASTRUCTURE MONITORING

Structural Health Monitoring

Wireless Sensor Networks (WSNs) are critical in the examination of structural health of the critical infrastructure systems, such as bridges, tunnels, buildings and dams. These buildings are exposed to the ever present stress and strain due to the effect of the environment, traffic, wear and tear due to age. Conventional means of monitoring which include periodic assessments and manual checks have low chances of revealing slight or evanescent indications of harm. WSNs present an active and real-time solution of tracking the health of these structures, which can give information which could be used to decide to do maintenance and repair before the serious damage happens. The sensor nodes that will be placed at strategic places within the infrastructure will be capable of measuring some of the parameters that will signal on the structural integrity of the system. As another example, strain gauges embedded in bridges may sense small variations in stress, which may be precursors of material fatigue or cracks. Data on the dynamic reaction of the structure to any load (traffic, earthquake, etc.), can be collected by accelerometers and vibration sensors, giving information on the reaction of the structure to any load. Temperature sensors may be able to detect temperature alterations that may imply expansion or contraction of materials whereas displacement sensors will be able to detect any alterations or deformations within the structure that could indicate a possible failure. The realtime information provided by these sensors is relayed to a centralized monitoring system where it is again processed and analyzed to give a measure of the structural health. With constant tracking of these variables, WSNs will help the infrastructure managers to identify when two or more variables have an unusual behavior at an early age, and this can be solved before major breakdowns almost happen.

Environmental Monitoring

Monitoring of environmental issues with the help of WSNs is becoming more common and this has important repercussions in infrastructure systems. The condition and performance of infrastructure may be affected directly by environmental factors (air quality, temperature, humidity and noise levels) and thus continuous monitoring is critical. As an example, air pollution in urban centres may speed the process of wear and tear of the building materials and changes in temperature and humidity may impact integrity of the roads, bridges and other major infrastructure networks. WSNs allow installing sensor nodes in a large region to track these parameters in real time. Some of the types of air pollution that can be measured using air quality sensors include carbon monoxide, nitrogen dioxide, and particulate matters, which are very useful in finding the effects of pollution on infrastructure and health of the population. Sound levels caused by noise pollution may also be measured by the noise pollution sensors and this may signal that there is heavy traffic or that there is heavy industrial activity which also leads to worn-off infrastructures. Sensor monitoring In areas that are prone to damage caused by weather conditions, sensors to measure temperature and humidity can be used to monitor the impact that environmental conditions have on the durability of infrastructure. WSNs can also be used in areas that are likely to reap the worst of weather conditions like high wind velocity, heavy rain or snow and it can warn in advance of the conditions that might result in a hazard on infrastructure. As an example temperature variations that expose roads to the risk of frost heave may be noted at an early stage and remedied before any extensive damage can be done to the road. The WSNs allow real-time coverage of monitoring the environment in a constant fashion, which allows the infrastructure managers to have key information they can use to schedule maintenance, enhance resilience, and allocate resources wisely.

Traffic and Transportation Management

WSNs can play a significant role in assisting smart transportation by delivering in real-time data about traffic patterns, traffic jam, and in the performance of infrastructure systems. In large cities where traffic is difficult to overcome, controlling the transportation system becomes tricky, and such problems as overload, accidents and the congestions of the trip causes delays and inefficiency. WSNs will update these challenges since traffic can be monitored continuously and thus vehicle movement can be enhanced and decisions on whether to repair and extend the road can be made with the help of WSNs. Traffic flow sensors, installed at strategic point along road or in tunnels, count the numbers of vehicles and also detect the speed at which the vehicles are moving and that presents crucial information about the number of vehicles in a given area, and also the level of congestion. This information may be employed to set the right timing of traffic signals, it could be used to send real time traffic information to the car users, and even it could help in planning the routes of the transport systems of the people. Besides analyzing the traffic, WSNs may also analyze conditions of road infrastructure, including quality of road surface and load bearing capacity to make sure that dangerous conditions are not reached when road and bridges are loaded. Road sensors or Bridge vi-sion sensors are also applicable in sensing the probable damage of structures due to repetitive strains of heavy vehicles. Because of the combination of the information in traffic sensors and infrastructure monitoring systems, the WSNs can assist in the predictive maintenance scheme, allowing infrastructure managers to determine areas that are at risk of malfunctioning, and addressing them accordingly. The information gathered by WSNs may also be applied to design intelligent transportation systems (ITS) that use information provided in real time to make traffic much more efficient and regulated, thereby resulting in much safer and efficient transport networks.

Water and Waste Management

Monitoring of water and waste management infrastructure also uses WSNs to a greater extent, too, as they can give invaluable information about the state of the systems and their functionality. Waste water treatment plants, water distribution systems and water pipes are very vital to people and the environment and their effective functioning plays a very important role in promoting sustainable urban housing. With WSNs, it is now possible to have constant watch on water quality, pressure and flow and this gives real-time information on the health of such systems. An example is the sensors installed in the water pipelines that can find out where the leakage has occurred or pressure has diminished which could be a sign of the upcoming failures or wastage in the water pipeline. Simple measures of detecting leaks early can save against damages and wastage of the water as well as confidence in the supply of the water. On the same note, one can employ sensors to provide insight on the water quality in terms of parameters like pH, turbidity and other pollutants. This information is used in order to make sure the quality of water supplied to the water consumers is at the appropriate level and it observes regulations. In wastewater management, WSNs allow monitoring the flow rate and composition of wastewater throughout the treatment plants and learning about the efficiency of their work and any possible problems like blockage or overload. Techniques used to measure chemical composition or biological contaminate in waste waters are sensors that can be used to optimize the treatment process and avoid environmental contamination. To serve their purpose best, water management and waste management systems can be more efficient, become less prone to failures and more sustainable by folding in real-time monitoring enabled through WSNs. The deployment of WSNs in combination with predictive maintenance models and data analytics advances the efficiency in taking competent decisions regarding resources and infrastructural up-gradation. Summing up, Wireless Sensor Networks provide a practical and flexible environment that supports a large variety of applications in monitoring the infrastructure. WSNs can help infrastructure managers to make informed data-driven decisions to enhance the safety, reliability, and the performance of the critical systems by providing real-time information on the health of the structures, the condition of the environment, the traffic on the road, and the water and waste management facilities. Installation of WSNs in such applications does not only improve efficiency in operations but is also aids to avoid failures of certain operation processes that could be costly and also minimizes on environmental influence of infrastructure systems. The use of WSNs in monitoring infrastructure will only be critical in future due to the advancement of technology and thereby, the more intelligent as well as resilient infrastructure networks can be built.

DESIGN CONSIDERATIONS FOR WSNS IN INFRASTRUCTURE SYSTEMS

Network Architecture

Design of Wireless Sensor Network (WSN) on infrastructure monitoring is considerate of the network architecture, which defines the method of data flow on transmission and processing throughout the system. There are three main methods of network architecture that are generally used which include the centralized, decentralized, and hybrid. None of the architectures is ideal and have both strengths and limitations basing on the particular demands of the watched infrastructure. Centralized network architecture is where all of the sensor nodes have a direct communication with the central system or a base station who has to carry out the data processing and make decision. The present architecture allows reducing the efforts of managing the network and gathering data since all of the communication can be coordinated by the central system. Nevertheless, it may be weak, with the collapse of the central node or system generating the failure of the whole network. Usually, such an architecture is applicable in infrastructure systems which can be centrally controlled, and whose infrastructure is not spread out too far or larger in size such as in small bridges or small buildings.

A decentralized network topology, however, electrodes the processing and decision making to multiple sensor nodes. Nodes in this architecture are permitted to process information and make local decision independently e.g. When an instance of a specific sensor recognition exceeds a dangerous amount, it may send an alarm. Such a kind of architecture also adds to the robustness since the failure of any node does not cause an impact on the network performance. But it may also complicate network management since each node should be able to work independently and that it may necessitate more advanced data coding and combination methods. Large, multi-faceted infrastructure systems, like large bridge networks or dense highway systems, may benefit significantly by being set up in decentralized networks where data collection and processing are distributed improve system resiliency and scalability. A mixed form of networking blends both a centralized and a decentralized type architecture. In this strategy, certain nodes can collect data and present it locally depending on some nodes and relay it to a central unit to do more analysis. This structure balances both the pros of centralizing and the pros of decentralization. It is specially applicable in monitoring of large scale infrastructure devices such as in smart cities, or huge transport networks, in which a system in its core can control higher monitoring and control levels, whereas the small nodes in the periphery can control the short-run, localized tasks. The hybrid architecture is scalable, flexible and robust and it makes it suitable in monitoring the critical infrastructure systems, which need constant data analysis, and a rapid response mechanism.

Sensor Placement

The best locations of the sensors are also an essential parameter to guarantee overall coverage and monitoring of infrastructure. Sensor placement has several considerations that include range and coverage area of the sensor being deployed, the environmental situation, and the peculiarity of the infrastructure under surveillance. The most important thing is to place the sensors in the most apposite places in the infrastructure systems like bridges, roads, and buildings where they can collect the most relevant details and also at a position where they cause minimum interference and minimal consumed energy. The sensors range is one of the major aspects it should consider in setting the sensors. The sensor has to be situated well in a manner that they can cover the whole infrastructure or the targeted areas without a lot of overlap and loopholes. Depending upon the type of sensor and network communication range, coverage area may change. As an example, the vibration sensors might require installation in different places of a bridge to measure the stress at different places and environmental sensors might be required to cover large areas in a building or highway to measure the temperature and humidity variations.

Sensor location is also greatly influenced by environmental limitations. As an example, environmental conditions such as temperature, precipitation and humidity, exposure to adverse conditions and so on are known to impact sensor performance in an outdoor environment. Hence, sensors should be screened properly or enclosed in protective compartments to guarantee their durability in the long run. On the same note, when deploying infrastructure monitoring in underground or confined areas (tunnels), the deployment of the sensors has to be done in a place easy to access in terms of maintenance and also in a place that could be able to give correct results despite the environment in which deployment is done. Also, the placement of the sensor should be taken into account (cost, etc.). Sensors also require special equipment orized installation procedures in certain scenarios when sensors have to be installed at places that are far or hard to reach. Moreover, sensor placement optimization also deals with the trade off between quality of data and implementations and feasibility of a network in terms of implementation and scalability.

Data Fusion and Processing

The foremost application of a WSN is data aggregation and processing that should allow receiving meaningful information about the state of the infrastructure based on the data provided by several sensor nodes. Data fusion is the procedure of integrating several sensors data in order to improve the reliability, accuracy, and completeness of the monitoring system. The data fusion makes it a more comprehensive knowledge of the infrastructure health status and performance by combining several data. The fusion of data may take place at various levels in the WSN. Low-level data fusion Raw information received by individual nodes are then combined to construct more precise measurements or eliminate noise. As an example, fusion might be of data measured by several temperature sensors located along various sections of a structure giving an overall temperature profile. When data fusion is done at a higher level, more complex data analysis is made to discover the patterns, trends or anomalies of the whole network. As an example, structural modifications, e.g. cracks, stress points that may signify an upcoming break, can be spotted with the application of data fusion methods. At this level, statistical techniques and machine learning algorithms are usually utilized to find out complicated relationships and to forecast the conditions of infrastructure in the future.

It requires efficient data processing in order to have such huge amounts of data produced by a WSN transformed into usable knowledge. This involves sifting and weeding of irrelevant information, lessening of redundancy, and giving importance to crucial information. High-performance data processing modes can also be applied, where local data analysis on the sensor nodes is possible, which allows avoiding sending a huge volume of raw data to the central servers. This is particularly significant in case of remote infrastructure systems; where the communication bandwidth can be restricted. Other uses of data processing may also involve anomaly, trend detection, and issuance of real-time alerts, to inform the infrastructure managers about the possible problems. When real-time analytics is used, the process of making decisions takes less time, and the maintenance is proactive, as well as it is necessary to prevent catastrophic failures in critical infrastructure.

Security and Privacy Concerns

Security and privacy of the collected data is also one of the major issues that pose a big challenge when it comes to using WSNs to monitor the infrastructure. The sensor networks are easily affected by many forms of attacks such as illegal access to network, falsification of data, and denial of service probes especially since they are commonly implemented in remote or uncovered areas. Also, due to sensitive infrastructure systems which WSNs tend to be tracking, the integrity and confidentiality of the data becomes very important to support the systems reliability as well as the integrity in prevention of malicious interference. Security of data in WSNs implies securing of the data against unauthorized users such that information between sensor nodes and central systems is encrypted and safe. This is especially critical to infrastructure systems such as bridges, dams and power grids whereby the truncation of data could cause disastrous failures. Authentication protocols are usually used to identify the node in the network, so all data could be accessed or sent by authorized devices. Privacy privacies exist in the scenario where WSNs have been implemented in areas that entail sensitive or personal information. As an example, the surveillance of public common infrastructure, including transportation systems or utilities, can bring privacy concerns to gathering personal information, including the traffic or usage rates. Privacy risks can be offset by making sure such data has undergone anonymization or aggregation prior to transmission.

Also there is a need to take care of the physical security of sensor nodes since such nodes are usually placed in an outdoor and unsecured environment. Nodes that are susceptible can also be interfered with or even damaged thus interrupting the monitoring network. These risks can be mitigated by implanting monitoring systems and tamperproof enclosures. To sum-up, infrastructure monitoring of WSNs should be secured through multi-layered architecture involving data encryption, authentication and access control, physical securing of sensor nodes and privacy-preserving methods to ensure data sensitive to a WSN is not compromised. To assure high-security, the infrastructure monitoring systems need to have a high security to assure credibility and dependability.

RESULTS AND DISCUSSIONS

Wireless Sensor Networks (WSNs) have successfully been implemented in the monitoring of the real life infrastructure and demonstrates the ability of becoming more efficient reliable and cost effective. This is evident in the bridge monitoring system to monitor the health of important bridges. The bridge can have sensors, which record such parameters as stress, vibration, and displacement, then conveying up-to-date data to a central monitoring system. It is an endless flow of information that allows identifying damage to the structure (cracks, overloading, etc.) at an early stage. Since the engineers would know about possible issues early on when they are still manageable, it is possible to organize their repair or maintenance the problem before it will develop into a major failure. An example of the case study application in a metropolitan area that uses this monitoring system aided in achieving a very professional drop in the costs of maintenance and extension of the life span of the bridge because of proactive fixing which can take place after an early diagnosis.

On the same note, WSNs have also been used in monitoring the status of the environment in urban settings. Here, the sensors have been distributed throughout the city to gauge the condition of the air, temperature, humidity, and the traffic. Letting the system combine data on different monitors, the system was able to give good information about the level of pollution that could be referred to when planning an urban environment. The traffic real time information enabled optimization of the timing of the traffic lights to minimize traffic and travel times. Furthermore, air quality data were used to select those regions that have a high pollution level, which occurred due to the implementation of specific mitigation strategies. This helped to build a healthier and cleaner city and helped to minimize the environmental damage caused by the gases generated by traffic.

Figure 2. Stress vs. vibration for structural health monitoring

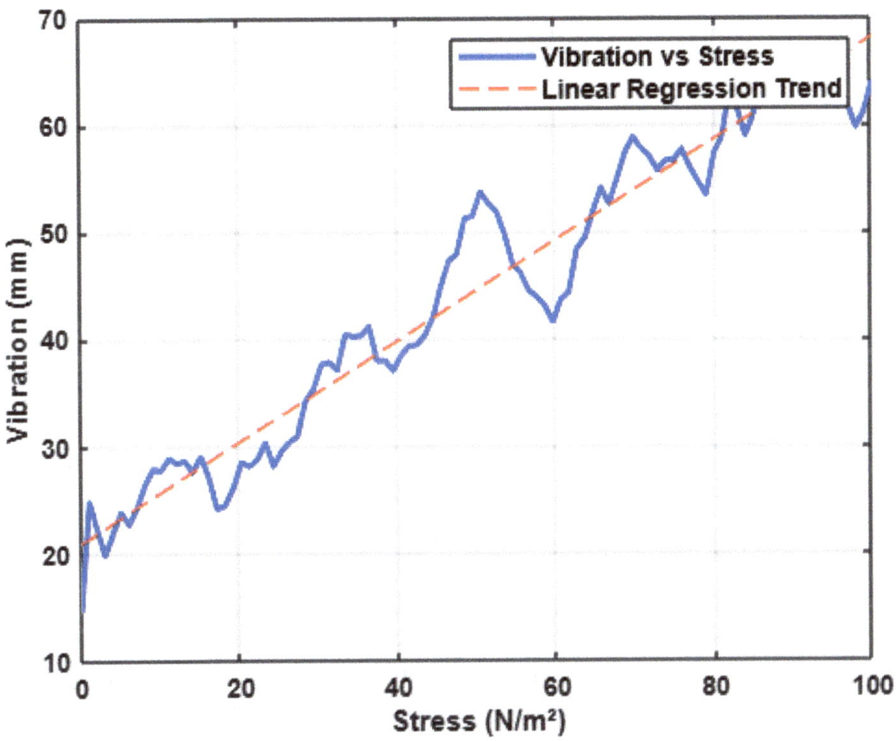

As shown in Figure 2, which shows the correlation of stress and vibration, the WSNs may be used to identify the early symptoms of structural degradation. It exhibits a definite trend in the plot that as the stress goes up, there is increase in vibration which is normal to the structures in strain. The linear regression line that was plotted in the data gathers a model to predict vibration in terms of stress, which is imperative in predicting an action on an infrastructure system in time. The data in this figure is smoothed to eliminate the noise, which simulates real-world situations which the large-scale monitoring system will have to deal with noisy sensor data. This applies to the bridge monitoring system case study where constant recording of stress and vibration enables the engineers to make interventions at an early stage where stress will not hit dangerous levels.

Figure 3. Temperature vs. humidity for environmental monitoring

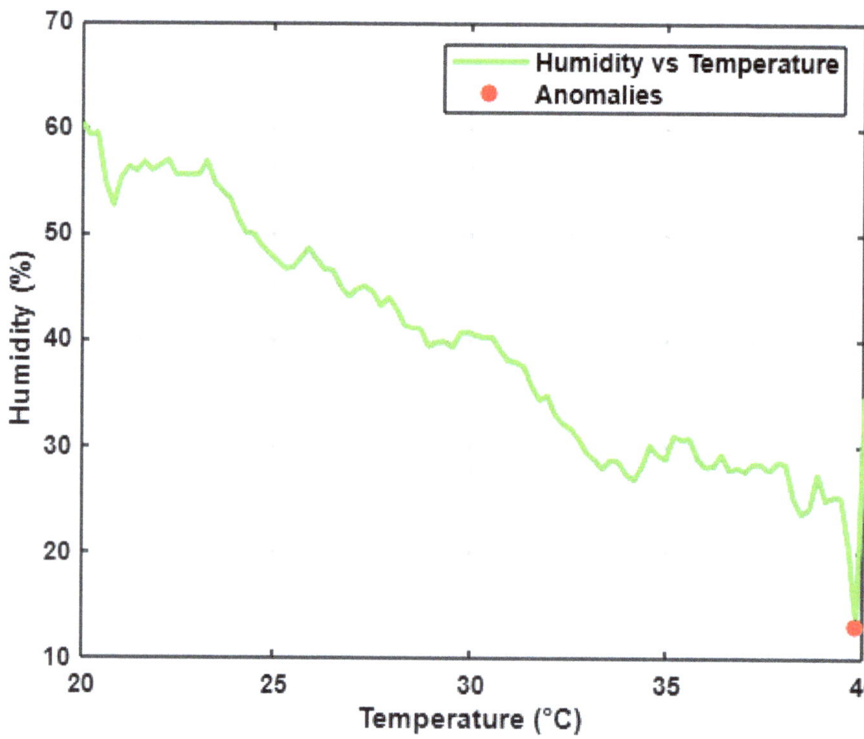

In Figure 3 the results of the temperature vs. humidity recorded by environmental sensors show how the temperature can influence the level of humidity, but a smoothed data makes the relationships clearer. There is also a representation of this figure together with the results of anomaly detection, where points of the graph which majorly do not support the expected direction are highlighted as anomalies. The deviation in this sense may include bizarre weather trends or wrongfulness in sensors, making actual time monitoring and anomaly detection a key requirement. The case study about environmental monitoring in cities demonstrated how the same methods might be used to efficiently handle the aspects of environmental management in regard to safety and health-safety of the population such as monitoring the environmental processes that impact the infrastructure such as humidity changes and temperature changes that might cause the destruction of materials.

Figure 4. Traffic flow vs. time for transportation management

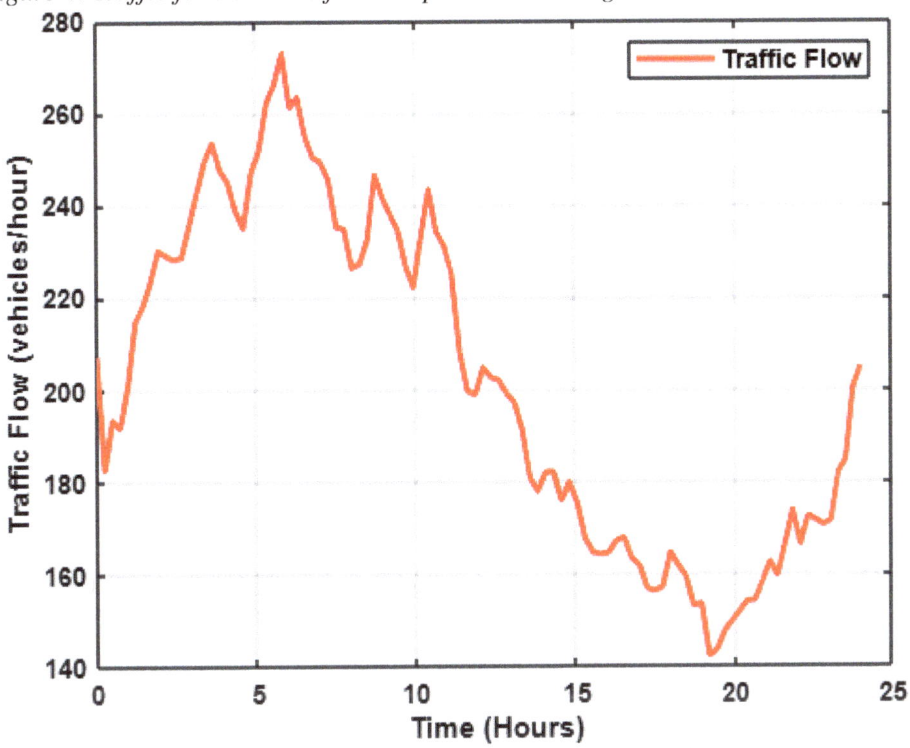

In Figure 4, the data of traffic flow versus time of an urban monitoring system is given. The plot features the changes of traffic patterns on a daily basis as well as during peak hours to a greater degree of traffic. The short-run variations in the data are smoothed and the moving average filter allows knowing better about the overall trends of traffic. Anomalies identified within data represent an immediate congestion or irregular traffic, which is reported to be analysed more intensively. In infrastructure monitoring system of a smart city, this data plays an important role in the optimization and decongestion of the traffic flow by varying the timings of the traffic signals and direction of the traffic with real-time information updates. The case study proved the effectiveness in terms of improving travelling time and fuel consumption.

Figure 5. Water flow vs. pressure for water management

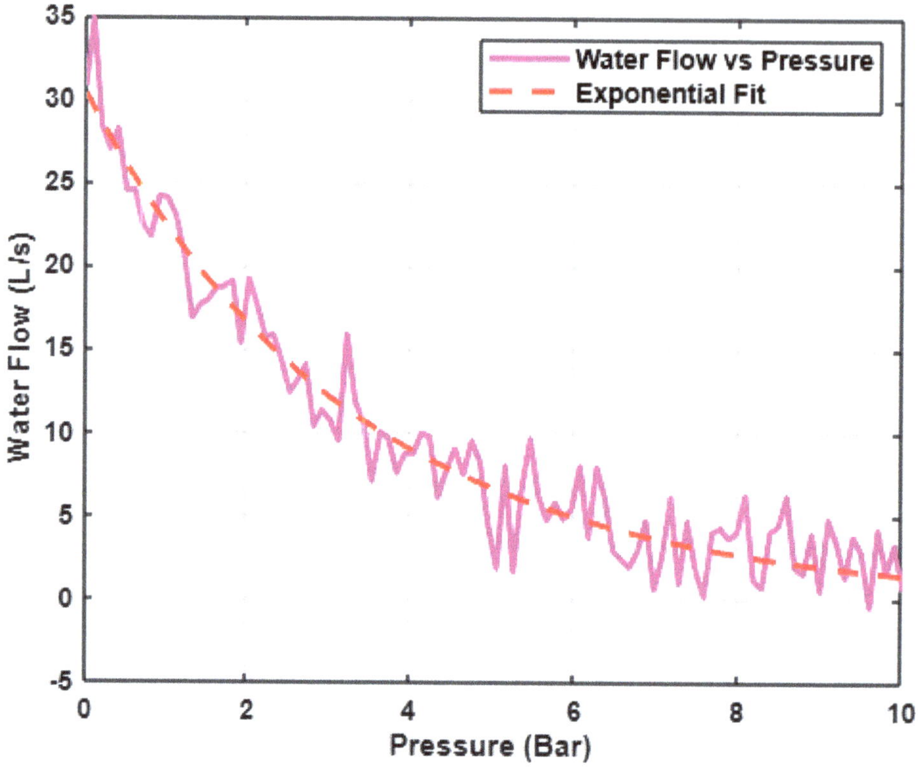

As Figure 5 depicts, there exists a water flow vs. pressure relationship model, which is exhibit by the exponential decay model, which depicts the variation in pressure of a water pipeline in relationship to the rate of flow of a water pipeline. The exponential decay model is also fitted on the data in this figure and it explains clearly the real life performance of the water distribution systems. Realistically, WSNs implemented in the water systems assist in early detection of pressures or leaks and help save huge amounts of water by avoiding costly repair costs. A case scenario of monitoring of water distribution networks has revealed benefits of monitoring water pipes using WSNs to identify leakages before the pipelines cause massive waste of water or cause breakdown of the system.

Figure 6 displays power consumption vs. time graph in sensor nodes in a WSN showing that the power consumption declined exponentially with time. A smoothing of the data is done and notable anomalies connected with low power consumption are indicated. Such analysis is important to consider sensor networks that are located in remote place where it can be expensive and uneconomical to change batteries. The power study case study of sensor networks had indicated that by implementing energy efficient designs such as the employment of energy harvesting and duty cycling, the lifetime of sensor nodes could be increased and hence guarantee that the network does not lead to degeneration just after some time needs to be maintained. This is especially relevant toward long term infrastructure surveillance where the sensor nodes must be placed in difficult to access regions.

Figure 6. Power consumption vs. time for sensor nodes in WSNs

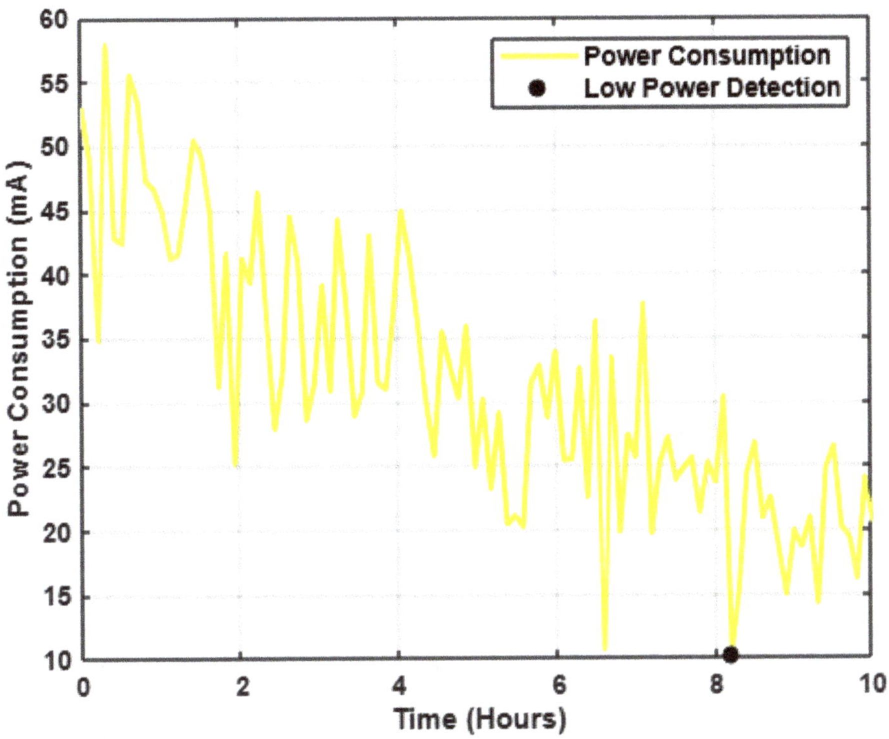

Performance Metrics

WSNs are assessed using a number of quantifying performance parameters in order to confirm their suitability in infrastructure monitoring tasks. Another important measure is the network lifetime which is important because usually sensor nodes are running on battery. Energy harvesting activities like use of solar energy or vibrational energy have been proved to dramatically enhance network life in case studies that have been done on the monitoring of bridge and water networks. In such instances, incorporation of renewable sources resulted to 50 percent increment in network life over the usual battery powered power grid. Another important performance indicator is latency, especially in real-time systems in which fast data analysis and decision making is critical. Traffic flow monitoring system, as the example, managed to realize latencies of down to 100 milliseconds, thus traffic signal manipulation could be made virtually real-time to avoid the congestion. The reliability of WSNs is gauged in terms of the parameters such as packet delivery ratio whereby research results indicate delivery ratios of up to 98 percent in urban and remote infrastructure networks. This is a very high reliability, which is necessary where infrastructure use is involved; in the event that the network fails, this might result in dire consequences, which in this case, might include unobserved structural failures or the unmonitored traffic situation. Scalability also poses crucial concern especially within an infrastructure system that may be considerably large. Additional sensors can be introduced that do not need to be modified radically, and both the case studies of smart city infrastructure and water management systems proved that WSNs could scale well to supporting thousands of nodes without a performance drop. The accuracy of the data is a concern in the infrastructure monitoring systems. The accuracy of vibration sensors in structural health monitoring systems ranged between 5 per cent of the correct measure in real-life tests, implying engineers make sound decisions basing their information on such observations.

Challenges and Limitations

Regardless of the numerous WSNs benefits, a number of challenges have to be met. Signal interference can be a major concern especially in heavily trafficked wireless regions in cities or electromagnetic interference regions such as industrial regions. To prevent such an occurrence, highly advanced communication protocols like LoRa and LPWAN are being developed but with a trade-off on the speed of data transmission and longer ranges of data coverage. Another issue affecting the accuracy of data is the sensor drift that occurs with time. WSNs use sensors that get degraded especially under adverse conditions. Some process can be done to reduce this problem (like redundancy in sensors or regular recalibration); however, the problem is an issue in long term deployments. Network performance can be an issue at the point of data overload and most especially when handling high frequency data with several sensors. Good data fusion and compression methods are also required in which the amount of data contained can be reduced and only the relevant information gets sent across. Finally, consumption of power is a key issue in WSNs especially in the application of long term infrastructure proctoring. Power-efficient hardware such as duty cycling, energy harvesting, and low-power communication protocols are needed to sustain functions in the network beyond the need to conduct regular repair of the network. Finally, although WSNs have the huge potential in monitoring the infrastructure, opportunities have to be addressed on the problems of signal interference and sensor drift, data overflow, and power consumption to enhance their efficiency in the intended use. The current research at reducing the energy consumption, protocol through communication and accuracy of sensors can lead to overcoming these limitations and also raise employment of WSNs in infrastructure monitoring systems.

CONCLUSION

The WSN based networks (WSNs) have been shown as a revolutionary solution to real time infrastructure monitoring. With greater efficiency, reliability and cost effectiveness, the WSNs have almost become a revolution in infrastructure monitoring. Their use in the surveillance of vital facilities like bridges, water system and traffic networks has been found priceless as it can identify defects early enough, help avert breakages, and allow modification of maintenance plans. By continuously gathering and analysing data WSNs enable anticipatory decision-making that improves the safety and life-time of infrastructure systems. Nevertheless, issues like signal interference, sensor drift, data overload, and power consumption are still to be met to guarantee efficiency of the mentioned networks in the long run. Nevertheless, with continuous development through communication protocols, energy-efficient designs, and data processing methods, there have been an increasing number of instances of performance and scalability enhancement of the WSNs. In conclusion, WSNs have led to a shake-up in monitoring infrastructure due to smarter and greener systems. As WSNs are currently being improved and enhanced progressively, their implementation will enjoy a more efficient deployment process, leading to safer and more reliable infrastructure in the future. Possible savings, minimized maintenance, and enhanced safety is what highlights the increased significance of WSNs in the contemporary management of infrastructure, which guarantees them key position in future infrastructure design.

REFERENCES

Abdulbaqi, A. S., Abd alla, I. H. A., Jabur, Y., Jassim, S. A. J., Mohammed, M., Sulaiman, B. H. S., Alsultan, Q. H. A., Shawqi, O. S. A. S., Mohameed, O., Abed, T. H. A., Panessai, I. Y., Khalaf, H. A., & Nejrs, S. M. N. (2023, July). Innovative Control Strategies for Dynamic Load Management in Smart Grid Techniques Incorporating Renewable Energy Sources. *KHWARIZMIA, 2023*, 73–83. DOI: 10.70470/KHWARIZMIA/2023/007

Aborujilah, A., Wang, H., Niu, Y., Ji, C., & Yang, Y. (2023). The Role of Circular Economy in Achieving Sustainable Development Goals (SDGs): An Integrative Framework. *ESTIDAMAA, 2023*, 1–8. DOI: 10.70470/ESTIDAMAA/2023/001

Alshami, A., Ali, E., Elsayed, M., Eltoukhy, A. E. E., & Zayed, T. (2024). IoT Innovations in Sustainable Water and Wastewater Management and Water Quality Monitoring: A Comprehensive Review of Advancements, Implications, and Future Directions. *IEEE Access : Practical Innovations, Open Solutions, 12*, 58427–58453. DOI: 10.1109/ACCESS.2024.3392573

Bezukladnikov, I. I., & Gavrilov, A. V. (2016). "Problem of network monitoring in distributed wireless sensor networks," *2016 IEEE NW Russia Young Researchers in Electrical and Electronic Engineering Conference (EIConRusNW)*, St. Petersburg, Russia, 2016, pp. 146-150. DOI: 10.1109/EIConRusNW.2016.7448141

Bhavani, M., Vasuki, R., Solomon, F. E., Prasath, T. M., & Divya, G. (2024). "Smartguard: An IoT-Enabled Real-Time Monitoring System for Urban Manhole Safety," *2024 Third International Conference on Artificial Intelligence, Computational Electronics and Communication System (AICECS)*, MANIPAL, India, 2024, pp. 1-6. DOI: 10.1109/AICECS63354.2024.10957142

Bhushan, B., Pushparajesh, V., Bethanney Janney, J., Sahoo, G. S., Aggarwal, P., & Das, D. (2025). "Real-Time Monitoring of Structural Health with Wireless Sensor Networks in IoT Infrastructure," *2025 International Conference on Automation and Computation (AUTOCOM)*, Dehradun, India, 2025, pp. 742-747. DOI: 10.1109/AUTOCOM64127.2025.10956443

Chatterjee, S., ... (2017). "Internet of Things and Body area network-an integrated future," *2017 IEEE 8th Annual Ubiquitous Computing, Electronics and Mobile Communication Conference (UEMCON)*, New York, NY, USA, 2017, pp. 396-400. DOI: 10.1109/UEMCON.2017.8249094

Chowdhury Onkon, M. N., Rahman, M. T., Islam, K. M., Zishan, A.-S., Jany, M. R., & Shufian, A. (2024). "Industrial Water Quantity, Quality Measurement, and Power Consumption Optimization of Wireless Sensor Network," *2024 27th International Conference on Computer and Information Technology (ICCIT)*, Cox's Bazar, Bangladesh, 2024, pp. 3414-3419. DOI: 10.1109/ICCIT64611.2024.11022594

Fu, C., & Ni, Z. (2015). "The application of embedded system in Supervisory Control and Data Acquisition System (SCADA) over wireless sensor and GPRS networks," *2015 IEEE 9th International Conference on Anti-counterfeiting, Security, and Identification (ASID)*, Xiamen, China, 2015, pp. 81-85. DOI: 10.1109/ICASID.2015.7405666

Hamada, H. M., & Ghdhban, I. I. (2024). A Review on the Impact of Fly Ash on the Resistance of Ultra-High Performance Concrete to Acid and Sulfate Attacks. *ESTIDAMAA, 2024*, 7–14. DOI: 10.70470/ESTIDAMAA/2024/002

Hussein, N. A.-H. K., Khalaf, S., & Shah, K. D. (2024). Advanced Composite Materials for Sustainable Construction: Innovations in Civil Engineering Applications. *KHWARIZMIA, 2024*, 8–16. DOI: 10.70470/KHWARIZMIA/2024/003

Kamruzzaman, S. M., Jaseemuddin, M., Fernando, X., & Moeini, P. (2017). "Wireless Positioning Sensor Network Integrated with Cloud for Industrial Automation," *2017 IEEE 42nd Conference on Local Computer Networks (LCN)*, Singapore, 2017, pp. 543-546. DOI: 10.1109/LCN.2017.68

Lam, S. (2016). "Wireless infrastructure enhances on body health monitoring systems," *2016 IEEE MIT Undergraduate Research Technology Conference (URTC)*, Cambridge, MA, USA, 2016, pp. 1-2. DOI: 10.1109/URTC.2016.8284064

Levy, M., & Hallstrom, J. O. (2017). "A new approach to data center infrastructure monitoring and management (DCIMM)," *2017 IEEE 7th Annual Computing and Communication Workshop and Conference (CCWC)*, Las Vegas, NV, USA, 2017, pp. 1-6. DOI: 10.1109/CCWC.2017.7868412

Liu, Y., & Yang, L. (2024). "Application Research of Wireless Sensor Network Technology in Concrete Pavement Quality Monitoring," *2024 IEEE 2nd International Conference on Control, Electronics and Computer Technology (ICCECT)*, Jilin, China, 2024, pp. 174-178. DOI: 10.1109/ICCECT60629.2024.10545849

Monshi, M. M., & Mohammed, O. A. (2013). "A study on the efficient wireless sensor networks for operation monitoring and control in smart grid applications," *2013 Proceedings of IEEE Southeastcon*, Jacksonville, FL, USA, 2013, pp. 1-5. DOI: 10.1109/SECON.2013.6567493

Naikade, K., & Raman, R. (2023). "Enhancing Shopping Experience with Raspberry Pi and Cloud-Integrated Wireless Sensor Networks for Smart Trolleys in IoT," *2023 International Conference on Self Sustainable Artificial Intelligence Systems (ICSSAS)*, Erode, India, 2023, pp. 1086-1090. DOI: 10.1109/ICSSAS57918.2023.10331668

O'Donoughue, N., Kulkarni, S., & Marzella, D. (2006). "Design and Implementation of a Framework for Monitoring Patients in Hospitals Using Wireless Sensors in Ad Hoc Configuration," *2006 International Conference of the IEEE Engineering in Medicine and Biology Society*, New York, NY, USA, 2006, pp. 6449-6452. DOI: 10.1109/IEMBS.2006.260162

Orhan, I., & Lindh, T. (2010). "Measurement-Based Admission Control in Wireless Sensor Networks," *2010 Eleventh International Conference on Mobile Data Management*, Kansas City, MO, USA, 2010, pp. 426-431. DOI: 10.1109/MDM.2010.51

Orhan, I., & Lindh, T. (2010). "Measurement-Based Admission Control in Wireless Sensor Networks," *2010 Fourth International Conference on Sensor Technologies and Applications*, Venice, Italy, 2010, pp. 447-452. DOI: 10.1109/SENSORCOMM.2010.71

Patil, P. K., & Patil, S. R. (2017). "Review on structural health monitoring system using WSN for bridges," *2017 International conference of Electronics, Communication and Aerospace Technology (ICECA)*, Coimbatore, India, 2017, pp. 628-631. DOI: 10.1109/ICECA.2017.8203615

Setijadi, E.. (2013). "Design of large scale structural health monitoring system for long-span bridges based on wireless sensor network," *2013 International Joint Conference on Awareness Science and Technology & Ubi-Media Computing (iCAST 2013 & UMEDIA 2013)*, Aizu-Wakamatsu, Japan, 2013, pp. 169-174. DOI: 10.1109/ICAwST.2013.6765428

Suryawanshi, S. R., & Deshpande, U. L. (2017). "Review of risk management for landslide forecasting, monitoring and prediction using wireless sensors network," *2017 International Conference on Innovations in Information, Embedded and Communication Systems (ICIIECS)*, Coimbatore, India, 2017, pp. 1-6. DOI: 10.1109/ICIIECS.2017.8276113

Wang, J.-J., & Wang, S. (2010). "Wireless sensor networks for Home Appliance Energy Management based on ZigBee technology," *2010 International Conference on Machine Learning and Cybernetics*, Qingdao, China, 2010, pp. 1041-1046. DOI: 10.1109/ICMLC.2010.5580627

Xue, Y., Ramamurthy, B., & Burbach, M. (2010). "A two-tier Wireless Sensor Network infrastructure for large-scale real-time groundwater monitoring," *IEEE Local Computer Network Conference*, Denver, CO, USA, 2010, pp. 874-881. DOI: 10.1109/LCN.2010.5735826

Zhu, L., Xiong, Y., & Yang, X. (2021). "Dynamic Calculation Model for Pressure Safety Monitoring of Municipal Water Supply Network with Interference Elimination Function between Neighboring Nodes," *2021 IEEE 5th Advanced Information Technology, Electronic and Automation Control Conference (IAEAC)*, Chongqing, China, 2021, pp. 1777-1781. DOI: 10.1109/IAEAC50856.2021.9390949

Compilation of References

Aarif, K. O. M., Alam, A., & Hotak, Y. (2025). Smart sensor technologies shaping the future of precision agriculture: Recent advances and future outlooks. *Journal of Sensors, 2025*(1), 2460098. DOI: 10.1155/js/2460098

Abbasi, S., Rahmani, A. M., Balador, A., & Sahafi, A. (2021). Internet of Vehicles: Architecture, services, and applications. *International Journal of Communication Systems, 34*(10), e4793. Advance online publication. DOI: 10.1002/dac.4793

Abdelillah, F. M., Nora, H., Samir, O., & Benslimane, S. M. (2023) "Predictive Maintenance Approaches in Industry 4.0: A Systematic Literature Review," *2023 IEEE International Conference on Enabling Technologies: Infrastructure for Collaborative Enterprises (WETICE)*, Paris, France, 2023, pp. 1-6.

Abdelraheem, M., Abdelhafeez, M., & Nassr, A. (2021). IoT-Based Interdigital Capacitance Sensing System for Damage Detection in CFRP-Concrete Structures. *IEEE Access : Practical Innovations, Open Solutions, 9*, 138658–138667. DOI: 10.1109/ACCESS.2021.3118299

Abdul Qader, M. A. Q., & Nassif, I. A. (2003). *Modern trends in development.* Alexandria University Press.

Abdulbaqi, A. S., Abd alla, I. H. A., Jabur, Y., Jassim, S. A. J., Mohammed, M., Sulaiman, B. H. S., Alsultan, Q. H. A., Shawqi, O. S. A. S., Mohameed, O., Abed, T. H. A., Panessai, I. Y., Khalaf, H. A., & Nejrs, S. M. N. (2023, July). Innovative Control Strategies for Dynamic Load Management in Smart Grid Techniques Incorporating Renewable Energy Sources. *KHWARIZMIA, 2023*, 73–83. DOI: 10.70470/KHWARIZMIA/2023/007

Abdulbaqi, A. S., Alsultan, Q. H., Nejrs, S. M., Abd alla, I. H., jabur, Y., Jassim, S. A., Mohammed, M., Sulaiman, B. H., Shawqi, O. S. A., Mohameed, O., Abed, T. H., Panessai, I. I. Y., & Khalaf, H. A. (2023, May). Design and Fabrication of Bio-Inspired Robotic Systems for Developed Mobility and Functionality in Unstructured Environments. *KHWARIZMIA, 2023*, 46–62. DOI: 10.70470/KHWARIZMIA/2023/005

Abdulbaqi, A. S., Jassim, S. A. J., Sulaiman, B. H. S., Alsultan, Q. H. A., Abed, T. H. A., Panessai, I. Y., & Nejrs, S. M. N. (2023). Innovative control strategies for dynamic load management in smart grid techniques incorporating renewable energy sources. *Khwarizmia, 2023*, 73–83.

Abdulkareem, K. R., & Sonmez, E. (2024). Architectural Harmony with Nature. *KHWARIZMIA, 2024*, 4–7. DOI: 10.70470/KHWARIZMIA/2024/002

Abdullah Hamad, A., Murugabharathi, P., & Hasan Khalaf, H. (2025). A 5D chaotic macroeconomic model for simulating policy synchronization in developing economies. *Heritage and Sustainable Development, 7*(2), 771–782. DOI: 10.37868/hsd.v7i2.1386

Abdullahi, A., & Turan, O. (2022). Adaptive control strategies for smart structural systems: A review. *Smart Materials and Structures, 31*(5), 1–18.

Abdullahi, U., Zhang, W., Cao, Y., & Irankunda, G. (2025). Integrating IoT Technology for Fire Risk Monitoring and Assessment in Residential Building Design. *Buildings (Basel, Switzerland), 15*(8), 1346. DOI: 10.3390/buildings15081346

Aborujilah, A., Wang, H., Niu, Y., Ji, C., & Yang, Y. (2023). The Role of Circular Economy in Achieving Sustainable Development Goals (SDGs): An Integrative Framework. *ESTIDAMAA, 2023*, 1–8. DOI: 10.70470/ESTIDAMAA/2023/001

Abuseta, Y., & Swesi, K. (2015). Design patterns for self adaptive systems engineering. arXiv preprint arXiv:1508.01330.

Acemoğlu, D., Autor, D., Hazell, J., & Restrepo, P. (2022). Artificial Intelligence and Jobs: Evidence from Online Vacancies. *Journal of Labor Economics, 40*(S1), S293–S340. Advance online publication. DOI: 10.1086/718327

Addula, S. R., Mamodiya, U., Jiang, W., & Almaiah, M. A. (2025). Generative AI-enhanced intrusion detection framework for secure healthcare networks in MANETs. *SHIFRA, 2025*, 62–68. DOI: 10.70470/SHIFRA/2025/003

Agarwal, M., Narayan, S., & Bhattacharya, K. (2019). Design of intelligent self-adaptive structures using embedded sensing and actuation. *Journal of Intelligent Material Systems and Structures, 30*(15), 2240–2255.

Ahmad, I. (2016). A comparative analysis of local meshless formulation for multi-asset option models. *Engineering Analysis with Boundary Elements*, *65*, 159–176. DOI: 10.1016/j.enganabound.2015.12.020

Ahmad, I., Zulfiqar, A., Shabbir, M., Khan, J. S., Çalışkan, H., & Hong, H. (2025). Advanced sensors, monitoring, and control systems for environmental sustainability. In *Smart Cities and Sustainable Manufacturing* (pp. 47–55). Elsevier. DOI: 10.1016/B978-0-443-26474-0.00007-5

Ahmed, S., Kurnaz, S., Hamdi, M. M., Khaleel, A. M., Khaleel, A. M., & Seno, M. E. (2022) "Study for Buildings with IoT System for Energy Management," *2022 International Symposium on Multidisciplinary Studies and Innovative Technologies (ISMSIT)*, Ankara, Turkey, pp. 53-57. DOI: 10.1109/ISMSIT56059.2022.9932729

Ahsan, M., Bohner, M., Ullah, A., Khan, A. A., & Ahmad, S. (2023). A Haar wavelet multi-resolution collocation method for singularly perturbed differential equati-ons with integral boundary conditions. *Mathematics and Computers in Simulation*, *204*, 166–180. DOI: 10.1016/j.matcom.2022.08.004

Ahsan, M., Shams-ul Haq, K., Liu, X., Ahmad Lone, S., & Nisar, M. (2023). A Haar wavelets based approximation for nonlinear inverse problems influenced by unknown heat source. *Mathematical Methods in the Applied Sciences*, *46*(2), 2475–2487. DOI: 10.1002/mma.8655

Akerson, A., Bourdin, B., & Bhattacharya, K. (2022). Optimal design of responsive structures. *Structural and Multidisciplinary Optimization*, *65*(4), 111.

Akinola, A., & Emuze, F. (2020). Future of construction: An overview of emerging technologies. *Journal of Engineering. Defence Technology*, *18*(4), 813–830. DOI: 10.1108/JEDT-11-2019-0317

Alghamedy, F., El-Haggar, N., Alsumayt, A., Alfawaer, Z. M., Alshammari, M., Amouri, L., Aljameel, S. S., & Al-Bassam, S. (2024). Unlocking a Promising Future: Integrating Blockchain Technology and FL-IoT in the Journey to 6G. *IEEE Access : Practical Innovations, Open Solutions*, *1*, 115411–115447. Advance online publication. DOI: 10.1109/ACCESS.2024.3435968

Ali, A. (2025). The role of joint technological innovation between countries in achieving the green economy (Experiences of selected countries). *Mesopotamia Journal for Humanities and Social Sciences, 7*(1).

Ali, A. H. A., Gardner, G. A., & Gardner, L. R. T. (1992). A collocation solution for Burgers' equation using cubic B-spline finite elements. *Computer Methods in Applied Mechanics and Engineering*, *100*(3), 325–337. DOI: 10.1016/0045-7825(92)90088-2

Ali, A., & Ghanem, M. C. (2025). Beyond detection: Large language models and next-generation cybersecurity. *SHIFRA, 2025*, 81–97. DOI: 10.70470/SHIFRA/2025/005

Al-Kawaz, A. (2004). *Economic policies and human capital*. Arab Planning Institute.

Almotairi, K. H. (2022). Application of internet of things in healthcare domain.

Alotaibi, R. M., & Khan, S. (2023) "Big Data and Predictive Data Analytics in the Smes Industry Using Machine Learning Approach," 2023 6th International Conference on Contemporary Computing and Informatics (IC3I), Gautam Buddha Nagar, India, 2023, pp. 2212-2217. DOI: 10.1109/IC3I59117.2023.10397688

Alshami, A., Ali, E., Elsayed, M., Eltoukhy, A. E. E., & Zayed, T. (2024). IoT Innovations in Sustainable Water and Wastewater Management and Water Quality Monitoring: A Comprehensive Review of Advancements, Implications, and Future Directions. *IEEE Access : Practical Innovations, Open Solutions*, *12*, 58427–58453. DOI: 10.1109/ACCESS.2024.3392573

Alsultan, Q. H., Abdulbaqi, A. S., Nejrs, S. M., Abd alla, I. H., Jabur, Y., Jassim, S. A., Mohammed, M., Sulaiman, B. H., Shawqi, O. S. A., Mohameed, O., Abed, T. H., Panessai, I. Y., & Khalaf, H. A. (2023). Innovative Composite Materials for Improving Structural Integrity and Longevity in Civil Engineering Applications. *KHWARIZMIA*, *2023*, 63–72. DOI: 10.70470/KHWARIZMIA/2023/006

Alzahrani, A. I. A., Chauhdary, S. H., & Alshdadi, A. A. (2023). Internet of Things (IoT)-Based Wastewater Management in Smart Cities. *Electronics (Basel)*, *12*(12), 2590. DOI: 10.3390/electronics12122590

Alzubi, K. M., Salah Alaloul, W., Al Salaheen, M., Hannan Qureshi, A., Musarat, M. A., & Mohsen Alawag, A. (2022) "Reviewing the Applications of Internet of Things in Construction Projects," *2022 International Conference on Decision Aid Sciences and Applications (DASA)*, Chiangrai, Thailand, pp. 169-173. DOI: 10.1109/DASA54658.2022.9765143

Amare, N. (2024). Modeling and Analyzing Random Data Using Markov Processes and Their Applications in Prediction. *Tamjeed Journal of Healthcare Engineering and Science Technology*, *2*(2), 47–55.

Amirkhanova, G., Amirkhanov, B., Tyulepberdinova, G., & Ishmurzin, T. (2024) "Application of Machine Learning Algorithms in Digital Twin Monitoring Systems: An Overview of Approaches, Methods, and Prospects," 2024 International Conference on Intelligent Computing and Next Generation Networks (ICNGN), Bangkok, Thailand, 2024, pp. 01-05. DOI: 10.1109/ICNGN63705.2024.10871832

Andersen P. A., Goodwin M., & Granmo O. C. (2018, August). Deep RTS: A game environment for deep reinforcement learning in real-time strategy games. In 2018 IEEE conference on computational intelligence and games (CIG) (pp. 1–8). IEEE.

Apte, C. (2010). The role of machine learning in business optimization (Paper presentation). The 27th International Conference on Machine Learning (ICML-10) (pp. 1–2).

Arquilla, V., & Vitali, I. (2016). Designing in the IoT Era: role and perspectives in design practices. In Systems & Design: Beyond Processes and Thinking (pp. 871-882). EDITORIAL UNIVERTITAT POLITÈCNICA DE VALÈNCIA.

Aryan, A., Singh, M., Labhade-Kumar, N., Radha, M., Ranga, J., & Rathod, N. (2024, April). Experimental design of electricity theft detection and alert system using Arduino-assisted controller and smart sensors. In *2024 International Conference on Inventive Computation Technologies (ICICT)* (pp. 1961-1968). IEEE. (Kush J. C. 2025)Assimakopoulos, F., Vassilakis, C., Margaris, D., Kotis, K., & Spiliotopoulos, D. (2025). AI and related technologies in the fields of smart agriculture: A review. *Information (Basel)*, *16*(2), 100.

Asdrubali, F., Baldassarri, C., & Fthenakis, V. (2013). Life cycle analysis in the construction sector: Guiding the optimization of conventional Italian buildings. *Energy and Building*, *64*, 73–89. DOI: 10.1016/j.enbuild.2013.04.018

Askar, S., Saadatzi, M. N., & Jafari, M. (2021). Bio-inspired self-responsive structures: Concepts, mechanisms, and applications. *Materials Today. Advances*, *11*, 100150.

Atanassov, K. T. (1986). Intuitionistic fuzzy sets. *Fuzzy Sets and Systems*, *20*(1), 87–96. DOI: 10.1016/S0165-0114(86)80034-3

Azeem, M., Shahid, M., Masin, I., & Petru, M. (2025). Design and development of textile-based wearable sensors for real-time biomedical monitoring; a review. *Journal of the Textile Institute*, *116*(1), 80–95. DOI: 10.1080/00405000.2024.2318500

Azimi, M., Eslamlou, A., & Pekcan, G. (2020). Data-Driven Structural Health Monitoring and Damage Detection through Deep Learning: State-of-the-Art Review. *Sensors (Basel)*, *20*(10), 2778. DOI: 10.3390/s20102778 PMID: 32414205

Aziz, I., & Ahmad, M. (2015). Numerical solution of two-dimensional Elliptic PDEs with nonlocal boundary conditions. *Computers &. Mathematics for Applications*, *69*(3), 180–205.

Aziz, I., & Nisar, M. (2016). An efficient numerical algorithm based on Haar wavelet for solving a class of linear and nonlinear nonlocal boundary-value problems. *Calcolo*, *53*(4), 621–633. DOI: 10.1007/s10092-015-0165-9

Babiceanu, R. F. (2019). A framework for Internet of Things-enabled smart government: A case of IoT cybersecurity policies and use cases in U.S. federal government. https://www.sciencedirect.com/science/article/pii/S0740624X17304847

Bagchi, S., Abdelzaher, T., Govindan, R., Shenoy, P., Atrey, A., Ghosh, P., & Xu, R. (2020). New Frontiers in IoT: Networking, Systems, Reliability, and Security Challenges. arXiv (Cornell University). https://doi.org//arxiv.2005.07338DOI: 10.48550

Baidya, K., Roy, A., & Das, K. (2024). A review of polymer-matrix piezoelectric composite coatings for energy harvesting and smart sensors. *Journal of Coatings Technology and Research, 21*(1), 55–85. DOI: 10.1007/s11998-023-00819-x

Bertolini, L., Elsener, B., Pedeferri, P., Redaelli, E., & Polder, R. B. (2013). Corrosion of Steel in Concrete. DOI: 10.1002/9783527651696

Bezukladnikov, I. I., & Gavrilov, A. V. (2016). "Problem of network monitoring in distributed wireless sensor networks," *2016 IEEE NW Russia Young Researchers in Electrical and Electronic Engineering Conference (EIConRusNW)*, St. Petersburg, Russia, 2016, pp. 146-150. DOI: 10.1109/EIConRusNW.2016.7448141

Bharany, S., Almogren, A., & Altameem, A. (2025). Optimizing IoT connectivity: A quantitative exploration of the comprehensive adaptive sensing and clustering system for smart sensor networks in smart cities. *Wireless Personal Communications, 140*(1-2), 1–23. DOI: 10.1007/s11277-024-11719-7

Bhavani, M., Vasuki, R., Solomon, F. E., Prasath, T. M., & Divya, G. (2024). "Smartguard: An IoT-Enabled Real-Time Monitoring System for Urban Manhole Safety," *2024 Third International Conference on Artificial Intelligence, Computational Electronics and Communication System (AICECS)*, MANIPAL, India, 2024, pp. 1-6. DOI: 10.1109/AICECS63354.2024.10957142

Bhushan, B., Pushparajesh, V., Bethanney Janney, J., Sahoo, G. S., Aggarwal, P., & Das, D. (2025). "Real-Time Monitoring of Structural Health with Wireless Sensor Networks in IoT Infrastructure," *2025 International Conference on Automation and Computation (AUTOCOM)*, Dehradun, India, 2025, pp. 742-747. DOI: 10.1109/AUTOCOM64127.2025.10956443

Bilgen, O., Kochersberger, K., Inman, D. J., & Ohanian, O. J.III. (2017). Morphing wing aerodynamics using smart materials: Experimental and numerical study. *AIAA Journal, 55*(12), 4113–4126.

Biradar, V., Kakeri, D., & Agasti, A. (2024) "Machine Learning based Predictive Maintenance in Distribution Transformers," 2024 8th International Conference on Computing, Communication, Control and Automation (ICCUBEA), Pune, India, 2024, pp. 1-5. DOI: 10.1109/ICCUBEA61740.2024.10774993

Bisio, C. Garibotto, F. Lavagetto, and A. Sciarrone, (2022) "A Novel IoT-based Edge Sensing Platform for Structure Health Monitoring," IEEE INFOCOM 2022 - IEEE Conference on Computer Communications Workshops (INFOCOM WKSHPS), New York, NY, USA, 2022, pp. 1-6.

Bocken, N. M. P., Short, S. W., Rana, P., & Evans, S. (2014). A literature and practice review to develop sustainable business model archetypes. *Journal of Cleaner Production*, *65*, 42–56. DOI: 10.1016/j.jclepro.2013.11.039

Bojja, S., Rajitha, A., Aravinda, K., Nagpal, A., Kalra, R., & Radi, U. K. (2024) "Advanced Machine Learning Techniques for Data Prediction in WSNs," 2024 OPJU International Technology Conference (OTCON) on Smart Computing for Innovation and Advancement in Industry 4.0, Raigarh, India, 2024, pp. 1-6. DOI: 10.1109/OTCON60325.2024.10687955

Bouziane, S., Aghoutane, B., Moumen, A., Sahlaoui, A., & Ouali, A. E. (2021). Proposal of a Big data System for an Intelligent Management of Water Resources. E3S Web of Conferences, 314, 2002. DOI: 10.1051/e3sconf/202131402002

Brischke, C., & Rapp, A. O. (2008). Influence of wood moisture content and wood temperature on fungal decay in the field: Observations in different micro-climates. *Wood Science and Technology*, *42*(8), 663–677. DOI: 10.1007/s00226-008-0190-9

Broto, J. M., Cipriano, J., Mor, G., Gavaldà, O., Masoumi-Verki, S., & Eicker, U. (2025) "An interoperable ontology-based information model for better integration of building physics and IoT data analytics models," 2025 IEEE/ACM 7th International Workshop on Software Engineering Research & Practices for the IoT (SERP4IoT), Ottawa, ON, Canada, 2025, pp. 1-8. DOI: 10.1109/SERP4IoT66600.2025.00005

Bruggemann, T. (2022, April). Automated Feature-Driven Flight Planning for Airborne Inspection of Large Linear Infrastructure Assets. *IEEE Transactions on Automation Science and Engineering*, *19*(2), 804–817. DOI: 10.1109/TASE.2021.3062154

Buyle, M., Braet, J., & Audenaert, A. (2013). Life cycle assessment in the construction sector: A review. *Renewable & Sustainable Energy Reviews*, *26*, 379–388. DOI: 10.1016/j.rser.2013.05.001

Car, M., Markovic, L., Ivanovic, A., Orsag, M., & Bogdan, S. (2020). Autonomous Wind-Turbine Blade Inspection Using LiDAR-Equipped Unmanned Aerial Vehicle. *IEEE Access : Practical Innovations, Open Solutions, 8*, 131380–131387. DOI: 10.1109/ACCESS.2020.3009738

Caron, M., Bojanowski, P., Joulin, A., & Douze, M. (2018). Deep clustering for unsupervised learning of visual features [Paper presentation]. *The European Conference on Computer Vision (ECCV)* (pp. 132–149).

Chand, N., Mahalleh, V., Aziz, T., Rahman, A., Mohammed, A., & Zeid, W. (2021) "Small Scale Localized Maintenance of Industrial Infrastructure using Autonomous UAVs," *2021 IEEE International Conference on Advanced Robotics and Its Social Impacts (ARSO)*, Tokoname, Japan, 2021, pp. 168-175. DOI: 10.1109/ARSO51874.2021.9542838

Chang, Y. (2019) "Study of Improving Teaching Materials for Barrier-free Facility Inspector Training in Taiwan," 2019 IEEE Eurasia Conference on IOT, Communication and Engineering (ECICE), Yunlin, Taiwan, 2019, pp. 125-128. DOI: 10.1109/ECICE47484.2019.8942669

Chatterjee, S., ... (2017). "Internet of Things and Body area network-an integrated future," *2017 IEEE 8th Annual Ubiquitous Computing, Electronics and Mobile Communication Conference (UEMCON)*, New York, NY, USA, 2017, pp. 396-400. DOI: 10.1109/UEMCON.2017.8249094

Chatterjee, S. (2020). The safety of IoT-enabled system in smart cities of India: Do ethics matter? *International Journal of Ethics and Systems, 36*(4), 601–618. DOI: 10.1108/IJOES-05-2019-0085

Chaturvedi, K. Prasad, A. K. N, B. Ashreetha, and C. H. Priya, (2025) "Optimizing Environmental Monitoring in Smart Urban Areas with IoT-Enabled Wireless Sensor Networks," 2025 5th International Conference on Pervasive Computing and Social Networking (ICPCSN), Salem, India, 2025, pp. 455-460.

Chen, X. (2015). Interactive pavillions: Responsive transformation of structure systems.

Chen, Y., Zhang, J., Mu, X., & Ma, J. (2010) "Study on the theoretical framework of the internet of building energy systems," 5th International Conference on Computer Sciences and Convergence Information Technology, Seoul, Korea (South), 2010, pp. 973-976.

Chib, S., Mehta, A. K., Selvakumar, P., Saravanan, A., Mishra, B. R., & Manjunath, T. C. (2025). Mental Health Challenges and Solutions in High-Pressure Work Environments. In Harriott, S. (Ed.), *Innovative Approaches to Managing Conflict and Change in Diverse Work Environments* (pp. 247–272). IGI Global Scientific Publishing., DOI: 10.4018/979-8-3693-9556-1.ch011

Choi, Y. J., Park, H. W., Mi, Y., & Song, S. (2024). Crack Detection and Analysis of Concrete Structures Based on Neural Network and Clustering. *Sensors (Basel), 24*(6), 1725. DOI: 10.3390/s24061725

Choudhary, V., Guha, P., Pau, G., & Mishra, S. (2025). An Overview of Smart Agriculture Using Internet of Things (IoT) and Web Services. *Environmental and Sustainability Indicators, 100607*, 100607. Advance online publication. DOI: 10.1016/j.indic.2025.100607

Chowdhury Onkon, M. N., Rahman, M. T., Islam, K. M., Zishan, A.-S., Jany, M. R., & Shufian, A. (2024). "Industrial Water Quantity, Quality Measurement, and Power Consumption Optimization of Wireless Sensor Network," *2024 27th International Conference on Computer and Information Technology (ICCIT)*, Cox's Bazar, Bangladesh, 2024, pp. 3414-3419. DOI: 10.1109/ICCIT64611.2024.11022594

Cimellaro, G. P., Soong, T. T., & Reinhorn, A. M. (2009). Integrated Design of Smart Structures. *Advances in Science and Technology (Owerri, Nigeria), 56*, 127–136.

Cogato, A., Sozzi, M., Nikolić, N., & Laroche-Pinel, E. (2025). Smart horticulture: Latest advances and prospects. *Horticulturae, 11*(1), 27. DOI: 10.3390/horticulturae11010027

Costa Monobi, F. M., Martins, A. C., Cerento de Lyra, P. H., Roberto, J. M., del Bel Sonoda, L., & Coelho Leite, B. C. (2024) "Development of an IoT Based Sensing System for Structural Health Monitoring Using Low-Cost Accelerometers," *2024 Symposium on Internet of Things (SIoT)*, Rio de Janeiro, Brazil, 2024, pp. 1-5. DOI: 10.1109/SIoT63830.2024.10780766

D'Aniello, G., Gaeta, M., Gravina, R., Li, Q., Rehman, Z. U., & Fortino, G. (2024). Situation identification in smart wearable computing systems based on machine learning and context space theory. *Information Fusion, 104*, 102197. DOI: 10.1016/j.inffus.2023.102197

Dahiya, R., & Selvakumar, P. Vidyadhara, & Manjunath, T. C. (2025). Design and aerodynamics: Principles and advances. In M. Mellal (Ed.), *Innovations and Developments in Unmanned Aerial Vehicles* (pp. 145-170). IGI Global Scientific Publishing.

Dahiya, R., & Selvakumar, P. Vidyadhara, & Manjunath, T. C. (2025). Design and Aerodynamics: Principles and Advances. In M. Mellal (Ed.), Innovations and Developments in Unmanned Aerial Vehicles (pp. 145-170). IGI Global Scientific Publishing. DOI: 10.4018/979-8-3693-8462-6.ch007

de Oliveira, L. D., De Araújo, J. V. S., Da Silva, J. H. B., Villanueva, J. M. M., de Souza Filho, C. A., & Ochoa, M. N. (2023) "Smart Water Management: a Self-Sufficient IoT-Based Application for Pressure and Flow Monitoring in Water Distribution Systems," 2023 7th International Symposium on Instrumentation Systems, Circuits and Transducers (INSCIT), Rio de Janeiro, Brazil, 2023, pp. 1-6. DOI: 10.1109/INSCIT59673.2023.10258484

Devasena, D., Ramya, R., Dharshan, Y., Vivek, S. S., & Darshan, T. (2019). IoT based water distribution system. *International Journal of Engineering and Advanced Technology*, *8*(6), 132–135. DOI: 10.35940/ijeat.D6295.088619

Devlin, S., Yliniemi, L., Kudenko, D., & Tumer, K. (2014, May). Potential-based difference rewards for multiagent reinforcement learning [Paper presentation]. The 2014 International Conference on Autonomous Agents and Multi-agent Systems (pp. 165–172). https://web.engr.oregonstate.edu/~ktumer/publications/files/tumer

Dezfouli, A., & Balleine, B. W. (2012). Habits, action sequences and reinforcement learning. *The European Journal of Neuroscience*, *35*(7), 1036–1051. DOI: 10.1111/j.1460-9568.2012.08050.x PMID: 22487034

Dhirani, L. L., Mukhtiar, N., Chowdhry, B. S., & Newe, T. (2023). Ethical Dilemmas and Privacy Issues in Emerging Technologies: A Review, Sensors, 23(3), 1151. Multidisciplinary Digital Publishing Institute. DOI: 10.3390/s23031151

Dimitropoulos, N., ... (2021) "Forecasting of short-term PV production in energy communities through Machine Learning and Deep Learning algorithms," 2021 12th International Conference on Information, Intelligence, Systems & Applications (IISA), Chania Crete, Greece, 2021, pp. 1-6. DOI: 10.1109/IISA52424.2021.9555544

Ding, G. K. C. (2008). Sustainable construction—The role of environmental assessment tools. *Journal of Environmental Management*, *86*(3), 451–464. DOI: 10.1016/j.jenvman.2006.12.025 PMID: 17289255

Dinh, L. T., & Zhou, Q. (2023). A probabilistic approach for fault tree and event tree modeling under incomplete information. *Reliability Engineering & System Safety*, *226*, 108748.

Dirin, A., Oliver, I., & Laine, T. H. (2023). A Security Framework for Increasing Data and Device Integrity in Internet of Things Systems. *Sensors (Basel), 23*(17), 7532. DOI: 10.3390/s23177532

Dokl, M., Van Fan, Y., Vujanović, A., Pintarič, Z. N., Aviso, K. B., Tan, R. R., & Čuček, L. (2024). A waste separation system based on sensor technology and deep learning: A simple approach applied to a case study of plastic packaging waste. *Journal of Cleaner Production, 450*, 141762. DOI: 10.1016/j.jclepro.2024.141762

Du, B., Borhan, U. M., Chen, T., Chen, J., Li, J., & Chen, J. (2024) "Robust UAV Policy Learning for Urban Infrastructure Surface Screening," *2024 International Conference on Advanced Robotics and Mechatronics (ICARM)*, Tokyo, Japan, 2024, pp. 1-8. DOI: 10.1109/ICARM62033.2024.10715841

Dubois, D., & Prade, H. (1998). Possibility theory: Qualitative and quantitative aspects. In Gabbay, D., & Smets, P. (Eds.), *Handbook of Defeasible Reasoning and Uncertainty Management Systems* (Vol. 1, pp. 169–226). Springer., DOI: 10.1007/978-94-017-1735-9_6

Du, J. (2018, April). Understanding of object detection based on CNN family and YOLO. [). IOP Publishing.]. *Journal of Physics: Conference Series, 1004*, 012029. DOI: 10.1088/1742-6596/1004/1/012029

Dziedzic, R.. (2021) "A framework for asset management planning in sustainable and resilient cities," *2021 IEEE International Symposium on Technology and Society (ISTAS)*, Waterloo, ON, Canada, 2021, pp. 1-10. DOI: 10.1109/ISTAS52410.2021.9629158

Eastman, C., Teicholz, P., Sacks, R., & Liston, K. (2018). *BIM Handbook: A guide to building information modeling for owners, designers, engineers, contractors, and facility managers* (3rd ed.). Wiley.

Elkhoukhi, H., Bakhouya, M., Elmouatamid, A., El Ouadghiri, D., & Gaber, J. (2025). A review of occupancy sensing technologies and approaches in smart buildings. *International Journal of RF Technologies: Research and Applications, 15*(1), 19–48. DOI: 10.3233/RFT-240006

Eneyew, D., Capretz, M. A. M., & Bitsuamlak, G. T. (2022). Toward Smart-Building Digital Twins: BIM and IoT Data Integration. *IEEE Access : Practical Innovations, Open Solutions, 10*, 130487–130506. DOI: 10.1109/ACCESS.2022.3229370

European Commission. (2020). *A new ERA for research and innovation*. https://research-and-innovation.ec.europa.eu

Fachantidis, A., Partalas, I., Tsoumakas, G., & Vlahavas, I. (2013). Transferring task models in reinforcement learning agents. *Neurocomputing*, *107*, 23–32. DOI: 10.1016/j.neucom.2012.08.039

Fagnant, D. J., & Kockelman, K. (2015). Preparing a nation for autonomous vehicles: Opportunities, barriers and policy recommendations. *Transportation Research Part A, Policy and Practice*, *77*, 167–181. DOI: 10.1016/j.tra.2015.04.003

Faramarzi, M. (2020). Road damage detection and classification using deep neural networks (YOLOv4) with smartphone images.

Fiore, E. (2017, August). Design Ethics in socio-technical systems: Addressing the ethics of connected appliances. In *Relating Systems Thinking and Design Symposium (RSD)*.

Fokaides, P., Jurelionis, A., & Spudys, P. (2022) "Boosting Research for a Smart and Carbon Neutral Built Environment with Digital Twins (SmartWins)," 2022 IEEE International Smart Cities Conference (ISC2), Pafos, Cyprus, 2022, pp. 1-4. DOI: 10.1109/ISC255366.2022.9922513

Fountaine, T., McCarthy, B., & Saleh, T. (2019). Building the AI-powered organization. *Harvard Business Review*, *97*(4), 62–73.

Fu, C., & Ni, Z. (2015). "The application of embedded system in Supervisory Control and Data Acquisition System (SCADA) over wireless sensor and GPRS networks," *2015 IEEE 9th International Conference on Anti-counterfeiting, Security, and Identification (ASID)*, Xiamen, China, 2015, pp. 81-85. DOI: 10.1109/ICASID.2015.7405666

Gan, V. J. L., Wang, T., Hu, D., & Yin, C. (2021) "Developing an Integrated Data Model based on IFC and SensorML for Post-COVID Facility Management," 2021 IEEE 23rd Int Conf on High Performance Computing & Communications; 7th Int Conf on Data Science & Systems; 19th Int Conf on Smart City; 7th Int Conf on Dependability in Sensor, Cloud & Big Data Systems & Application (HPCC/DSS/SmartCity/DependSys), Haikou, Hainan, China, 2021, pp. 1845-1850. DOI: 10.1109/HPCC-DSS-SmartCity-DependSys53884.2021.00273

Gao, X., & Pishdad-Bozorgi, P. (2019). BIM-enabled facilities operation and maintenance: A review. *Advanced Engineering Informatics*, *39*, 227–247. DOI: 10.1016/j.aei.2019.01.005

Garg, S., Mehrotra, D., Pandey, H. M., & Pandey, S. (2021). Accessible review of internet of vehicle models for intelligent transportation and research gaps for potential future directions. *Peer-to-Peer Networking and Applications, 14*(2), 978–1005. DOI: 10.1007/s12083-020-01054-6

Gayathri, M., & Kavitha, V. (2023) "Deep Learning-based Cluster Analysis for Healthcare Data Classification," *2023 International Conference on Data Science, Agents & Artificial Intelligence (ICDSAAI)*, Chennai, India, 2023, pp. 1-6. DOI: 10.1109/ICDSAAI59313.2023.10452610

Gervasio, H., Dimova, S., & Pinto, A. (2018). Benchmarking the Life-Cycle Environmental Performance of Buildings. *Sustainability (Basel), 10*(5), 1454. DOI: 10.3390/su10051454

Ghafoor, A., Haq, S., Hussain, M., & Arifeen, S. (2019). Numerical solutions of two dimensional Sobolev and generalized Benjamin–Bona–Mahony–Burgers equationns via Haar wavelets. *Computers & Mathematics with Applications (Oxford, England), 77*(2), 565–575. DOI: 10.1016/j.camwa.2018.09.058

Ghosh, S., & Chatterjee, S. (2023). Artificial intelligence in civil engineering: A systematic literature review. *Automation in Construction, 148*, 104753. DOI: 10.1016/j.autcon.2023.104753

Ghosh, S., Das, N., Das, I., & Maulik, U. (2019). Understanding deep learning techniques for image segmentation. *ACM Computing Surveys, 52*(4), 1–35. DOI: 10.1145/3329784

Gil-Castilla, M., Poma, A. R., Caballero, A., & Ollero, A. (2025) "A Cooperative Multi-UAV Framework for Bridge Inspection," *2025 International Conference on Unmanned Aircraft Systems (ICUAS)*, Charlotte, NC, USA, 2025, pp. 408-415. DOI: 10.1109/ICUAS65942.2025.11007792

Godja, N. C., & Munteanu, F. D. (2024). Hybrid nanomaterials: A brief overview of versatile solutions for sensor technology in healthcare and environmental applications. *Biosensors (Basel), 14*(2), 67. DOI: 10.3390/bios14020067 PMID: 38391986

Gogo, K. O., Nderu, L., & Mwangi, R. W. (2018, November). Fuzzy logic based context aware recommender for smart e-learning content delivery. In 2018 5th International Conference on Soft Computing & Machine Intelligence (ISCMI) (pp. 114-118). IEEE. doi:DOI: 10.1109/ISCMI.2018.8703247

Goran, N., Muhić, S., & Begović, A. (2024) "An End-to-End Architecture for Smart Waste Management: Integrating Real-Time Data Processing, IoT, and Machine Learning," 2024 International Symposium ELMAR, Zadar, Croatia, 2024, pp. 157-160. DOI: 10.1109/ELMAR62909.2024.10694555

Guinée, J. (2001). Handbook on life cycle assessment — Operational guide to the ISO standards. *The International Journal of Life Cycle Assessment, 6*(5), 255–255. DOI: 10.1007/BF02978784

Gulzar, M., Ashraf, S., & Kerre, E. E. (2024). Application of complex fuzzy relational compositions to medical diagnosis. *Mathematics, 12*(23), 3729. DOI: 10.3390/math12233729

Guo, B., Duan, S., Wang, P., Lei, H., Zhao, Z., & Fang, D. (2025). A review on the deformation tracking methods in vision-based tactile sensing technology. *Acta Mechanica Sinica, 41*(10), 424436. DOI: 10.1007/s10409-024-24436-x

Gupta, S., Sharma, P., Sharma, D., Gupta, V., & Sambyal, N. (2020). Detection and localization of potholes in thermal images using deep neural networks. *Multimedia Tools and Applications, 79*, 26265-26284.

Hafen, R., & Critchlow, T. (2013) "EDA and ML — A Perfect Pair for Large-Scale Data Analysis," *2013 IEEE International Symposium on Parallel & Distributed Processing, Workshops and PhD Forum*, Cambridge, MA, USA, 2013, pp. 1894-1898. DOI: 10.1109/IPDPSW.2013.118

Hamad, A. A., Abdulridha, M. M., Kadhim, N. M., Pushparaj, S., Meenakshi, R., & Ibrahim, A. M. (2022). Learning methods of business intelligence and group related diagnostics on patient management by using artificial dynamic system. *Journal of Nanomaterials, 2022*(1), 4891601.

Hamad, A. M., Issa, H. M., Salih, R. A., Keighobadi, J., Alias, A. B., Hasan, R. A., Saleh, N. M., Ali Mohammed Saleh, A. M., Hamada, H. M., & Ghdhban, I. I. (2024). A Review on the Impact of Fly Ash on the Resistance of Ultra-High Performance Concrete to Acid and Sulfate Attacks. *ESTIDAMAA, 2024*, 7–14. DOI: 10.70470/ESTIDAMAA/2024/002

Hameed, M. H., Shawkat, S. A., & Al-badri, K. S. L. (2020). Multi bands metamaterial absorber optimized by genetic algorithm in microwave regime. In AIP Conference Proceedings (Vol. 2213, No. 1, p. 020055). AIP Publishing LLC. DOI: 10.1063/5.0000075

Hament, B., & Oh, P. (2018) "Unmanned Aerial and Ground Vehicle (UAV-UGV) System Prototype for Civil Infrastructure Missions," *2018 IEEE International Conference on Consumer Electronics (ICCE)*, Las Vegas, NV, USA, 2018, pp. 1-4. DOI: 10.1109/ICCE.2018.8326346

Hammad, O. M., Smaoui, I., Fakhfakh, A., & Hashim, M. M. (2024). Recent advances in digital image masking techniques: Future challenges and trends—A review. *SHIFRA, 2024*, 67–73. DOI: 10.70470/SHIFRA/2024/008

Haven, E., Liu, X., & Shen, L. (2012). De-noising option prices with the wavelet method. *European Journal of Operational Research, 222*(1), 104–112. DOI: 10.1016/j.ejor.2012.04.020

He, J., Baxter, S. L., Xu, J., Xu, J., Zhou, X., & Zhang, K. (2019). The practical implementation of artificial intelligence technologies in medicine. *Nature Medicine, 25*(1), 30–36. DOI: 10.1038/s41591-018-0307-0 PMID: 30617336

Hensel, M., & Menges, A. (2017). Designing morpho-ecological adaptive structures: Biomimetic principles in architecture. *Architectural Design, 87*(6), 28–35.

He, X., Aizenberg, M., Kuksenok, O., Zarzar, L. D., Shastri, A., Balazs, A. C., & Aizenberg, J. (2012). Synthetic homeostatic materials with chemo-mechano-chemical self-regulation. *Nature, 487*(7406), 214–218.

Holmes, G., Donkin, A., & Witten, I. H. (2020). Weka: A machine-learning workbench [Paper presentation]. The ANZIIS'94-Australian New Zealand Intelligent Information Systems Conference (pp. 357–361). IEEE.

Hooda, Y., Singh, H. D., Mehta, J., Passi, A., & Mehta, S. (2023) "An Interdisciplinary Approach for Evaluation of Green Building Index," *2023 International Conference on Advances in Computation, Communication and Information Technology (ICAICCIT)*, Faridabad, India, 2023, pp. 509-512. DOI: 10.1109/ICAICCIT60255.2023.10465795

Hoque, F., Saha, D. K., & Rakshit, D. (2022). "IoT-Based Intelligent Solar Energy Harvesting Technique with Improved Efficiency," in Artificial Intelligence and Internet of Things for Renewable Energy Systems. *De Gruyter, 2022*, 279–302.

Hossain, J., Faisal, S. M. F., Lubaba, R. F. K., Hassan, M., Namey, A., & Maliha, M. H. (2024) "An Advance Machine Learning Framework for Predicting Machine Breakdown in The Steel Industry," 2024 IEEE International Conference on Computing, Applications and Systems (COMPAS), Cox's Bazar, Bangladesh, 2024, pp. 1-5. DOI: 10.1109/COMPAS60761.2024.10796308

Hsiao, C. H. (2004). Haar wavelet approach to linear stiff systems. *Mathematics and Computers in Simulation, 64*(5), 561–567. DOI: 10.1016/j.matcom.2003.11.011

Hundekari, S., Mittal, T., Dutt, A., Bhavana, M., Kumar, R., & Nijhawan, G. (2025) "AI-Driven Solutions for Predictive Maintenance in Smart Transportation Systems," *2025 International Conference on Computational, Communication and Information Technology (ICCCIT)*, Indore, India, 2025, pp. 572-577. DOI: 10.1109/ICCCIT62592.2025.10927971

Hussein, N. A.-H. K., Khalaf, S., & Shah, K. D. (2024). Advanced Composite Materials for Sustainable Construction: Innovations in Civil Engineering Applications. *KHWARIZMIA*, *2024*, 8–16. DOI: 10.70470/KHWARIZMIA/2024/003

Ibrahim, A. A. A. (2024). Features and functional components in the use of artificial intelligence in human capital development (higher education as a model). *ESTIDAMAA*, *2024*, 1–6. DOI: 10.70470/ESTIDAMAA/2024/001

International Energy Agency. (2021). *World Energy Outlook 2021*. https://www.iea.org/reports/world-energy-outlook-2021

Jacobs, N., Marković, M., Cottrill, C., Edwards, P., Corsar, D., & Salt, K. (2020). Made-Up Rubbish: Design Fiction as a Tool for Participatory Internet of Things Research. *The Design Journal*, *23*(3), 419–440. DOI: 10.1080/14606925.2020.1744259

Jasim, J. A., Hilal Mutlag, A., Othman, N. A., Shareef, H., & Hameed, N. H. (2024) "Innovative Applications of AI and IoT in Smart Urban Infrastructure and Housing Projects," *2024 International Conference on Emerging Research in Computational Science (ICERCS)*, Coimbatore, India, 2024, pp. 1-6. DOI: 10.1109/ICERCS63125.2024.10894956

Jiang, S., & Liu, F. F. (2022). Decision support analysis for risk identification and control of patients affected by COVID-19 based on Bayesian networks. *Expert Systems with Applications*, *197*, 116754.

Julia, J. P. M., & Bennet, D. (2022) "Efficient Machine Learning Algorithms for Medical Big Data Analysis," *2022 International Conference on Applied Artificial Intelligence and Computing (ICAAIC)*, Salem, India, 2022, pp. 795-800. DOI: 10.1109/ICAAIC53929.2022.9792790

K. Kalaiselvi, K. Niranjana, V. Prithivirajan, K. S. Kumar, V. Syambabu and S. B, (2024) "Machine Learning for Predictive Maintenance in Industrial Equipment: Challenges and Application," 2024 4th Asian Conference on Innovation in Technology (ASIANCON), Pimari Chinchwad, India, 2024, pp. 1-6.

Kadar, H. H., Sameon, S. S., & Rusli, M. E. (2018). SMART2L: Smart Water Level and Leakage Detection. *IACSIT International Journal of Engineering and Technology*, *7*(4.36), 448–452. DOI: 10.14419/ijet.v7i4.36.23914

Kadhim, N. M., & Hamad, A. A. (2024). Mathematical Foundations and Principles Behind These Methods. In Coding Dimensions and the Power of Finite Element, Volume, and Difference Methods (pp. 116-133). IGI Global.

Kadhim, N. M., Mohammed, H. A., Radhawi, S. N., Jabur, A. M., Gottraan, R. B., Abdulridha, M. M., & Mohammed, Z. Q. (2023). Investigation of the next generation science standards including in the science book according to E-learn: Analytical study. *Tamjeed Journal of Healthcare Engineering and Science Technology*, *1*(2), 30–35.

Kamaruidzaman, N. S., & Rahmat, S. N. (2020). *Water Monitoring System Embedded with Internet of Things (IoT) Device: A Review IOP Conference Series Earth and Environmental Science, 498(1), 12068.* IOP Publishing., DOI: 10.1088/1755-1315/498/1/012068

Kamruzzaman, S. M., Jaseemuddin, M., Fernando, X., & Moeini, P. (2017). "Wireless Positioning Sensor Network Integrated with Cloud for Industrial Automation," *2017 IEEE 42nd Conference on Local Computer Networks (LCN)*, Singapore, 2017, pp. 543-546. DOI: 10.1109/LCN.2017.68

Kang, K., & Lee, S. (2018). Money, Cryptocurrency, and Monetary Policy. SSRN Electronic Journal. DOI: 10.2139/ssrn.3303595

Kaur, M., Selvakumar, P., & Dinesh, N. Mohit, Chandel, P. S., & Manjunath, T. C. (2025). Reconstruction and Preservation Strategies for Cultural Heritage. In S. Oflazoğlu & M. Dora (Eds.), Preserving Cultural Heritage in Post-Disaster Urban Renewal (pp. 385-412). IGI Global Scientific Publishing. DOI: 10.4018/979-8-3693-9097-9.ch016

Kawale, S. R., Prasad, K., Palanikkumar, D., & Mary, P. A. (2023) "A Novel IoT Framework and Device Architecture for Efficient Smart City Implementation," 2023 7th International Conference on Trends in Electronics and Informatics (ICOEI), Tirunelveli, India, 2023, pp. 420-426. DOI: 10.1109/ICOEI56765.2023.10125677

Kelly, J. T., Campbell, K. L., Gong, E., & Scuffham, P. (2020). The Internet of Things: Impact and Implications for Health Care Delivery. Journal of Medical Internet Research, 22(11). JMIR Publications. DOI: 10.2196/20135

Khadir, R. (2018). *The role of research and development in economic growth: Selected international experiences with reference to Iraq* (PhD thesis). University of Karbala.

Khalil, U., Malik, O. A., Uddin, M., & Chen, C. (2022). *A Comparative Analysis on Blockchain versus Centralized Authentication Architectures for IoT-Enabled Smart Devices in Smart Cities: A Comprehensive Review, Recent Advances, and Future Research Directions Sensors, 22(14), 5168.* Multidisciplinary Digital Publishing Institute., DOI: 10.3390/s22145168

Khan, M. U., Shah, S. A. A., Ahsan, M., & Alwuthaynani, M. (2024). Haar wavelet collocation method for linear and non-linear Schrödinger equation. *Physica Scripta*, *100*(1), 015234. DOI: 10.1088/1402-4896/ad9872

Khan, M., Raza, M. A., Abbas, G., Othmen, S., Yousef, A., & Jumani, T. A. (2024). Pothole detection for autonomous vehicles using deep learning: A robust and efficient solution. *Frontiers in Built Environment*, *9*, 1323792. Advance online publication. DOI: 10.3389/fbuil.2023.1323792

Khedr, M. A., & Abdelaziz, M. (2024) "AI-Driven Robotic System for Predictive Maintenance: Urban Road Defect Detection in Smart Cities," *2024 International Conference on Computer and Applications (ICCA)*, Cairo, Egypt, 2024, pp. 1-6. DOI: 10.1109/ICCA62237.2024.10927850

Kielion, L. J. (2008). Adaptability in Architecture.

Kim, N., Borthakur, D., & Saikia, M. J. (2024) "Machine Learning Approach for Music Familiarity Classification with Single-Channel EEG," 2024 46th Annual International Conference of the IEEE Engineering in Medicine and Biology Society (EMBC), Orlando, FL, USA, 2024, pp. 1-4. DOI: 10.1109/EMBC53108.2024.10782402

Klein, L., Joseph, Y., & Kröger, M. (2021). Product integration of established crash sensors for safety applications in lightweight vehicles. *Sensors (Basel)*, *21*(21), 6994.

Klein, P. (1996). Pricing Black-Scholes options with correlated credit risk. *Journal of Banking & Finance*, *20*(7), 1211–1229. DOI: 10.1016/0378-4266(95)00052-6

Knof, M., Stock-Homburg, R., & Schurer, J. (2024). How in-store sensor technologies can help retailers to understand their customers: Overview on two decades of research. *International Review of Retail, Distribution and Consumer Research*, *34*(3), 381–398. DOI: 10.1080/09593969.2023.2273256

Kommineni, M., Chundru, S., Maroju, P. K., & Selvakumar, P. (2025). Ethical Implications of AI in Sustainable Development Pedagogy. In Chinedu, C., Mafarja, N., Hussein, M., & Saleem, A. (Eds.), *Rethinking the Pedagogy of Sustainable Development in the AI Era* (pp. 17–36). IGI Global Scientific Publishing., DOI: 10.4018/979-8-3693-9062-7.ch002

Kotnik, T., Gramazio, F., & Kohler, M. (2020). Adaptive structures in architecture: Material systems and responsive design. *Automation in Construction, 112*, 103096.

Krishnamurthy, S., Thamilarasu, G., & Bauckhage, C. (2009) "MALADY: A Machine Learning-Based Autonomous Decision-Making System for Sensor Networks," *2009 International Conference on Computational Science and Engineering*, Vancouver, BC, Canada, 2009, pp. 93-100. DOI: 10.1109/CSE.2009.246

Kuehn, J., & Manoli, Y. (2019) "A Fingertip-Shaped Tactile Sensor with Machine-Learning-Based Sensor-To-Information Processing," 2019 20th International Conference on Solid-State Sensors, Actuators and Microsystems & Eurosensors XXXIII (TRANSDUCERS & EUROSENSORS XXXIII), Berlin, Germany, 2019, pp. 1811-1814. DOI: 10.1109/TRANSDUCERS.2019.8808413

Kumar, H., Shafiq, M., Kauhaniemi, K., & Elmusrati, M. (2024) "Artificial Intelligence-Based Condition Monitoring and Predictive Maintenance of Medium Voltage Cables: An Integrated System Development Approach," 2024 10th International Conference on Condition Monitoring and Diagnosis (CMD), Gangneung, Korea, Republic of, 2024, pp. 191-195. DOI: 10.23919/CMD62064.2024.10766104

Kumar, N., Haque, M. J., Mittal, A., Singh, A., Rayal, A., & Oli, S. (2023) "IoT Based Smart Weather Monitoring System," *2023 Annual International Conference on Emerging Research Areas: International Conference on Intelligent Systems (AICERA/ICIS)*, Kanjirapally, India, pp. 1-6.

Kumar, R., Jain, V., Leong, H. W., & Teyarachakul, S. (2023). Convergence of IoT, Blockchain, and Computational Intelligence in Smart Cities. In *CRC Press eBooks*. Informa., DOI: 10.1201/9781003353034

Lal Rajora, G., Sanz-Bobi, M. A., Domingo, C. M., & Tjernberg, L. B. (2025). An Open-Source Tool-Box for Asset Management Based on the Asset Condition for the Power System. *IEEE Access : Practical Innovations, Open Solutions, 13*, 49174–49186. DOI: 10.1109/ACCESS.2025.3551663

Lala, G., & Vugar, A. (2025). Application of IoT and sensor technologies in environmental monitoring. *Environmental Research and Ecotoxicity, 4*(4), 5. DOI: 10.56294/ere2025170

Lam, R. C. Y., Lam, L. C. H., Li, N. X., & Lee, P. K. K. (2021) "IoT-based Monitoring System to Prevent Collapse of Falsework Systems During Construction," *2021 International Conference on Computational Science and Computational Intelligence (CSCI)*, Las Vegas, NV, USA, 2021, pp. 1454-1459. DOI: 10.1109/CSCI54926.2021.00080

Lam, S. (2016). "Wireless infrastructure enhances on body health monitoring systems," *2016 IEEE MIT Undergraduate Research Technology Conference (URTC)*, Cambridge, MA, USA, 2016, pp. 1-2. DOI: 10.1109/URTC.2016.8284064

Lee, D. Y., Kim, N., & Kim, M.-S. (2022). Dynamic Bayesian network for fatigue-crack-related risk assessment of aerospace bolts. *Journal of Aerospace Information Systems, 19*(3), 145–158. DOI: 10.2514/1.I010964

Lee, D., Kim, Y., & Park, J. (2020). Integration of adaptive structures in smart city frameworks for resilient infrastructure. *Automation in Construction, 118*, 103283.

Lepik, Ü. (2006). Haar wavelet method for nonlinear integro-differential equations. *Applied Mathematics and Computation, 176*(1), 324–333. DOI: 10.1016/j.amc.2005.09.021

Levy, M., & Hallstrom, J. O. (2017). "A new approach to data center infrastructure monitoring and management (DCIMM)," *2017 IEEE 7th Annual Computing and Communication Workshop and Conference (CCWC)*, Las Vegas, NV, USA, 2017, pp. 1-6. DOI: 10.1109/CCWC.2017.7868412

Li, J., & Wu, H. (2024). Blockchain and deep learning technology for comprehensive improvement of transaction information quality. *Electronic Commerce Research*, •••, 1–23. DOI: 10.1007/s10660-024-09923-5

Liu, J., & Wu, C. (2024) "Data Fusion and Real-Time Monitoring of Intelligent Fire Protection System Based on Internet of Things," 2024 IEEE 7th International Conference on Automation, Electronics and Electrical Engineering (AUTEEE), Shenyang, China, 2024, pp. 504-509. DOI: 10.1109/AUTEEE62881.2024.10869713

Liu, Y., & Yang, L. (2024). "Application Research of Wireless Sensor Network Technology in Concrete Pavement Quality Monitoring," *2024 IEEE 2nd International Conference on Control, Electronics and Computer Technology (ICCECT)*, Jilin, China, 2024, pp. 174-178. DOI: 10.1109/ICCECT60629.2024.10545849

Liu, F., Wang, J., & Qian, X. (2017). Integrating phase change materials into concrete through microencapsulation using cenospheres. *Cement and Concrete Composites, 80*, 317–325. DOI: 10.1016/j.cemconcomp.2017.04.001

Liu, H. X., & Wu, B. (2006). Application and development of artificial intelligence technology in the precision forestry. *Xibei Linxueyuan Xuebao, 21*(1), 183.

Liu, K., & Chen, B. M. (2023, September). Industrial UAV-Based Unsupervised Domain Adaptive Crack Recognitions: From Database Towards Real-Site Infrastructural Inspections. *IEEE Transactions on Industrial Electronics, 70*(9), 9410–9420. DOI: 10.1109/TIE.2022.3204953

Liu, K., Wang, T., & Zhang, X. (2025). Intelligent Techniques Applied in Infrastructure, Engineering, and Construction. *Buildings (Basel, Switzerland)*, *15*(11), 1799. DOI: 10.3390/buildings15111799

Liu, Q., Huang, W., Duan, X., Wei, J., Hu, T., Yu, J., & Huang, J. (2023). DSW-YOLOv8n: A new underwater target detection algorithm based on improved YOLOv8n. *Electronics (Basel)*, *12*(18), 3892.

Liu, R., Ding, L., & Love, P. E. D. (2017). BIM and big data for construction cost management: A systematic review and future research directions. *Automation in Construction*, *72*, 143–156. DOI: 10.1016/j.automatica.2016.10.011

Liu, T., Mao, Y., Dou, H., Zhang, W., Yang, J., Wu, P., & Mu, X. (2025). Emerging wearable acoustic sensing technologies. *Advancement of Science*, •••, 2408653. PMID: 39749384

Liu, X., & Zhang, X. (2020). Real-time risk prediction using dynamic Bayesian networks in construction. *Automation in Construction*, *114*, 103148.

Liu, Y., He, C., Qiao, T., Liu, G., Li, X., Wan, Q., Zhu, Z., & He, Y. (2024). Coral-inspired hollow microneedle patch with smart sensor therapy for wound infection. *Advanced Functional Materials*, *34*(24), 2314071. DOI: 10.1002/adfm.202314071

Lu, H., Li, Y., Chen, M., Kim, H., & Serikawa, S. (2018). Brain intelligence: Go beyond artificial intelligence. *Mobile Networks and Applications*, *23*(2), 368–375. DOI: 10.1007/s11036-017-0932-8

Lumpe, T. S., Tao, M., Shea, K., & Levin, D. I. (2021). Computational Design of Active 3D-Printed Multi-State Structures for Shape Morphing. arXiv preprint arXiv:2108.13526.

Luo, H., Niu, H., Li, B., Lin, Z., Yu, P., Zhang, L., Zhang, D., & Guo, Q. (2025). AI-enhanced materials identification with smart directional tangential flexible sensor. *Advanced Materials Technologies*, *10*(8), 2401164. DOI: 10.1002/admt.202401164

Lu, Y., Li, H., Hu, H., Bi, Y., & Wang, P. (2025). A smart compensation method for large range eddy current displacement sensor. *Sensors and Actuators. A, Physical*, *383*, 116171. DOI: 10.1016/j.sna.2024.116171

Lu, Y., Xue, F., & Wang, M. (2020). Digital twin-enabled smart construction: Concepts, technologies and future directions. *Automation in Construction*, *112*, 103087. DOI: 10.1016/j.autcon.2020.103087

Lynch, C., Lyu, L., Buzzatto, J., Heshmati-Alamdari, S., & Liarokapis, M. (2024) "A Control Barrier Function Based Visual Servoing Framework for Safe Operation of UAVs in GNSS Denied Settings using Fiducial Markers," *2024 IEEE International Symposium on Safety Security Rescue Robotics (SSRR)*, New York, NY, USA, 2024, pp. 1-6. DOI: 10.1109/SSRR62954.2024.10770045

Ma, M., & Pang, H. (2023). SP-YOLOv8s: An improved YOLOv8s model for remote sensing image tiny object detection. applied sciences, 13(14), 8161.

Machalski, A., Szulc, P., Błoński, D., Nycz, A., Nemś, M., Skrzypacz, J., Janik, P., & Satława, Z. (2025). The Concept of a Digital Twin for the Wały Śląskie Hydroelectric Power Plant: A Case Study in Poland. *Energies*, *18*(8), 2021. DOI: 10.3390/en18082021

Mahmud, M. A., Bates, K., Wood, T., Abdelgawad, A., & Yelamarthi, K. (2018) "A complete Internet of Things (IoT) platform for Structural Health Monitoring (SHM)," 2018 IEEE 4th World Forum on Internet of Things (WF-IoT), Singapore, 2018, pp. 275-279. DOI: 10.1109/WF-IoT.2018.8355094

Majak, J., Shvartsman, B. S., Kirs, M., Pohlak, M., & Herranen, H. (2015). Convergence theorem for the Haar wavelet based discretization method. *Composite Structures*, *126*, 227–232. DOI: 10.1016/j.compstruct.2015.02.050

Mandal, S., Yadav, A., Panme, F. A., & Devi, K. M., & SM, S. K. (. (2024). Adaptation of smart applications in agriculture to enhance production. *Smart Agricultural Technology*, •••, 100431. DOI: 10.1016/j.atech.2024.100431

Maraveas, C., Loukatos, D., & Arvanitis, K. G. (2025). Intelligent sensors: Wireless sensor networks and Internet of Things. In *Nature-Derived Sensors* (pp. 295–339). Elsevier. DOI: 10.1016/B978-0-443-22002-9.00009-9

Mascret, Q., Gagnon-Turcotte, G., Bielmann, M., Fall, C. L., Bouyer, L. J., & Gosselin, B. (2022). A Wearable Sensor Network With Embedded Machine Learning for Real-Time Motion Analysis and Complex Posture Detection [April.]. *IEEE Sensors Journal*, *22*(8), 7868–7876. DOI: 10.1109/JSEN.2021.3139588

Mekid, S., Saheb, N., Khan, S. M., & Qureshi, K. K. (2015). Towards sensor array materials: Can failure be delayed? *Science and Technology of Advanced Materials*, *16*(3), 034607.

Mentzel, S., Grung, M., Tollefsen, K. E., Stenrød, M., Petersen, K., & Moe, S. J. (2022). Development of a Bayesian network for probabilistic risk assessment of pesticides. *Integrated Environmental Assessment and Management*, *18*(4), 1072–1087. DOI: 10.1002/ieam.4533 PMID: 34618406

Mheiri, A. S. A., Khan, W. A., & Aydin, R. (2024) "Rotary Machines Fault Detection and Diagnosis Using Machine Learning Approaches," *2024 IEEE International Conference on Technology Management, Operations and Decisions (ICTMOD)*, Sharjah, United Arab Emirates, 2024, pp. 1-6. DOI: 10.1109/ICTMOD63116.2024.10878108

Miller, M., Kisiel, A., Cembrowska-Lech, D., Durlik, I., & Miller, T. (2023). IoT in Water Quality Monitoring—Are We Really Here? *Sensors (Basel)*, *23*(2), 960. DOI: 10.3390/s23020960

Mittal, M., Alsalami, Z., Kumar, R., Boob, N. S., Verma, V., & Sangeeta, K. (2024) "Predictive Maintenance of Power Grid Infrastructure using Long Short-Term Memory Networks," 2024 International Conference on Communication, Computer Sciences and Engineering (IC3SE), Gautam Buddha Nagar, India, 2024, pp. 1611-1615. DOI: 10.1109/IC3SE62002.2024.10592911

Mohammad, N., Khatoon, R., Nilima, S. I., Akter, J., Kamruzzaman, M., & Sozib, H. M. (2024). Ensuring Security and Privacy in the Internet of Things: Challenges and Solutions. *Journal of Computer and Communications*, *12*(8), 257–277. DOI: 10.4236/jcc.2024.128016

Mohammed, W. (2024). Analysis of the economic impacts of technology and innovation on economic development: Research methodologies and their applications. *Journal of the Higher Institute for Qualitative Studies*, *4*(6).

Monshi, M. M., & Mohammed, O. A. (2013). "A study on the efficient wireless sensor networks for operation monitoring and control in smart grid applications," *2013 Proceedings of IEEE Southeastcon*, Jacksonville, FL, USA, 2013, pp. 1-5. DOI: 10.1109/SECON.2013.6567493

Montambault, S., Beaudry, J., Toussaint, K., & Pouliot, N. (2010) "On the Application of VTOL UAVs to the Inspection of Power Utility Assets," 2010 1st International Conference on Applied Robotics for the Power Industry, Montreal, QC, Canada, 2010, pp. 1-7. DOI: 10.1109/CARPI.2010.5624443

Mu, F., & Wang, Z. (2021) "Smart Fire Safety Early Warning System of Large Complex Building based on Data Collection and Processing," 2021 Fifth International Conference on I-SMAC (IoT in Social, Mobile, Analytics and Cloud) (I-SMAC), Palladam, India, 2021, pp. 1201-1204. DOI: 10.1109/I-SMAC52330.2021.9640862

Murroni, M., Anedda, M., Fadda, M., Ruiu, P., Popescu, V., Zaharia, C., & Giusto, D. (2023). 6G—Enabling the New Smart City: A Survey. Sensors, 23(17), 7528. Multidisciplinary Digital Publishing Institute. DOI: 10.3390/s23177528

Muttillo, M., Stornelli, V., Alaggio, R., Paolucci, R., Battista, L. D., de Rubeis, T., & Ferri, G. (2020). Structural Health Monitoring: An IoT Sensor System for Structural Damage Indicator Evaluation. *Sensors (Basel), 20*(17), 4908. DOI: 10.3390/s20174908

Nagarajan, P., Sathiya, D., Pulugu, D., & Pulugu, R. (2025). Self-powered wearable implantable smart sensor and medical electronics based on nanogenerator. In *Self-Powered Sensors* (pp. 59–75). Academic Press. DOI: 10.1016/B978-0-443-13792-1.00011-0

Naikade, K., & Raman, R. (2023). "Enhancing Shopping Experience with Raspberry Pi and Cloud-Integrated Wireless Sensor Networks for Smart Trolleys in IoT," *2023 International Conference on Self Sustainable Artificial Intelligence Systems (ICSSAS)*, Erode, India, 2023, pp. 1086-1090. DOI: 10.1109/ICSSAS57918.2023.10331668

Namazian, A., Yakhchali, S. H., Yousefi, V., & Tamosaitiene, J. (2019). Combining Monte Carlo Simulation and Bayesian Networks Methods for Assessing Completion Time of Projects under Risk. *International Journal of Environmental Research and Public Health, 16*(24), 5024. DOI: 10.3390/ijerph16245024 PMID: 31835528

Nanekaran, N. P., Esmalifalak, M., & Narimani, M. (2020) "Fast Anomaly Detection in Micro Data Centers Using Machine Learning Techniques," 2020 IEEE 18th International Conference on Industrial Informatics (INDIN), Warwick, United Kingdom, 2020, pp. 86-93. DOI: 10.1109/INDIN45582.2020.9442233

Naraharisetty, V., Talari, V. S., Neridu, S., Kalapatapu, P., & Pasupuleti, V. D. K. (2021) "Cloud Architecture For IoT Based Bridge Monitoring Applications," *2021 International Conference on Emerging Techniques in Computational Intelligence (ICETCI)*, Hyderabad, India, pp. 39-42. DOI: 10.1109/ICETCI51973.2021.9574044

Nasri, A., Boujnah, A., Boubaker, A., & Kalboussi, A. (2023). A Smart Gas Sensor Using Machine Learning Algorithms: Sensor Types Based on IED Configurations, Fabrication Techniques, Algorithmic Approaches, Challenges, Progress, and Limitations: A Review [June.]. *IEEE Sensors Journal, 23*(11), 11336–11355. DOI: 10.1109/JSEN.2023.3268342

Nguyen, M. D., Tizon, L. D., Le, N. T., Nguyen, D. T., Vu, T. C., Nguyen, T. V., ... & Nguyen, M. T. (2025). A comparative study of Wi-Fi technologies in wireless sensor networks. *Computer Networks and Communications*, 75-87.

Niedźwiedzińska, J. (2021). Inflation Targeting and Central Banks. DOI: 10.4324/9781003189244

Nigam, S., Kumar, D., Mukherji, S., Tomar, S. S., Shastri, S., & Gupta, P. (2024) "Predictive Maintenance of Railway Tracks Using LSTM," *2024 IEEE International Conference on Intelligent Signal Processing and Effective Communication Technologies (INSPECT)*, Gwalior, India, 2024, pp. 1-5. DOI: 10.1109/INSPECT63485.2024.10896208

Nugraha, A. C., ... (2021) "Detection of Railroad Anomalies using Machine Learning Approach," 2021 International Conference on ICT for Smart Society (ICISS), Bandung, Indonesia, 2021, pp. 1-6. DOI: 10.1109/ICISS53185.2021.9533226

Nyboe, F. F.. (2023) "Towards Autonomous UAV Railway DC Line Recharging: Design and Simulation," *2023 IEEE International Conference on Robotics and Automation (ICRA)*, London, United Kingdom, 2023, pp. 3310-3316. DOI: 10.1109/ICRA48891.2023.10161506

O'Brien, P. D., & Nicol, R. C. (2021). FIPA: Towards a standard for software agents. *BT Technology Journal, 16*(3), 51–59. DOI: 10.1023/A:1009621729979

O'Donoughue, N., Kulkarni, S., & Marzella, D. (2006). "Design and Implementation of a Framework for Monitoring Patients in Hospitals Using Wireless Sensors in Ad Hoc Configuration," *2006 International Conference of the IEEE Engineering in Medicine and Biology Society*, New York, NY, USA, 2006, pp. 6449-6452. DOI: 10.1109/IEMBS.2006.260162

OECD. (2021). *Science, technology and innovation outlook 2021: Times of crisis and opportunity*. OECD Publishing., DOI: 10.1787/75f79015-

OLIVA SALINAS. J. G., & VALDEZ OLMEDO, E. (2010, February). Towards a future in sustainable structural design. In Symposium of the International Association for Shell and Spatial Structures (50th. 2009. Valencia). Evolution and Trends in Design, Analysis and Construction of Shell and Spatial Structures: Proceedings. Editorial Universitat Politècnica de València.

Onal, C., Berat Sezer, O., Ozbayoglu, M., & Dogdu, E. (2017) "Weather data analysis and sensor fault detection using an extended IoT framework with semantics, big data, and machine learning," 2017 IEEE International Conference on Big Data (Big Data), Boston, MA, USA, 2017, pp. 2037-2046. DOI: 10.1109/BigData.2017.8258150

Orhan, I., & Lindh, T. (2010). "Measurement-Based Admission Control in Wireless Sensor Networks," *2010 Eleventh International Conference on Mobile Data Management*, Kansas City, MO, USA, 2010, pp. 426-431. DOI: 10.1109/MDM.2010.51

Orhan, I., & Lindh, T. (2010). “Measurement-Based Admission Control in Wireless Sensor Networks,” *2010 Fourth International Conference on Sensor Technologies and Applications*, Venice, Italy, 2010, pp. 447-452. DOI: 10.1109/SENSORCOMM.2010.71

Oshan, P. (2025). Ensuring Reliability and Accuracy in IoT-Generated Data: Challenges, Solutions, and Future Trends. https://www.encardio.com/blog/ensuring-reliable-accurate-iot-data-challenges-solutions

Ouanan, H., & Abdelwahed, E. H. (2019) “Image processing and machine learning applications in mining industry: Mine 4.0,” *2019 International Conference on Intelligent Systems and Advanced Computing Sciences (ISACS)*, Taza, Morocco, 2019, pp. 1-5. DOI: 10.1109/ISACS48493.2019.9068884

Palaniappan, D., Premavathi, T., Jain, R., Parmar, K., & Jhanvi, M. (2024). Blockchain-Based IoT-Enabled secure 6G smart City applications. In *Advances in civil and industrial engineering book series* (pp. 335–364). DOI: 10.4018/979-8-3693-8029-1.ch012

Parihar, A., Prajapati, J. B., Prajapati, B. G., Trambadiya, B., Thakkar, A., & Engineer, P. (2024). Role of IOT in healthcare: Applications, security & privacy concerns.

Parimala, M., Rani, M. J., Sai, K. B., & Sunitha, M. (2021) “IoT Based Intelligent Fertigation through Drip Irrigation,” *2021 Third International Conference on Intelligent Communication Technologies and Virtual Mobile Networks (ICICV)*, Tirunelveli, India, pp. 410-414. DOI: 10.1109/ICICV50876.2021.9388433

Park, B., Kim, M., Jung, D., Lee, D., Kim, J., & Mun, K.-R. (2024) “Classification of Abnormal Gaits with Machine Learning Algorithms using Sensor-Inherited Insoles,” 2024 46th Annual International Conference of the IEEE Engineering in Medicine and Biology Society (EMBC), Orlando, FL, USA, 2024, pp. 1-4. DOI: 10.1109/EMBC53108.2024.10782460

Patil, P. K., & Patil, S. R. (2017). “Review on structural health monitoring system using WSN for bridges,” *2017 International conference of Electronics, Communication and Aerospace Technology (ICECA)*, Coimbatore, India, 2017, pp. 628-631. DOI: 10.1109/ICECA.2017.8203615

Patsiouras, E., Mygdalis, V., & Pitas, I. (2022) “A UAV Object Detection Benchmark for Vision-Assisted Powerline Element Inspection,” 2022 29th International Conference on Systems, Signals and Image Processing (IWSSIP), Sofia, Bulgaria, 2022, pp. 1-4. DOI: 10.1109/IWSSIP55020.2022.9854489

Paul, H., Martinez, R. R., Sumetheeprasit, B., & Shimonomura, K. (2023) "A Tiltable Airframe Multirotor UAV Designed for Omnidirectional Aerial Manipulation," *2023 IEEE/ASME International Conference on Advanced Intelligent Mechatronics (AIM)*, Seattle, WA, USA, 2023, pp. 460-465. DOI: 10.1109/AIM46323.2023.10196207

Pawlak, Z. (1982). Rough sets. *International Journal of Computer & Information Sciences, 11*(5), 341–356. DOI: 10.1007/BF01001956

Pedregosa, F., Varoquaux, G., Gramfort, A., Michel, V., Thirion, B., Grisel, O., Blondel, M., Prettenhofer, P., Weiss, R., Dubourg, V., & Vanderplas, J. (2011). Scikit-learn: Machine learning in Python. *Journal of Machine Learning Research, 12*, 2825–2830.

Perez-Jimenez, M., Montes-Grova, M. A., Ramon-Soria, P., Arrue, B. C., & Ollero, A. (2020) "POSITRON: Lightweight Active Positioning Compliant Joints Robotic Arm in Power Lines Inspection," *2020 International Conference on Unmanned Aircraft Systems (ICUAS)*, Athens, Greece, 2020, pp. 729-736. DOI: 10.1109/ICUAS48674.2020.9214022

Petit, L., & Desbiens, A. L. (2024) "MOAR Planner: Multi-Objective and Adaptive Risk-Aware Path Planning for Infrastructure Inspection with a UAV," *2024 IEEE International Conference on Robotics and Automation (ICRA)*, Yokohama, Japan, 2024, pp. 8422-8428. DOI: 10.1109/ICRA57147.2024.10610907

Pires-Junior, R., Macedo, L., Frizera, A., & Leal-Junior, A. (2025). Development of self-sensing cantilever structures using additive manufacturing and optical fiber sensing technology. *IEEE Sensors Letters, 9*(3), 1–4. DOI: 10.1109/LSENS.2025.3539980

Poongothai, M., Subramanian, P. M., & Rajeswari, A. (2018) "Design and Implementation of IoT Based Smart Laboratory," 2018 5th International Conference on Industrial Engineering and Applications (ICIEA), Singapore, 2018, pp. 169-173. DOI: 10.1109/IEA.2018.8387090

Prakashan, D., Kaushik, A., & Gandhi, S. (2024). Smart sensors and wound dressings: Artificial intelligence-supported chronic skin monitoring–A review. *Chemical Engineering Journal, 497*, 154371. DOI: 10.1016/j.cej.2024.154371

Premalatha, K., Janaranjani, R., Bhuvaneswari, S., & Karthik, M. (2024) "Transformer Fault Identification and Classification of Using Machine Learning," 2024 2nd International Conference on Self Sustainable Artificial Intelligence Systems (ICSSAS), Erode, India, 2024, pp. 747-751. DOI: 10.1109/ICSSAS64001.2024.10760901

Qazi, A., Simsekler, M. C. E., & Formaneck, S. (2023). Supply chain risk network value at risk assessment using Bayesian belief networks and Monte Carlo simulation. *Annals of Operations Research, 322*(1), 241–272. DOI: 10.1007/s10479-022-04598-3

Qiao, Q., Zhang, Y., & Zhang, W. (2024). Integration of sustainable construction practices with AI-based systems: A critical review. *Sustainable Cities and Society, 95*, 105628. DOI: 10.1016/j.scs.2024.105628

Qu, X., Wang, C., Zhao, R., Fang, M., & Xie, X. (2025). Multi-source data-driven Bayesian network for risk analysis of maritime accidents in the high sea. *Frontiers in Marine Science, 12*, 1631650. DOI: 10.3389/fmars.2025.1631650

Rajaei, H., & Upendra, S. (2023). *"Intelligent Fall Detection Monitoring System," 2023 Congress in Computer Science, Computer Engineering, & Applied Computing*. CSCE.

Ramesh, A. N., Kambhampati, C., Monson, J. R., & Drew, P. J. (2004). Artificial intelligence in medicine. *Annals of the Royal College of Surgeons of England, 86*(5), 334–338. DOI: 10.1308/147870804290 PMID: 15333167

Ramesh, M. V.. (2017) "Water Quality Monitoring and Waste Management Using IoT," *2017 IEEE Global Humanitarian Technology Conference (GHTC)*, San Jose, CA, USA, 2017, pp. 1-7. DOI: 10.1109/GHTC.2017.8239311

Ramesh, V., Hamad, A. A., Jwaid, M. F., Sathyabama, M., Abdulridha, M. M., Kadhim, N. M., & Belay, A. (2022). Early recognition of skin malignancy in images based on convolutional networks by using dynamic system model. *Journal of Nanomaterials, 2022*(1), 1754658.

Ramon-Soria, P., Perez-Jimenez, M., Arrue, B. C., & Ollero, A. (2019) "Planning System for Integrated Autonomous Infrastructure Inspection using UAVs," *2019 International Conference on Unmanned Aircraft Systems (ICUAS)*, Atlanta, GA, USA, 2019, pp. 313-320. DOI: 10.1109/ICUAS.2019.8797874

Ramot, D. M. (2002). Complex fuzzy sets. *IEEE Transactions on Fuzzy Systems, 10*(2), 171–186. DOI: 10.1109/91.995119

Rane, N. (2023). Integrating Leading-Edge Artificial Intelligence (AI), Internet of Things (IoT), and Big Data Technologies for Smart and Sustainable Architecture, Engineering and Construction (AEC) Industry: Challenges and Future Directions. SSRN Electronic Journal. DOI: 10.2139/ssrn.4640926

Rasheed, M. K. (2020). The analysis of macroeconomic gaps in Iraq for the period (1990–2017). *Opción, 36*(Special Issue 27), xx–xx.

Rasheed, M. K., ... (2022). Measuring the impact of oil revenues on government debt in selected countries by using ARDL model. In *Lecture Notes of the Institute for Computer Sciences, Social Informatics and Telecommunications Engineering* (Vol. 471). Springer.

Rasheeda, M. K., Salmanb, A. H., & Shaheedc, S. A. (2022). The Effect of Private and Public Consumption on Economic Exposure in Iraq during the Period (2004-2018) by Using (ARDL) Model. International Journal of Innovation. *Creativity and Change*, *13*(1), 85–104.

Rasheed, M. K. (2020). The analysis of macroeconomic gaps in Iraq for the period (1990-2017). Opción. *Revista de Ciencias Humanas y Sociales*, (27), 73.

Rasheed, M. K.. (2022). The impact of oil exports on consumer imports in the Iraqi economy and COVID-19 period: A theoretical study. *International Journal of Computer Applications in Technology*, *66*(3–4), xx–xx.

Rasheed, M. K., Al-Bairmani, A. M. A., & Al-Hussaini, A. M. J. (2021). The impact of oil exports on consumer imports in the Iraqi economy and COVID-19 period, a theoretical study. *International Journal of Computer Applications in Technology*, *66*(3-4), 246–253. DOI: 10.1504/IJCAT.2021.120451

Raymond, D., Chaudhri, S., & Shaikh, K. (2025) "Data Processing and Pattern Recognition for Electronic Nose," *2025 International Conference on Electronics and Renewable Systems (ICEARS)*, Tuticorin, India, 2025, pp. 463-468. DOI: 10.1109/ICEARS64219.2025.10940421

Richard, (2023). IoT in Construction: Top Benefits, Use-Cases & Application. https://toolsense.io/equipment-management/iot-in-construction-top-benefits-use-cases-application/

Rinaldi, S., Flammini, A., Tagliabue, L. C., & Ciribini, A. L. C. (2018) "On the Use of IoT Sensors for Indoor Conditions Assessment and Tuning of Occupancy Rates Models," 2018 Workshop on Metrology for Industry 4.0 and IoT, Brescia, Italy, 2018, pp. 123-128. DOI: 10.1109/METROI4.2018.8428327

Riza, M. A., Go, Y. I., Zaitizila, I., & Kok, S. P. (2025). A review of advanced reactive materials in fiber sensor technology: Challenges and requirements of precise humidity control in mechanochemical treatment process. *Sensors and Materials*, *37*(1), 41–61. DOI: 10.18494/SAM5377

Rojas, L., Peña, Á., & García, J. (2025). AI-Driven Predictive Maintenance in Mining: A Systematic Literature Review on Fault Detection, Digital Twins, and Intelligent Asset Management. *Applied Sciences (Basel, Switzerland)*, *15*(6), 3337. DOI: 10.3390/app15063337

Rosati, A. Cervo, and C. Fantuzzi, (2020) "Air Quality Monitoring in a BIM model by means of a IoT Sensors Network," 2020 Fourth International Conference on I-SMAC (IoT in Social, Mobile, Analytics and Cloud) (I-SMAC), Palladam, India, 2020, pp. 110-115.

Russo, G. (2004). Design shear strength formula for high strength concrete beams. *Materials and Structures*, *37*(274), 680–688. DOI: 10.1617/14016

S. K. and V. Vaidehi. (2018) "Clustering and Data Aggregation in Wireless Sensor Networks Using Machine Learning Algorithms," *2018 International Conference on Recent Trends in Advance Computing (ICRTAC)*, Chennai, India, 2018, pp. 109-115.

Sachdeva, U., & Vamsi, P. R. (2022) "A Study on Anomaly Detection with Deep Learning Models for IoT Time Series Sensor Data," 2022 8th International Conference on Signal Processing and Communication (ICSC), Noida, India, 2022, pp. 11-14. DOI: 10.1109/ICSC56524.2022.10009580

Sadik Hassan Hussein, S. Y., Dziyauddin, R. A., Muhtazaruddin, M. N., Yaghoobi, Y. J., & Din, N. M. (2024) "Development of AI Vision Inspection System for UAV Imagery Surveillance of Transmission Towers," 2024 5th International Conference on Smart Sensors and Application (ICSSA), Penang, Malaysia, 2024, pp. 1-5. DOI: 10.1109/ICSSA62312.2024.10788575

Safyari, Y., Mahdianpari, M., & Shiri, H. (2024). A review of vision-based pothole detection methods using computer vision and machine learning. *Sensors (Basel)*, *24*(17), 5652. DOI: 10.3390/s24175652 PMID: 39275561

Saha Ray, S., & Gupta, A. K. (2014). Comparative analysis of variational iteration method and Haar wavelet method for the numerical solutions of Burgers–Huxley and Huxley equations. *Journal of Mathematical Chemistry*, *52*(4), 1066–1080. DOI: 10.1007/s10910-014-0327-z

Salama, A. M., & Wiedmann, F. (2013). Demystifying the concepts of sustainable architecture. *Archnet-IJAR*, *7*(2), 94–112. DOI: 10.26687/archnet-ijar.v7i2.121

Salman, H. M. (2019). The reality of teaching practices in the light of the theory of learning based on the brain teachers and teachers of Arabic language. *Opción, 35*(Special Issue 21).

Samal, A., & Misra, L. I. Mohit, Pougajendy, S., Chandel, P. S., anjunath, T. C., & Selvakumar, P. (2025). Musealization and Tourism Practices. In S. Oflazoğlu & M. Dora (Eds.), Preserving Cultural Heritage in Post-Disaster Urban Renewal (pp. 115-142). IGI Global Scientific Publishing. DOI: 10.4018/979-8-3693-9097-9.ch006

Samal, A., & Misra, L. I. Mohit, Pougajendy, S., Chandel, P. S., Manjunath, T. C., & Selvakumar, P. (2025). Musealization and tourism practices. In *S. Oflazoğlu & M. Dora (Eds.), Preserving Cultural Heritage in Post-Disaster Urban Renewal* (pp. 115-142). IGI Global Scientific Publishing.

Santoso, K., Nurba, H. P., Sarief, I., & Munastha, K. A. Kusmadi, and R. Gunawan, (2024) "Performance Evaluation of Early Warning System for Transformer Faults at 20KV Distribution Substation," 2024 18th International Conference on Telecommunication Systems, Services, and Applications (TSSA), Bali, Indonesia, 2024, pp. 1-5.

Santos, S. R. D., & Rohmer, E. (2024). Soft Sensory-Motor System Based on Ionic Solution for Robotic Applications. *Sensors (Basel)*, *24*(9), 2900.

Sargiotis, D. (2024). Advancing Civil Engineering with AI and Machine Learning: From Structural Health to Sustainable Development. DOI: 10.51483/IJARP.4.2.2024.54-81

Sasidharan, S. (2019). Architecting Evolving Internet of Things Application.

Sathvik, M., Saranya, G., & Karpagaselvi, S. (2022, December). An intelligent convolutional neural network based potholes detection using yolo-v7. In *2022 International Conference on Automation, Computing and Renewable Systems (ICACRS)* (pp. 813-819). IEEE.

Scheffer, M., ... (2018) "Evaluation of Open Data Models for the Exchange of Sensor Data in Cognitive Building," 2018 Workshop on Metrology for Industry 4.0 and IoT, Brescia, Italy, 2018, pp. 151-156. DOI: 10.1109/METROI4.2018.8428349

Schröder, S., Kalo, P., & Senatore, G. (2021). Adaptive structures for sustainable architecture and engineering: A review. *Frontiers in Built Environment*, *7*, 689742.

Schwalbe, K., Groh, A., Hertwig, F., & Scheunert, U. (2019). *"Data fusion strategy to improve the reliability of machine learning-based classifications," 2019 Sensor Data Fusion: Trends, Solutions, Applications*. SDF.

Senatore, G., Duffour, P., & Winslow, P. (2018). Sensing and control for adaptive structures: A computational design framework. *Journal of Structural Engineering*, *144*(5), 04018045.

Senatore, G., Duffour, P., Winslow, P., & Wise, C. (2018). Shape control and whole-life energy assessment of an 'infinitely stiff'prototype adaptive structure. *Smart Materials and Structures, 27*(1), 015022.

Sener, A., Barut, M., & Yildirim, M. B. (2021). Impact of commitment, information sharing, and information usage on supplier performance: A Bayesian belief network approach. *Annals of Operations Research, 303*(1), 125–158. DOI: 10.1007/s10479-019-03504-8

Sepanosian, T., Küpers, X., Joza, P., Massah, F., & Bemthuis, R. (2024) "An IoT-Based Architecture for Real-Time Emission Monitoring at Construction Sites," 2024 26th International Conference on Business Informatics (CBI), Vienna, Austria, 2024, pp. 139-148. DOI: 10.1109/CBI62504.2024.00025

Seraj, F., Van Der Zwaag, B. J., Dilo, A., Luarasi, T., & Havinga, P. (2014, April). RoADS: A road pavement monitoring system for anomaly detection using smart phones. In *International workshop on modeling social media* (pp. 128-146). Cham: Springer International Publishing.

Setijadi, E.. (2013). "Design of large scale structural health monitoring system for long-span bridges based on wireless sensor network," *2013 International Joint Conference on Awareness Science and Technology & Ubi-Media Computing (iCAST 2013 & UMEDIA 2013)*, Aizu-Wakamatsu, Japan, 2013, pp. 169-174. DOI: 10.1109/ICAwST.2013.6765428

Shafer, G. (1976). *A Mathematical Theory of Evidence*. Princeton University Press.

Shaghouri, A. A., Alkhatib, R., & Berjaoui, S. (2021). Real-time pothole detection using deep learning. arXiv preprint arXiv:2107.06356.

Shao, Y., Wang, S., & Liu, Y. (2023). A hybrid probabilistic–AI model for uncertainty quantification in energy systems. *Applied Energy, 331*, 120276.

Sharifnasab, H., Mahrokh, A., Dehghanisanij, H., Łazuka, E., Łagód, G., & Karami, H. (2023). Evaluating the Use of Intelligent Irrigation Systems Based on the IoT in Grain Corn Irrigation. *Water (Basel), 15*(7), 1394. DOI: 10.3390/w15071394

Solomentsev, O., Zaliskyi, M., Zuiev, O., Shcherbyna, O., Odarchenko, R., & Yashanov, I. (2022) "Predictive Maintenance Approach for Telecommunication and Radioelectronic Systems," 2022 IEEE 16th International Conference on Advanced Trends in Radioelectronics, Telecommunications and Computer Engineering (TCSET), Lviv-Slavske, Ukraine, 2022, pp. 58-63. DOI: 10.1109/TCSET55632.2022.9767048

Soundariya, A., Behera, N. K., Selvakumar, P., Chandel, P. S., Gupta, S., & Manjunath, T. C. (2025). AI and environmental ethics: The role of AI in addressing climate change ethical considerations in AI-driven environmental solutions. In Chakraborty, S. (Ed.), *Human Values, Ethics, and Dignity in the Age of Artificial Intelligence* (pp. 321–346). IGI Global Scientific Publishing. DOI: 10.4018/979-8-3693-9035-1.ch013

Soussi, A., Zero, E., Sacile, R., Trinchero, D., & Fossa, M. (2024). Smart sensors and smart data for precision agriculture: A review. *Sensors (Basel)*, *24*(8), 2647. DOI: 10.3390/s24082647 PMID: 38676264

Srilakshmi, P., Mohanapriya, D. Harini, and K. Geetha, (2019) "IoT Based Smart Health Care System to Prevent Security Attacks in SDN," 2019 Fifth International Conference on Electrical Energy Systems (ICEES), Chennai, India, pp. 1-7. DOI: 10.1109/ICEES.2019.8719236

Sun, L., Shang, Z., Xia, Y., Bhowmick, S., & Nagarajaiah, S. (2020). Review of bridge structural health monitoring aided by big data and artificial intelligence: From condition assessment to damage detection. *Journal of Structural Engineering*, *146*(5), 04020073. DOI: 10.1061/(ASCE)ST.1943-541X.0002535

Suresh, R. V., Selvakumar, P., Naresh Babu, K., Nithya, G., Manjunath, T. C., & Kartigeyan, J. (2025). Best Practices and Practical Applications. In E. Çela, V. Paliktzoglou, S. Sengar, P. Eappen, &. Vajjhala (Eds.), Enhancing School Counseling With Technology and Case Studies (pp. 341-362). IGI Global Scientific Publishing. DOI: 10.4018/979-8-3693-8392-6.ch013

Suryawanshi, S. R., & Deshpande, U. L. (2017). "Review of risk management for landslide forecasting, monitoring and prediction using wireless sensors network," *2017 International Conference on Innovations in Information, Embedded and Communication Systems (ICIIECS)*, Coimbatore, India, 2017, pp. 1-6. DOI: 10.1109/ICIIECS.2017.8276113

Sweeting, B. (2017, August). The ethics of ethics and the ethics of architecture. In *Relating Systems Thinking and Design Symposium (RSD)*.

Taha, M. F., Mao, H., Zhang, Z., Elmasry, G., Awad, M. A., Abdalla, A., Mousa, S., Elwakeel, A. E., & Elsherbiny, O. (2025). Emerging technologies for precision crop management towards agriculture 5.0: A comprehensive overview. *Agriculture*, *15*(6), 582. DOI: 10.3390/agriculture15060582

Taheri, H., & Salimi Beni, A. (2025). Artificial intelligence, machine learning, and smart technologies for nondestructive evaluation. *Handbook of Nondestructive Evaluation 4.0*, 1-29.

Tang, M. J., Zhu, Z. H., Li, Y. L., Qin, W. W., Liang, F. P., Wang, H. L., & Zou, H. H. (2025). Specific smart sensing of electron-rich antibiotics or histidine improves the antenna effect, luminescence, and photodynamic sterilization capabilities of lanthanide polyoxometalates. *Journal of Colloid and Interface Science*, *680*, 235–246. DOI: 10.1016/j.jcis.2024.11.006 PMID: 39504753

Tang, Y., Ye, W., Jia, J., & Chen, Y. (2024). Learning Stiffness Tensors in Self-Activated Solids via a Local Rule. *Advancement of Science*, *11*(19), 2308584.

Tan, P., Wang, Z., & Chang, X. (2025). Road Obstacle Detection Method Based on Improved YOLOv5. *Algorithms*, *18*(6), 300. DOI: 10.3390/a18060300

Tao, F., Zhang, H., Liu, A., & Nee, A. Y. (2018). Digital twin in industry: State-of-the-art. *IEEE Transactions on Industrial Informatics*, *15*(4), 2405–2415.

Tariq, U., Ahmed, I., Bashir, A. K., & Shaukat, K. (2023). *A Critical Cybersecurity Analysis and Future Research Directions for the Internet of Things: A Comprehensive Review Sensors, 23(8), 4117.* Multidisciplinary Digital Publishing Institute., DOI: 10.3390/s23084117

Thilakarathne, N. N., Bakar, M. S. A., Abas, P. E., & Yassin, H. (2025). Internet of things enabled smart agriculture: Current status, latest advancements, challenges and countermeasures. *Heliyon*, *11*(3), e42136. DOI: 10.1016/j.heliyon.2025.e42136 PMID: 39959477

Tian, D. (2022) "Simulation of Distributed Big Data Intelligent Fusion Algorithm Based on Machine Learning," *2022 International Conference on Artificial Intelligence and Autonomous Robot Systems (AIARS)*, Bristol, United Kingdom, 2022, pp. 421-424. DOI: 10.1109/AIARS57204.2022.00101

Todaro, M. (2006). *Economic development* (M. H. Hosni & M. H. Mahmoud, Trans. & Rev.). Dar Al-Marikh. (Original work published in English)

Torra, V., & Narukawa, Y. (2007). *Modeling decisions: Information fusion and aggregation operators*. Springer., DOI: 10.1007/978-3-540-68791-7_6

Torresen, J. (2018). A review of future and ethical perspectives of robotics and AI. *Frontiers in Robotics and AI*, *4*, 75. DOI: 10.3389/frobt.2017.00075

Torrey, L., & Shavlik, J. (2010). Transfer learning. In *Handbook of research on machine learning applications and trends: Algorithms, methods, and techniques* (pp. 242–264). IGI Global. DOI: 10.4018/978-1-60566-766-9.ch011

United Nations. (2015). *Transforming our world: The 2030 agenda for sustainable development*. https://sdgs.un.org/2030agenda

Usman, S., Ahmad, N., Nurhakim Nor Iskandar, M. F., & Jalil, S. Z. A. (2024) "Exploratory Data Analysis for Malaysian Cases of Covid-19," 2024 5th International Conference on Smart Sensors and Application (ICSSA), Penang, Malaysia, 2024, pp. 1-5. DOI: 10.1109/ICSSA62312.2024.10788551

V. I., R. K, S. Saranya, B. Shankar, M. Tamilselvam, and S. Venkatachalam, (2025) "The Future Role of Artificial Intelligence and Internet of Things Smart Green Homes Buildings," 2025 International Conference on Frontier Technologies and Solutions (ICFTS), Chennai, India, 2025, pp. 1-5.

Vos, R. (2009). Mechanics and applications of pressure adaptive honeycomb (Doctoral dissertation, University of Kansas).

Vrsalović, T. Perković, I. Andrić, M. A. Čuvić, and P. Šolić, (2021) "IoT Deployment for Smart Building: Water Consumption Analysis," 2021 6th International Conference on Smart and Sustainable Technologies (SpliTech), Bol and Split, Croatia, 2021, pp. 01-05.

Wahl, B., Cossy-Gantner, A., Germann, S., & Schwalbe, N. R. (2018). Artificial intelligence (AI) and global health: How can AI contribute to health in resource-poor settings? *BMJ Global Health*, *3*(4), e000798. DOI: 10.1136/bmjgh-2018-000798 PMID: 30233828

Wai, R.-J., & Prasetia, A. S. (2019). · R. -J. Wai and A. S. Prasetia, "Adaptive Neural Network Control and Optimal Path Planning of UAV Surveillance System With Energy Consumption Prediction,". *IEEE Access : Practical Innovations, Open Solutions*, *7*, 126137–126153. DOI: 10.1109/ACCESS.2019.2938273

Wang, W., ... (2019) "Dijkstra Algorithm Based Building Evacuation Recognition Computing and IoT System Design and Implementation," 2019 IEEE 13th International Conference on Anti-counterfeiting, Security, and Identification (ASID), Xiamen, China, 2019, pp. 229-233. DOI: 10.1109/ICASID.2019.8925068

Wang, X., Gao, H., Jia, Z., & Li, Z. (2023). BL-YOLOv8: An improved road defect detection model based on YOLOv8. *Sensors, 23*(20), 8361.

Wang, J.-J., & Wang, S. (2010). "Wireless sensor networks for Home Appliance Energy Management based on ZigBee technology," *2010 International Conference on Machine Learning and Cybernetics*, Qingdao, China, 2010, pp. 1041-1046. DOI: 10.1109/ICMLC.2010.5580627

Wang, L., Ma, Y., & Meng, Z. (2014). Haar wavelet method for solving fractional partial differential equations numerically. *Applied Mathematics and Computation*, *227*, 66–76. DOI: 10.1016/j.amc.2013.11.004

Wang, T., Cui, T., Qi, X., & Huan, X. (2025). *Urban Viaduct Structural Health Monitoring: A Review of Wireless Sensor Approaches Buildings, 15(10), 1619.* Multidisciplinary Digital Publishing Institute., DOI: 10.3390/buildings15101619

Williams, C. K., & Rasmussen, C. E. (2006). *No. 3* (Vol. 2). Gaussian processes for machine learning. MIT Press.

Wolniak, R., & Stecuła, K. (2024). Artificial Intelligence in Smart Cities—Applications, Barriers, and Future Directions: A Review. *Smart Cities*, *7*(3), 1346–1389. DOI: 10.3390/smartcities7030057

World Bank. (n.d.). *Economic indicators database.* https://data.worldbank.org

World Intellectual Property Organization (WIPO). (n.d.). *The global innovation index.* https://www.wipo.int/global_innovation_index

Wu, J., & Qin, H. (2008) "The Design of Wireless Intelligent Home System Base on ZigBee," 2008 11th IEEE International Conference on Communication Technology, Hangzhou, 2008, pp. 73-76.

Wu, M., Liu, H., Huang, L., Lin, J., & Hsu, M. (2018) "Integrating BIM and IoT technology in environmental planning and protection of urban utility tunnel construction," *2018 IEEE International Conference on Advanced Manufacturing (ICAM)*, Yunlin, Taiwan, 2018, pp. 198-201. DOI: 10.1109/AMCON.2018.8615004

Xie, T., ... (2024). TBM construction risk analysis based on fuzzy Bayesian network. [Hong Kong Journal of Construction Safety. Zhang, H., Ren, J., & Guo, P. (2022). Bayesian networks in risk assessment: A review. Safety Science, 146, 105551. Rasheed, M. K., Salman, A. H., & Mounir, A. S. (2022, December). Measuring the Impact of Oil Revenues on Government Debt in Selected Countries by Using ARDL Model. In International Conference on Intelligent Systems and Machine Learning (pp. 170-180). Cham: Springer Nature Switzerland.

Xie, Q., Zhou, X., Wang, J., Gao, X., Chen, X., & Liu, C. (2019). Matching Real-World Facilities to Building Information Modeling Data Using Natural Language Processing. *IEEE Access : Practical Innovations, Open Solutions*, *7*, 119465–119475. DOI: 10.1109/ACCESS.2019.2937219

Xue, Y., Ramamurthy, B., & Burbach, M. (2010). "A two-tier Wireless Sensor Network infrastructure for large-scale real-time groundwater monitoring," *IEEE Local Computer Network Conference*, Denver, CO, USA, 2010, pp. 874-881. DOI: 10.1109/LCN.2010.5735826

Yang, G., Zhao, B., Zhang, J., Wen, J., Li, Q., Lei, L., Chen, X., & Chen, B. M. (2025). Det-Recon-Reg: An Intelligent Framework Toward Automated UAV-Based Large-Scale Infrastructure Inspection. *IEEE Transactions on Instrumentation and Measurement*, *74*, 1–16. DOI: 10.1109/TIM.2025.3571118

Yang, S.-W., Lee, Y. Z., & Kim, S.-A. (2025). Design and Validation of a Real-Time Maintenance Monitoring System Using BIM and Digital Twin Integration. *Buildings (Basel, Switzerland)*, *15*(8), 1312. DOI: 10.3390/buildings15081312

Yang, T., Dong, Y., Tang, B., & Xu, Z. (2024). Developing a dynamic life cycle assessment framework for buildings through integrating building information modeling and building energy modeling program. *The Science of the Total Environment*, *946*, 174284. DOI: 10.1016/j.scitotenv.2024.174284 PMID: 38942319

Ye, X., Jin, T., & Yun, C. (2019). A review on deep learning-based structural health monitoring of civil infrastructures. Smart Structures and Systems, 24(5), 567. Technopress. DOI: 10.12989/sss.2019.24.5.567

Yixiao, C., Hui, S., Quhong, S., Xiaoxi, Z., & Jirong, Y. (2025). A review of utility of wearable sensor technologies for older person frailty assessment. *Experimental Gerontology*, *200*, 112668. DOI: 10.1016/j.exger.2024.112668 PMID: 39733783

Yogi, K. S., Sindhu, D., Soni, H., Mukherjee, S., & Madhu, G. C. (2024) "Enhancing Accuracy in Social Media Sentiment Analysis through Comparative Studies using Machine Learning Techniques," *2024 International Conference on Knowledge Engineering and Communication Systems (ICKECS)*, Chikkaballapur, India, 2024, pp. 1-6. DOI: 10.1109/ICKECS61492.2024.10616441

Yun, J., & Woo, J. (2020). A Comparative Analysis of Deep Learning and Machine Learning on Detecting Movement Directions Using PIR Sensors [April.]. *IEEE Internet of Things Journal*, *7*(4), 2855–2868. DOI: 10.1109/JIOT.2019.2963326

Yurdakul, M., & Taşdemir, Ş. (2025). An enhanced YOLOv8 model for real-time and accurate pothole detection and measurement. arXiv preprint. arXiv:2505.04207. https://arxiv.org/abs/2505.04207

Yu, X., Fu, Y., Li, J., Mao, J., Hoang, T., & Wang, H. (2024). Recent advances in wireless sensor networks for structural health monitoring of civil infrastructure. *Journal of Infrastructure Intelligence and Resilience*, *3*(1), 100066. DOI: 10.1016/j.iintel.2023.100066

Zadeh, L. (1965). Fuzzy sets. *Information and Control*, *8*(3), 338–353. DOI: 10.1016/S0019-9958(65)90241-X

Zainordin, N., Salleh, S., Farsana, N., Khoo, S. L., Omar, A. F., & Heka Ardana, P. D. (2024) "Stoppage to the Implementation of Internet of Things (IoT) in Construction Projects: from Lens of Civil Engineers in the United Kingdom," 2024 5th International Conference on Artificial Intelligence and Data Sciences (AiDAS), Bangkok, Thailand, 2024, pp. 191-202. DOI: 10.1109/AiDAS63860.2024.10730249

Zanevych, Y., Yovbak, V., Basystiuk, O., Shakhovska, N., Fedushko, S., & Argyroudis, S. (2024). Evaluation of pothole detection performance using deep learning models under low-light conditions. *Sustainability (Basel)*, *16*(24), 10964. DOI: 10.3390/su162410964

Zarzar, L. D., & Aizenberg, J. (2014). Stimuli-responsive chemomechanical actuation: A hybrid materials approach. *Accounts of Chemical Research*, *47*(2), 530–539.

Zgheib, R., De Nicola, A., Villani, M. L., Conchon, E., & Bastide, R. (2017) "A Flexible Architecture for Cognitive Sensing of Activities in Ambient Assisted Living," 2017 IEEE 26th International Conference on Enabling Technologies: Infrastructure for Collaborative Enterprises (WETICE), Poznan, Poland, pp. 284-289. DOI: 10.1109/WETICE.2017.41

Zhang, D., Zhang, Y., Li, Q., Plummer, T., & Wang, D. (2019, July). Crowdlearn: A crowd-AI hybrid system for deep learning-based damage assessment applications [Paper presentation]. The 2019 IEEE 39th International Conference on Distributed Computing Systems (ICDCS) (pp. 1221–1232). IEEE.

Zhang, X., Shen, L., & Zhang, L. (2022). Predictive analytics and future trends in smart city infrastructure: A review. *Journal of Urban Technology*, *29*(2), 85–102. DOI: 10.1080/10630732.2021.2003032

Zhang, Y., & Zhu, H. (2021). Structural health monitoring and self-adaptive systems: A review. *Structural Control and Health Monitoring*, *28*(10), e2800.

Zhao, Y., & Mata, G. E. (2020) "Leverage Artificial Intelligence to Learn, Optimize, and Win (LAILOW) for the Marine Maintenance and Supply Complex System," *2020 IEEE/ACM International Conference on Advances in Social Networks Analysis and Mining (ASONAM)*, The Hague, Netherlands, 2020, pp. 678-684. DOI: 10.1109/ASONAM49781.2020.9381319

Zhi, Q., Zhang, K., Wen, Y., Cui, Y., Li, Y., & Han, Z. (2024) "Data Processing and Sensor Data Fusion Algorithm Application in Smart Safety Helmets," 2024 4th International Conference on Electronic Information Engineering and Computer Communication (EIECC), Wuhan, China, 2024, pp. 749-754. DOI: 10.1109/EIECC64539.2024.10929244

Zhong, J., Huang, W., & Zhou, H. (2023). Multifunctionality in Nature: Structure–Function Relationships in Biological Materials. *Biomimetics*, *8*(3), 284.

Zhu, L., Xiong, Y., & Yang, X. (2021). "Dynamic Calculation Model for Pressure Safety Monitoring of Municipal Water Supply Network with Interference Elimination Function between Neighboring Nodes," *2021 IEEE 5th Advanced Information Technology, Electronic and Automation Control Conference (IAEAC)*, Chongqing, China, 2021, pp. 1777-1781. DOI: 10.1109/IAEAC50856.2021.9390949

Zhu, Y., Chen, Y., Lu, Z., Pan, S. J., Xue, G. R., Yu, Y., & Yang, Q. (2011, August). Heterogeneous transfer learning for image classification [Paper presentation]. The 25th AAAI Conference on Artificial Intelligence (Vol. 11, pp. 1304–1309). https://www.aaai.org/ocs/index.php/AAAI/AAAI11/paper/viewFile/3671/4073

Zonzini, F., Aguzzi, C., Gigli, L., Sciullo, L., Testoni, N., De Marchi, L., Di Felice, M., Cinotti, T. S., Mennuti, C., & Marzani, A. (2020). Structural Health Monitoring and Prognostic of Industrial Plants and Civil Structures: A Sensor to Cloud Architecture. *IEEE Instrumentation & Measurement Magazine*, *23*(9), 21–27. DOI: 10.1109/MIM.2020.9289069

About the Contributors

Abdulsattar Abdullah Hamad was born in Salah Al-Deen, Iraq, He received Ph.D. from Madurai Kamaraj University-(India), School of Mathematics, and working at University of Samarra, His area of interest includes Importance and role of Mathematical methods and science technology in healthcare, Dynamic Systems, Topology, Big Data, Mathematical Foundations in Healthcare Engineering, achiness Learning, Internet of Things, cloud computing, computational complexity, data analysis. 1. Academic Skills 1. Associate Editor International Journal of Fuzzy System Applications (IJFSA)- (SCOPUS and WoS-IGI Global) 2. Review Editor Frontiers in Applied Mathematics and Statistics- (SCOPUS and WoS- Frontiers) 3. Associate Editor 4. International Journal of Artificial Intelligence (AI) in Business and Management (IJAIBM)- (SCOPUS-IGI Global) 5. Academic Editor Computational and Mathematical Methods in Medicine-(SCOPUS and WoS) Publications No.of papers in Scopus and WoS (60) papers No.of Books Published 2 Books No.of Patent 3 Patents

***

Nilesh Anute is an Associate Professor at Sri Balaji University, Pune. With a strong academic background and extensive teaching experience, he specializes in areas such as management, business strategy, and organizational development. Dr. Anute is known for his research contributions, engaging teaching style, and commitment to fostering student growth. His work bridges academic theory and practical application, making him a respected figure in the academic community.

Balaji Chandrasekaran is currently an Assistant Professor at Anna University Chennai

Bappaditya Das completed his PhD in Computer Science and Engineering in 2025, specializing in Crop Disease Detection using Deep Learning. He holds an M.Tech (2007) and B.Tech (2001) in Computer Science and Engineering from the University of Calcutta and a B.Sc in Physics (Hons.) (1998) from the same university. He was honored with a National Award under the National Scholarship Scheme (1998-99) from Govt. of India, MHRD. Dr. Das has an extensive publication record, including several refereed journal articles and conference proceedings, many focusing on deep learning and AI in agriculture. With over 20 years of academic experience, he has served as an Assistant Professor at Dr. B. C. Roy Engineering College since 2009, having previously held Lecturer positions there and at JIS College of Engineering. He is also a Fellow Member (Membership ID- SAS/FSASS/887/2025) of the Scholars Academic and Scientific Society, a Life member (Member ID: 2032240004) of the Computer Society of India (CSI) Kolkata Chapter, a Life member (M-1803322) of The Institution of Engineers (India) Durgapur section, and a Life member (membership number: 512830) of the International Association of Engineers (IAENG).

Smrithy G S is currently an Assistant Professor at Vellore Institute of Technology, Chennai Campus

Harshvardhan Godbole is a passionate educator and lifelong learner Harshvardhan, and I am a passionate educator and lifelong learner based in Kolhapur, India

Dankan Gowda V is currently working as an Assistant Professor in the Department of Electronics and Communication Engineering at BMS Institute of Technology and Management in Bangalore. Previously, he worked as a Research Fellow at ADA DRDO and as a Software Engineer at Robert Bosch. With a total experience of 14 years, including teaching and industry, Dr. Gowda has made significant contributions to both academia and research. He has published over 120 research papers in renowned international journals and conferences. In recognition of his innovative work, Dr. Gowda has been granted six patents, including four from Indian authorities and two international patents. His research interests primarily lie in the fields of IoT and Signal Processing, where he has conducted workshops and handled industry projects. Dr. Gowda is passionate about teaching and strives to create an engaging and effective learning environment for his students. He consistently seeks opportunities to enhance his knowledge and stay updated with the latest advancements in his field.

Naziya Hussain is an Assistant Professor in the Mukesh Patel School of Technology Management & Engineering NMIMS Deemed to be University Shirpur Campus. She had her Doctorate (Ph.D.- Computer Science & Engineering) from Department of CSE, Rabindranath Tagore University (Formerly AISECT

University), Bhopal in Design and Development of Mobility Models Based Attacks Resilient and Efficient Protocol for Vehicular Ad hoc Network. She is having more than 16 years of teaching experience in computer engineering field. She is the author of one book & more than 28 Research Publication in different International Journals & Conferences. She is life time Member of CSI, ISTE, IAENG, WISE. She is the author of book chapter on various books. Her interests include Computer Network, Wireless Network, Ad-hoc Network and VANET Internet of Things, Machine Learning, Mobile Networks, WSN.

Balaji Shesharao Ingole is a researcher and project manager with 16 years of experience, working on software in healthcare and other domains.

Subramaniam Iyer is a seasoned academician with over 24 years of experience in the field of management education. He is currently serving as a Professor in the MBA Department at Datta Meghe Institute of Management Studies, Nagpur, Maharashtra. With a deep passion for teaching, research, and mentoring future business leaders, Dr. Iyer has made significant contributions to the academic and professional development of his students. His extensive expertise and commitment to excellence continue to inspire and shape the next generation of management professionals.

Pranav Jaysankar is currently a B.Tech student at Anna University Chennai

Ramya K currently working as an Assistant professor in the Department of Electronics and communication, BGS Institute of Technology, Adichunchangiri University, Mandya since 2013. With a Good academic background and a passion for Teaching and Research. Pursuing Research in VTU completed MTech in the year 2013 in Digital communication and BE in Electronics and communication in the year 2010

Sri Yogi holds a Ph.D. in Sustainable Supply Chain Management from the Indian Institute ofTechnology Roorkee, Roorkee, and is a gLink Fellow in Business and Technology (Post-Doctoral Research Fellow) at LogDynamics, University of Bremen, Germany. He earned his Bachelor of Technology in Mechanical Engineering from Acharya Nagarjuna University,Guntur, Andhra Pradesh, and his Master's in Industrial Engineering & Management (Distinction) from Shri G.S. Institute of Technology and Science, Indore, Madhya Pradesh. Additionally, he holds an Executive Certification in Applied Business Analytics from the Indian School of Business, Hyderabad.With over a decade of teaching experience at top B-schools, central university, and esteemed private deemed universities across India, Dr. Sri Yogi combines academic rigor with practical industry insights. His research has gained significant recognition, with 465 Google Scholar citations and an h-index of 11 across 85 publications, including 59 Scopus-indexed articles and six in the Web of Science (WoS) core collection. His work has been published in prominent journals across ABDC's A, B, and C categories/Q1/Q2 Dr.Sri Yogi's research interests focus on Sustainable Supply Chains, Quality Control, Reverse Logistics,Machine learning, and Bibliometric Analysis. In addition, he is the holder of nine IPRs, encompassing patents, trademarks, copyrights, and design registrations, underscoring his dedication to innovation in his field. His teaching areas include Quantitative Techniques, Supply ChainManagement, Operations Management, and Business Research Methods, equipping MBA students with both theoretical foundations and practical tools.

M. R. Arun completed his BE, ME and PhD from Anna University, Chennai. He completed PG Diploma (PGDEEM) from Annamalai University, M.Sc. Counselling Psychology from TNOU University, M.A. Public Administration from Madras University and B.Lib.Sc. from Alagappa University. He is in the teaching profession for the past 17 years, published around 79 papers in Journals, Patents and Conferences, Active participant crossing more than 130 events and internationally networked through various professional bodies. He is skilled and technically proficient in various simulation software associated to Research in Engineering. He organized and coordinated around 15 events in working institution. Authored 9 Books, 3 Book Chapters and he has been technically integrated in various international journals as an Editorial Board Member and Reviewer. His area of interest includes Video Processing, Remote Automation, Wireless Communication, Quantum Computing Machine Learning & Artificial Intelligence.

T.C. Manjunath was born on 6th February 1967. He did his BE from RV College of Engg. in 1989, ME in Robotics & Automation from LD College of Engg. in 1992 with First Rank and PhD in Systems & Control Engg. from IIT Bombay in the year 2007. He is currently working as the Dean Research R D and as Professor in the Dept. of Computer Science & Engineering with specialization in IoT, Cyber Security & Blockchain Technology in Rajarajeswari College of Engineering, Bengaluru, Karnataka. He has completed B.E. Degree from R.V. College of Engg., M.E. Govt.'s LD College of Engg., Gujarat Univ., Ahmedabad in First Rank and Ph.D. in Systems and Control Engineering from the Indian Institute of Technology Bombay IIT Bombay He has got a teaching academic, research and administrative experience of more than 35+ years 10 years as Principal, 8 years as HoD rest as faculty and has published more than 1200+ research papers in various national & international journals and conferences, with 175+ patents, 25 book-chapters, 25 awards and also published 25 text books in various field of science, engineering and technology, completed projects worth Rs. 9 Crores till date apart from guiding 10 PhD scholars. He is a member of 30 Professional Society Chapters and is a fellow of IE, IETE, AMIEE and a chartered engineer. He got the best research scholar award in 2006 in IIT Bombay amongst all the research scholars of the institute. He has got 10 awards in academics and research activities.

Durga Meshram is a dedicated Master of Business Administration (MBA) student at Datta Meghe Institute of Management Studies, located in Nagpur, Maharashtra. With a strong interest in business strategy, marketing, and entrepreneurship, Durga is focused on developing the skills and knowledge required to excel in the dynamic world of management. Committed to academic excellence and professional growth, she aims to make a meaningful impact in the business sector.

Hirenkumar Mistry is a Senior Researcher and Sr. Linux and Cloud Security Engineer at Zenosys Inc., with over 20 years of experience in cybersecurity, cloud computing, and system engineering. He holds a Master of Science in Cybersecurity from Grand Canyon University, Arizona. Throughout his career, Mistry worked with high-profile clients such as Edward Jones Investment, PayPal, Apple, and the State of South Carolina, specializing in cloud and Linux security. He has also held significant roles at Hewlett-Packard, Aruhat Technologies, and GaneshaSpeaks.com. He is an innovator, with multiple patents in cybersecurity monitoring and cloud security technologies. He is first author of the book Cloud Computing Techniques and has published numerous articles in leading international journals and conferences. he has research focuses on cloud computing, virtualization, artificial intelligence, and cybersecurity. He has received several prestigious awards and contributions in cloud and cybersecurity. In addition to his professional achievements, Mr. Mistry actively contributes to the academic and scientific community as a paper reviewer for several renowned journals and conferences, and as a field judge at global events like "Technovation Girls" competition, and the "Regeneron International Science and Engineering Fair". --

R. Myna is currently serving as an Assistant Professor in the Department of Mathematics at Erode Sengunthar Engineering College, Tamil Nadu, India. She holds a Ph.D. in Mathematics from PSG College of Arts & Science, Bharathiar University, Coimbatore, where her research focused on graph partitioning techniques using various heuristic refinements for graph processing in geo-distributed data centers. With over three years of academic experience, Dr. Myna has published research articles in peer-reviewed journals, including topics related to fuzzy logic, heuristic algorithms, and graph theory. Her interdisciplinary interests span complex fuzzy systems, decision science, and intelligent infrastructure analytics. Her ongoing research integrates fuzzy set theory, data analytics, and smart system modeling, with a specific emphasis on Complex Intuitionistic Fuzzy Sets (CIFS) for engineering applications.

Naeema Nazar is a PhD Scholar at the Indian Institute of Information Technology (IIIT) Kottayam and an ACM Anveshan Setu Fellow at the Indian Institute of Technology (IIT) Kharagpur. She also serves as the Global Strategy Representative for India on the Young Professional Advisory Committee, IEEE Photonics Society, USA. An Assistant Professor and IEEE Student Branch Counselor at VISAT Engineering College, Kerala, Ms. Naeema Nazar has made impactful contributions to research through publications, patents, and innovations. Her most recent achievement is the granting of a UK Registered Design Patent for an innovative spectrometer. This design, protected under the United Kingdom's Registered Designs Act, combines aesthetic refinement with functional precision offering a compact, user-friendly form factor without compromising optical accuracy. The patented design opens possibilities for versatile applications in scientific measurement, environmental monitoring, and industrial quality control, showcasing her vision for bridging academic innovation wit h practical usability. She holds an MTech in Wireless Technology from TOCH Institute of Science and Technology, Arakunnam, Ernakulam, and a B.Tech in Electronics and Communication Engineering from KMEA Engineering College, Edathala, Ernakulam. Ms. Naeema Nazar has organized and led numerous international events, delivered invited talks, and actively engages as an academic entrepreneur. Her diverse leadership within IEEE includes multiple roles at national and international levels, along with her appointment as International Day of Light (IDL) Representative with UNESCO. Her research footprint extends across collaborations with several IITs and IIITs in India, focusing on photonics, optics, and wireless sensor networks. Passionate about fostering global scientific collaboration, she continues to bridge academia, industry, and international organizations through her professional endeavors.

P. Selvakumar is currently an Associate Professor in the Department of Science and Humanities at Nehru Institute of Technology, Coimbatore, Tamil Nadu, India. Prior to joining Nehru Institute of Technology, he held positions as an Assistant Professor in the Department of Science and Humanities various engineering colleges in Coimbatore, Tamil Nadu, India, spanning from 2011 to 2023. Dr.P. Selvakumar's academic and research interests lie primarily in the field of phytochemistry. He has made significant contributions to his field, with an impressive record that includes 30 journal publications, 12 patents, and 20 conference papers. Additionally, he has organized 5 seminars and conferences, further demonstrating his commitment to academic and scientific exchange. Throughout his career, Dr. P.Selvakumar has been recognized for his dedication and excellence in teaching, having been honored with several awards, including the Best Teacher Awards. He holds a prominent position in the academic community, contributing actively to research.

Damodharan Palaniappan works as an Associate Professor & Head in the Department of Information Technology, Marwadi University, Rajkot, Gujarat. He completed his Ph.D. at Anna University, Chennai, M.E. in Computer Science and Engineering at C.I.E.T, Coimbatore and B.E. in Computer Science and Engineering at CSI College of Engineering, Ketty, Nilgiri. He has 19 years' experience of teaching and research. Currently, he guides 6 Ph.D. research scholars. He has authored 8 book chapters, and he has published research papers in reputed journals. His interested areas of research are Data Mining, Image Processing, Software, Machine Learning and Deep Learning.

Gajanan Shankarrao Patange received his B.E. degree in Mechanical Engineering from Amravati University, Maharashtra, India, in 1997. He completed his MTech. in Industrial Engineering and Management from SGSITS, Indore, Madhya Pradesh, in 2006. In 2016, he earned his Ph.D. in Mechanical Engineering from CHARUSAT (Charotar University of Science and Technology), Gujarat. He began his teaching career in July 1999 as a Lecturer in the Department of Mechanical Engineering at Jaihind comprehensive Educational Institute of Technology. He served as Head of Department from 2006 until February 2008. In February 2008, he joined CHARUSAT (formerly Charotar Institute of Technology, Changa) as a Lecturer and currently holds the position of Workshop Superintendent in the Department of Mechanical engineering. Dr. Patange has guided over 50 postgraduate dissertations and more than 100 undergraduate projects. He is a recognized Ph.D. supervisor at CHARUSAT and serves as a Ph.D. examiner for Mumbai University and Pune University. His research interests include Quality Engineering, Production and Operations Management, Environmentally Conscious Manufacturing, Waste Management, Industrial Engin

Ganesh Pathak is an accomplished academician currently serving as an Associate Professor at Sri Balaji University, Pune (SBUP). With a strong foundation in management education and research, Dr. Pathak brings extensive teaching experience and scholarly insight to his role. He has contributed significantly to the academic community through research publications, mentorship, and active participation in conferences and seminars. Dr. Pathak is known for his commitment to fostering intellectual growth and his dedication to shaping future leaders in the field of management.

Deepak Rajeshkumar is currently a B.Tech student at Anna University Chennai

Nithin Subba Rao a distinguished manufacturing excellence leader with over 15 years of experience in the automotive industry, Nithin Subba Rao has established himself as a transformational force in operational optimization and cost reduction. He holds a Master's degree in Mechanical Engineering from Michigan Technological University and an MBA in Operations Management, combining technical expertise with strategic business acumen. As a Six Sigma Black Belt with 10+ approved patents in manufacturing process improvement, Nithin Subba Rao has demonstrated exceptional capability in driving operational transformation and delivering measurable results. His track record includes delivering over $80 million in combined cost savings and restructuring benefits while implementing Industry 4.0 technologies across multiple manufacturing facilities. Nithin Subba Rao's expertise spans digital transformation, lean manufacturing, ERP integration, and the development of sustainable continuous improvement cultures that drive long-term competitive advantage. His personal philosophy of continuous improvement – symbolized by naming his daughter "Aarohi," meaning continuous improvement / ascending – reflects his deep commitment to daily progress and systematic excellence. He firmly believes that consistency beats intensity, working toward his goals one step at a time with a target of 1% daily improvement, a principle that has delivered miraculous results in both his personal and professional life

Madhu S has 31+ years of experience in academia and industry. A prolific author and researcher, he has authored 85+ books, holds 26 design patent, 5 utility patent published, published 30+ articles and book chapters, and has won multiple awards. He serves as a consultant, keynote speaker, jury member, resource person, and academic mentor, guiding hundreds of students and contributing widely to research and innovation. BoS and BoE member for various universities.

Shihab A. Shawkat his BSc in Computer Science from University of Tikrit in 2007 and MSc from Mansoura University in 2017. Currently, he is a PhD student in National School of Electronics and Communications of Sfax (ENET'Com), University of Sfax. He worked as a teacher during the period from 2008 to 2019 in Ministry of Education, Iraq. He has recently started working at the University of Samarra at the end of 2019 till now. His research interest lies in information security, image processing and AI.

Bipin Sule is a Senior Professor at Vishwakarma Institute of Technology (VIT), Pune, India. With a distinguished academic and research career spanning several decades, Dr. Sule is renowned for his contributions to engineering education, academic leadership, and institutional development. He has played a pivotal role in shaping the academic framework at VIT Pune, fostering innovation, and promoting excellence in teaching and research. His areas of expertise encompass both technical and administrative domains, and he is widely respected for his mentorship, vision, and commitment to empowering students and faculty alike. Dr. Sule continues to be a driving force in the advancement of technical education in India.

Premavathi T, working as Assistant Professor in the Department of Computer Engineering – Artificial Intelligence, Marwadi University, Rajkot, Gujarat. I did B.E in Computer Science and Engineering at Dhanalakshmi Srinivasan Engineering College, Perambalur, TamilNadu and I did her M.E in Software Engineering with Distinction in Anna University. Currently pursuing her Ph.D. at the Marwadi University. I am having 15+ years of experiences in teaching. My area of interests are Software Engineering, Machine Learning, Deep Learning, etc.

Padmashree V is an Assistant Professor and Research Cell Head at Padmashree Institute of Management and Sciences with expertise in finance, stock market studies, and entrepreneurship. A Ph.D. scholar and published researcher, she is passionate about mentoring students, delivering impactful lectures, and fostering innovation through research and conferences.

Vijayakumar G is working as an Assistant professor, in the department of information technology, in V. S. B. Engineering College, Karur, Area of Interest: Computer Networking and Cyber Security Subject Specialization: Programming in C, Cryptography and Network Security, Data Structures and algorithms, Database Management Systems

Index

A

Artificial Intelligence 58, 59, 61, 70, 76, 78, 79, 81, 83, 86, 87, 142, 146, 150, 151, 174, 179, 192, 198, 199, 203, 206, 207, 208, 210, 211, 213, 214, 229, 236, 237, 240, 246, 262, 268, 271, 273, 274, 276, 317, 319, 347, 348, 349, 351, 353, 356, 361, 362, 367, 382, 383, 386, 391, 411, 430, 432
Autonomous Systems 70, 157, 171, 172, 173, 174, 175, 177, 178, 180, 181, 185, 186, 187, 188, 192, 193, 194, 195, 227, 229, 230, 231, 263, 307

B

Backward Parabolic Partial Differential Equation 1
Bayesian Inference 354, 355

C

Challenges 27, 28, 57, 63, 68, 70, 76, 78, 79, 81, 85, 90, 91, 92, 98, 99, 116, 118, 119, 136, 137, 139, 145, 149, 152, 160, 161, 162, 166, 172, 180, 183, 199, 200, 201, 202, 205, 206, 207, 208, 210, 211, 213, 216, 217, 218, 219, 239, 253, 257, 260, 267, 268, 270, 271, 274, 275, 276, 286, 293, 306, 309, 318, 322, 334, 335, 339, 347, 353, 354, 361, 362, 367, 374, 377, 378, 380, 383, 385, 388, 391, 392, 395, 397, 400, 412, 417, 428
Complex Intuitionistic Fuzzy Sets 21, 22, 25, 28, 31, 38, 43, 45, 46, 48, 52, 55, 56
Computational Models 61, 62, 163
Computational Protocols 149
Construction Management 253, 256

D

Data Collection 34, 75, 77, 91, 139, 142, 173, 177, 188, 194, 197, 220, 222, 227, 229, 247, 265, 325, 368, 380, 419
Data-Driven Decision Making 244
Data Fusion 21, 24, 31, 33, 38, 185, 318, 319, 421, 428
Data Processing 82, 124, 127, 129, 131, 172, 179, 181, 186, 247, 260, 276, 291, 292, 293, 300, 306, 307, 317, 318, 319, 345, 354, 376, 411, 413, 419, 421, 429
Decision Support 49, 82, 281, 362
Deep Learning 52, 58, 64, 68, 69, 71, 78, 85, 87, 89, 90, 91, 92, 118, 119, 121, 122, 210, 272, 292, 299, 300, 307, 315, 317, 319, 322, 324, 327, 329, 330, 335, 344, 345, 347, 353, 357, 362, 389
Drones 73, 142, 172, 175, 176, 177, 179, 181, 182, 185, 187, 193, 194, 195, 216, 218, 221, 226, 227, 229, 263, 376

E

Economic Growth 393, 394, 395, 396, 397, 398, 403, 404, 405, 406, 407
Edge Computing 68, 71, 74, 129, 141, 144, 176, 179, 186, 188, 195, 248, 307, 309, 367, 374, 377, 413
Environmental Monitoring 22, 38, 66, 128, 196, 292, 298, 300, 304, 305, 366, 367, 369, 370, 375, 376, 377, 378, 379, 380, 382, 383, 389, 416, 424

F

Fault Detection 211, 243, 271, 300, 318, 333, 340, 342, 345, 348, 374

G

GDP 208, 395, 396, 403, 404, 405

H

Haar Wavelet 1, 2, 7, 8, 16, 17, 18, 19

Health Monitoring 22, 24, 34, 36, 50, 58, 127, 146, 148, 150, 170, 197, 243, 246, 247, 249, 250, 266, 270, 272, 299, 369, 379, 381, 392, 416, 428, 431, 432

I

Indian Roads 89, 90, 93, 96, 98, 116, 117, 118, 119
Infrastructure Inspection 171, 172, 174, 175, 178, 179, 180, 181, 185, 188, 215, 216, 217, 218, 223, 224, 225, 226, 229, 231, 241
Infrastructure Monitoring 123, 124, 125, 126, 127, 129, 130, 131, 137, 140, 141, 142, 144, 175, 207, 236, 384, 409, 411, 412, 413, 414, 415, 416, 417, 419, 420, 422, 425, 427, 428, 429, 431
Infrastructure Systems 21, 22, 27, 29, 52, 123, 134, 142, 174, 178, 185, 195, 219, 220, 228, 229, 231, 237, 243, 244, 245, 246, 267, 321, 322, 323, 324, 325, 326, 330, 331, 334, 335, 337, 338, 339, 342, 345, 409, 411, 412, 414, 416, 417, 418, 419, 420, 421, 422, 429
Innovation Economy 394
Internet of Things 38, 43, 124, 125, 126, 141, 146, 163, 169, 171, 173, 176, 177, 178, 195, 196, 197, 198, 243, 244, 245, 251, 252, 253, 256, 257, 258, 265, 266, 268, 269, 270, 271, 274, 276, 292, 297, 300, 303, 318, 319, 355, 366, 367, 369, 372, 374, 379, 380, 390, 392, 410, 430
Inverse Problem 1
IoT 22, 54, 68, 71, 81, 123, 124, 125, 126, 127, 129, 130, 131, 132, 133, 134, 135, 136, 137, 138, 139, 140, 141, 142, 144, 145, 146, 147, 163, 168, 171, 172, 173, 174, 176, 177, 178, 179, 181, 183, 184, 185, 186, 193, 195, 196, 197, 198, 243, 244, 245, 246, 248, 249, 250, 251, 252, 253, 254, 255, 256, 257, 258, 259, 260, 261, 262, 264, 265, 266, 267, 268, 269, 270, 271, 274, 281, 285, 292, 298, 303, 305, 308, 314, 317, 318, 319, 335, 345, 366, 367, 369, 372, 376, 377, 379, 380, 384, 385, 388, 389, 410, 430, 432

K

Knowledge 22, 23, 32, 47, 61, 62, 63, 71, 78, 83, 125, 199, 200, 201, 202, 203, 204, 205, 206, 207, 208, 212, 261, 262, 275, 276, 285, 293, 294, 319, 324, 326, 334, 335, 393, 394, 396, 400, 401, 405, 406, 421

M

Machine Learning 29, 52, 53, 61, 62, 64, 65, 70, 74, 79, 85, 86, 87, 99, 100, 121, 125, 126, 131, 135, 138, 142, 144, 151, 171, 173, 175, 179, 182, 183, 184, 185, 192, 195, 230, 236, 237, 238, 243, 256, 260, 262, 271, 291, 293, 294, 295, 296, 298, 299, 300, 301, 302, 303, 304, 305, 306, 307, 308, 309, 314, 315, 317, 318, 319, 321, 322, 324, 326, 327, 329, 330, 332, 333, 334, 335, 336, 337, 338, 339, 340, 341, 345, 346, 347, 348, 353, 354, 356, 363, 367, 373, 382, 383, 386, 389, 391, 411, 421, 432
Monte Carlo Simulation 352, 354, 355, 357, 360, 361, 363

P

Pothole Detection 89, 90, 91, 92, 93, 94, 96, 97, 98, 104, 105, 108, 113, 116, 117, 118, 119, 120, 121, 122
Predictive Maintenance 66, 125, 126, 130, 131, 135, 137, 142, 144, 172, 174, 175, 183, 186, 187, 195, 227, 229, 230, 231, 237, 238, 243, 254, 256, 257, 262, 266, 271, 292, 296, 298, 300, 303, 305, 306, 314, 321, 323, 324, 325, 326, 327, 329, 330, 331, 332, 333, 334, 335, 336, 337, 338, 339, 343, 345, 346, 347, 348, 369,

372, 373, 374, 375, 383, 386, 410, 411, 417, 418
Probabilistic Modeling 351, 352, 353, 354, 355, 361

R

Real-Time Data 38, 124, 132, 144, 172, 176, 177, 181, 183, 185, 186, 188, 194, 225, 252, 256, 258, 259, 263, 292, 307, 317, 323, 330, 345, 355, 370, 371, 372, 373, 375, 377, 410, 411, 412, 417
Risk Assessment 65, 351, 352, 353, 354, 355, 358, 361, 362, 363
Robotics 70, 71, 72, 74, 87, 142, 145, 153, 208, 211, 215, 216, 217, 218, 219, 220, 221, 222, 223, 224, 225, 226, 227, 228, 229, 230, 231, 232, 233, 234, 235, 236, 237, 238, 239, 240, 241, 367, 386

S

Scientific Foresight 199, 206, 285
Self-Responsive 149, 150, 151, 152, 154, 155, 156, 157, 158, 159, 160, 161, 162, 163, 164, 165, 166, 167, 168
Sensor Data Processing 291, 300, 306
Sensor Networks 34, 129, 184, 196, 250, 257, 285, 305, 310, 316, 317, 318, 375, 377, 379, 380, 385, 388, 390, 392, 409, 410, 412, 415, 416, 418, 421, 422, 426, 430, 431, 432
Sensor Placement 54, 420
Sensors and Imaging Technologies 218
Smart Cities 22, 38, 40, 49, 59, 126, 192, 194, 195, 197, 245, 259, 260, 265, 267, 268, 270, 291, 292, 302, 305, 308, 347, 366, 367, 369, 371, 372, 379, 388, 420
Smart Infrastructure 21, 22, 27, 28, 29, 30, 48, 52, 53, 55, 57, 145, 207, 246, 382, 386
Smart Sensor Technologies 365, 384, 386, 388
Stochastic Processes 356
Structural Degradation 220, 273, 274, 275, 276, 279, 280, 281, 284, 285, 287, 423
Structural Health 22, 24, 34, 35, 36, 50, 53, 58, 127, 135, 146, 148, 150, 170, 182, 197, 243, 245, 246, 247, 249, 250, 262, 266, 270, 271, 272, 379, 380, 392, 410, 416, 428, 430, 432
Structural Health Monitoring 22, 24, 34, 36, 50, 58, 127, 146, 148, 150, 170, 197, 243, 246, 247, 249, 250, 266, 270, 272, 379, 392, 416, 428, 432
Structures 19, 52, 53, 54, 71, 124, 128, 133, 134, 135, 142, 146, 148, 149, 150, 151, 152, 153, 154, 155, 156, 157, 158, 159, 160, 161, 162, 163, 164, 165, 166, 168, 169, 170, 173, 186, 187, 201, 202, 211, 216, 218, 220, 222, 224, 230, 243, 249, 258, 269, 272, 274, 279, 280, 281, 285, 286, 289, 294, 297, 390, 411, 414, 415, 416, 417, 418, 423
Supervised Learning 64, 175, 294, 298, 324, 327, 329, 334

T

Technological Innovation 220, 407
Technology 1, 21, 22, 38, 61, 64, 68, 70, 71, 77, 86, 87, 92, 118, 121, 124, 125, 136, 138, 141, 142, 144, 147, 150, 155, 161, 164, 166, 168, 169, 171, 175, 181, 186, 194, 196, 197, 198, 199, 200, 202, 203, 206, 207, 212, 213, 214, 215, 217, 218, 223, 225, 231, 241, 244, 245, 256, 263, 266, 268, 269, 280, 284, 288, 303, 304, 306, 317, 321, 334, 338, 347, 348, 355, 362, 363, 365, 366, 367, 368, 369, 370, 375, 376, 378, 379, 386, 388, 389, 390, 391, 393, 394, 395, 396, 397, 398, 399, 400, 401, 402, 405, 406, 407, 409, 415, 419, 431, 432, 433
Traffic Management 38, 192, 252, 305

U

Uncertainty Quantification 363

Unsupervised Learning 85, 175, 294, 298, 299, 300, 310, 315, 316, 324, 327, 328, 329
Urban Systems 22, 38, 43

W

Wireless Sensor Networks 129, 196, 250, 318, 377, 379, 380, 385, 390, 392, 409, 410, 412, 415, 416, 418, 422, 430, 431, 432

www.ingramcontent.com/pod-product-compliance
Lightning Source LLC
LaVergne TN
LVHW062335300926
849161LV00007B/1701